6th National Conference on Processing and Characterization of Materials (NCPCM 2016)

IOP Conference Series: Materials Science and Engineering
Volume 178

Rourkela, India
9 – 10 December 2016

Editors:

B.K. Mishra
Ashish Garg
Anindya Basu
Manojit Ghosh
Partha Sarathi De

ISBN: 978-1-5108-3716-4
ISSN: 1757-8981

Printed from e-media with permission by:

Curran Associates, Inc.
57 Morehouse Lane
Red Hook, NY 12571

Some format issues inherent in the e-media version may also appear in this print version.

Printed by Curran Associates, Inc. (2017)

For permission requests, please contact the Institute of Physics at the address below.

Institute of Physics
Dirac House, Temple Back
Bristol BS1 6BE UK

Phone: 44 1 17 929 7481
Fax: 44 1 17 920 0979

techtracking@iop.org

Additional copies of this publication are available from:

Curran Associates, Inc.
57 Morehouse Lane
Red Hook, NY 12571 USA
Phone: 845-758-0400
Fax: 845-758-2633
Email: curran@proceedings.com
Web: www.proceedings.com

Table of contents

Volume 178

National Conference on Processing and Characterization of Materials
9–10 December 2016, NIT Rourkela, Orissa, India

Accepted papers received: 8 February 2017
Published online: 1 March 2017

Preface

011001
OPEN ACCESS
National Conference on Processing and Characterization of Materials

011002
OPEN ACCESS
Peer review statement

Papers

012001
OPEN ACCESS
Growth, structure and photoluminescence properties of SnO_2-ZnO (TZO) micro cubes

S. Samal, A. B. Panda and B.S. Acharya.....1

012002
OPEN ACCESS
Designing and synthesis of a polymer matrix piezoelectric composite for energy harvesting

Asutya Kumar Biswal, Satyabati Das and Amritendu Roy.....8

012003
OPEN ACCESS
Characterization of iron ore pellets with dextrin added organic binders under different time and temperature conditions

Anand babu kotta, Swapan kumar karak and M. Kumar.....15

012004
OPEN ACCESS
Dry Sliding Wear Behaviour of Micro oven Treated Fly Ash Reinforced Epoxy Composite using ExtendedTaguchi Approach Optimization

A Pattanaik, S K Bhyan, S C Mishra, S. S. Patro and Ajit Behera.....27

012005
OPEN ACCESS
Simulation Study of Stress and Deformation Behaviour of Debonded Laminated Structure

C K Hirwani, H Mittal, S K Panda, S S Mahapatra, S K Mandal and A K De.....38

012006
OPEN ACCESS
Tensile behavior of MWCNT enhanced glass fiber reinforced polymeric composites at various crosshead speeds

K K Mahato, D K Rathore, R K Prusty, K Dutta and B C Ray.....49

012007
OPEN ACCESS
Structural and Dielectric Properties of Bismuth Doped Cobalt Nano Ferrites Prepared by Sol-Gel Auto Combustion Method

Krutika L. Routray and Dhrubananda Behera.....56

012008
OPEN ACCESS
Investigating the Magneto Electric Coupling of [90 wt% $Na_{0.5}Bi_{0.5}TiO_3$ (NBT)-10 wt% $BaFe_{12}O_{19}$ (BaM)] novel multiferroiccomposite system by increasing of BaM grain size

Ranjit Pattanayak, Subhajit Raut, Sourav Kuila, Mallam Chandrasekhar and Simanchala Panigrahi.....67

012009
OPEN ACCESS
Effect of rapid thermal annealing on the structural and electrical properties of RF sputtered CCTO thin film

N Tripathy, K C Das, S P Ghosh, G Bose and J P Kar.....76

012010
OPEN ACCESS
Effect of Moisture Absorption on the Mechanical Properties of Ceramic Filled Jute/Epoxy Hybrid Composites

Priyadarshi Tapas Ranjan Swain and Sandhyarani Biswas.....80

012011
OPEN ACCESS
Fabrication and characterization of the Al6063/5%ZrO_2/5%Al_2O_3 composite

Munmun Bhaumik and Kalipada Maity.....89

012012
OPEN ACCESS
Processing and characterization of epoxy composites reinforced with short human hair

Bishnu Prasad Nanda and Alok Satapathy.....96

012013
OPEN ACCESS
Effect of Hydrothermal immersion and Hygrothermal Conditioning on Mechanical Properties of GRE Composite

A.P. Chakraverty,, U. K. Mohanty, S. C. Mishra and B. B. Biswal.....102

012014
OPEN ACCESS
Dielectric and conduction behaviour of H_2SO_4 doped conducting Polyaniline

J. Mohanty, P. Behera, S. R. Mishra, T. Badapanda and S. Anwar.....114

012015
OPEN ACCESS
Fabrication of nano ZrO_2 dispersed novel $W_{79}Ni_{10}Ti_5Nb_5$ alloy by mechanical alloying and pressureless sintering

R. R. Sahoo, A. Patra and S. K. Karak.....122

012016
OPEN ACCESS
Elastic behavior of CNT-reinforced polymer composites with discontinuities in CNT configurations

Puneet Kumar and Dr. J. Srinivas.....132

012017
OPEN ACCESS
Impact of Annealing Temperature on the Phase of CZTS with the Variation in Surface Morphological Changes and Extraction of Optical Bandgap

J Raiguru, BVRS Subramanyam, K Sa, I Alam, S Das, J Mukherjee, P C Mahakul, B Subudhi and P Mahanandia.....143

012018
OPEN ACCESS
Sea water corrosion behavior of plasma sprayed abradable coatings

M Parida, S.P. Nanda and S.C. Mishra.....151

012019
OPEN ACCESS
Mechanical and wear characteristics of epoxy composites filled with industrial wastes: A comparative study

A Purohit and A Satapathy.....157

012020
OPEN ACCESS
Fabrication of flexible piezoelectric PMN-PT based composite films for energy harvesting

Satyabati Das, Asutya Kumar Biswal and Amritendu Roy.....164

012021
OPEN ACCESS
Synthesis of copper doped Zinc oxide nanowires with enhanced ultraviolet photoresponse behavior

S P Ghosh, K C Das, N Tripathy, A Moharana, A Adhikari, G Bose, D H Kim, T I Lee, J M Myoung and J P Kar.....170

012022
OPEN ACCESS
Biochar, a substitute for fossil fuels

Akankshya Supriya, R Samantray and S C Mishra.....175

012023
OPEN ACCESS
Evidence of Magnetoelectric coupling in $Bi_{2(1-x)}Ho_{2x}Fe_4O_9$ (x=0, 0.01) multiferroic ceramics

S R Mohapatra, A K Singh and S D Kaushik.....182

012024
OPEN ACCESS
Structural and electrical characteristics of solution processed P3HT-carbon nanotube composite

Prakash Chandra Mahakul and Pitamber Mahanandia.....187

012025
OPEN ACCESS
Wear behavior of Al-Si alloy based metal matrix composite reinforced with TiB2

J.K. Sahoo, S.K. Sahoo, H. Sutar and B. Sarangi.....194

012026
OPEN ACCESS
Effect of magnetostrictive material layer on the stress and deformation behaviour of laminated structure

S D Suman, C K Hirwani, A Chaturvedi and S K Panda.....204

012027
OPEN ACCESS
Degradation in Mechanical and Thermal Properties of Partially Aligned CNT/Epoxy Composites due to Seawater Absorption

Sunirmal Saha and Smrutisikha Bal.....215

012028
OPEN ACCESS
Modeling of microstructure evolution in direct metal laser sintering: A phase field approach

Jyotirmoy Nandy, Hrushikesh Sarangi and Seshadev Sahoo.....226

012029
OPEN ACCESS
Development of Ceramic Coating on Metal Substrate using Industrial Waste and Ore Minerals

S K Bhuyan, T K Thiyagarajan and S C Mishra.....234

012030
OPEN ACCESS
Study of process parameter on mist lubrication of Titanium (Grade 5) alloy

Kalipada Maity and Swastik Pradhan.....241

012031
OPEN ACCESS
Flexural Strength of Functionally Graded Nanotube Reinforced Sandwich Spherical Panel

Trupti R Mahapatra, Kulmani Mehar, Subrata K Panda, S Dewangan and Sushmita Dash.....263

012032
OPEN ACCESS
Dielectric, Ferroelectric and Piezoelectric study of BNT-BT solid solutions around the MPB region

T. Badapanda, S. Sahoo and P. Nayak.....271

012033
OPEN ACCESS
Selection of Levels of Dressing Process Parameters by Using TOPSIS Technique for Surface Roughness of En-31 Work piece in CNC Cylindrical Grinding Machine

Sanjay S. Patil and Yogesh J. Bhalerao.....279

012034
OPEN ACCESS
Effect of WC and Si additions on the mechanical properties of ZrB_2-WC-Si based cermet

R.R. Rout, B. Mishra, N. Sahoo and S.K. Karak.....285

IOP Conf. Series: Materials Science and Engineering **178** (2017) 011001 doi:10.1088/1757-899X/178/1/011001

PREFACE

This volume contains selected full length technical papers amongst fifty oral presentations made in the 6th National Conference on Processing and Characterization of Materials (NCPCM 2016), NIT Rourkela, Rourkela, Odisha, India, December 9 – 10, 2016.

The first conference of the NCPCM series was held at the same place in December 2011. Seeing the enthusiasm of the participants, it was decided to organize such conference in Rourkela every year. The basic idea was to establish a periodical national forum for multi–scale approaches in processing and characterization of materials in the eastern part of India. The conference NCPCM 2016 has successfully carried the tradition of previous conferences; more than fifty participants from twenty different organizations across India have registered. The conference was consisted of seven technical sessions of about fifty contributory talks along with four keynote lectures. Out of these, thirty four best peer – reviewed contributions are published in this volume of *IOP Conference Series: Materials Science and Engineering*.

We would like to thank all the contributors, members of the organizing committee, session chairs as well as colleagues and students who helped with the preparation of the conference and, particularly, with the preparation of this volume. We convey our heartiest gratitude to the sponsors and advertisers for their contribution.

January 2017 Editorial Board, NCPCM 2016

Processing & Characterization
of Materials
9th - 10th, December 2016
PCM

IOP Conf. Series: Materials Science and Engineering **178** (2017) 011001 doi:10.1088/1757-899X/178/1/011001

IOP Conf. Series: Materials Science and Engineering **178** (2017) 011001 doi:10.1088/1757-899X/178/1/011001

Our Sponsors

- Council of Scientific and Industrial Research (CSIR), India
- Board of Research in Nuclear Sciences (BRNS), India

LIST OF COMMITTEES

Editorial Board

Prof. B. K. Mishra, IMMT Bhubaneswar, India

Prof. Ashish Garg, IIT Kanpur, India

Prof. Anindya Basu, NIT Rourkela, India

Prof. Manojit Ghosh, IIEST Shibpur, India

Prof. Partha Sarathi De, IIT Bhubaneswar, India

Organizing Committee (NIT Rourkela, India)

Patron: Prof. A. Biswas, Director, NIT Rourkela

Chairman:	Prof. S. C. Mishra, Head of the Department, Metallurgical and Materials Engineering, NIT Rourkela
Convener:	Prof. D. Chaira
Co-convener:	Prof. S. K. Karak
Treasurer:	Prof. A. Basu

Technical Committee (NIT Rourkela, India)

Prof. S.N. Alam

Prof. D. Nagarajan

Prof. A. Patra

Local Organizing Committee (NIT Rourkela, India)

Prof. K. Dutta

Prof. S. K. Sahoo

Mr. S. Pradhan

Mr. S. Hemberam

Mr. R. Pattnaik

IOP Conf. Series: Materials Science and Engineering **178** (2017) 011002 doi:10.1088/1757-899X/178/1/011002

Peer review statement

All papers published in this volume of *IOP Conference Series: Materials Science and Engineering* have been peer reviewed through processes administered by the proceedings Editors. Reviews were conducted by expert referees to the professional and scientific standards expected of a proceedings journal published by IOP Publishing.

Growth, structure and photoluminescence properties of SnO_2-ZnO (TZO) micro cubes

S. Samal[1], A. B. Panda[2*], B.S. Acharya[2]

1. Department of Materials Science, Indian Institute of Technology, Kanpur-208016
2. Department of Physics, CVRCE,Bidyanagar,Bhubaneswar-752054,INDIA

*A. B. Panda; e-mail id: atalabit@gmail.com

Abstract: Structural investigation from X-ray diffraction patterns of TZO revealed the presence of SnO_2 (cassiterite), ZnO (wurtzite) and $ZnSnO_3$ and $ZnSnO_4$ phases. The optical absorption spectra recorded for these powders dispersed in ethanol show the nature of the composite nanopowders to be semiconducting one having an optical absorption band around 280nm. Electron microscopy confirms a change in shape and size of the particles with a change in Sn/Zn ratio. The amorphous nature increases while going from Sn/Zn ratio of 0.87 to 3.75 and cubes of size ~ 2μm, rhombohedral and irregular shaped particles are also observed. The prepared tin zinc oxide samples (TZO) show excitation bands at 241, 287 and 351 nm and common emission bands at 448, 485, 462 and 531 nm for all excitation energies. The photoluminescence and other studies reveal the role played by composition in changing the shape.

Key words: SnO_2-ZnO (TZO), nano cubes, photoluminescence, Nano Composite oxide

1. Introduction

Out of all transition metal oxides, few oxides like, ZnO, In_2O_3, SnO_2 and TiO_2 have received considerable attention for their use in optoelectronic devices, sensors and catalytic application [1-5]. SnO_2 doped with Fluorine or Indium oxide has been extensively used as conducting media for determining the electrical properties of many solid state devices. There have been considerable studies on SnO_2 thin films for sensor application also [6-7]. The key to understand the nano SnO_2 surface properties is the dual valance state of Sn i.e., +2 and +4. This property of capturing electron and hole to become Sn^{2+} and Sn^{4+} introduces changes in the surface properties by changing the electronic structure thus making surface states to lie deep inside the band gap and lowering the work function. Since reduction of Sn^{4+} to Sn^{2+} depends on oxygen chemical potential of the system, systematic investigation of nano SnO_2 prepared through soft chemical solution route is desirable for understanding the basic mechanism of such reduction process. In addition, ZnO is a wide band gap (3.37 eV) n-type semiconductor with large exciton energy of 60 meV having interesting technological properties for applications in photovoltaic and functional coatings [4, 6]. Moreover both SnO_2 and ZnO exhibt diverse group of growth morphologies such as nanocombs, nanorings, nanohelixes/nanosprings, nanobelts, nanowires and nanocages at lower dimensions [4]. Due to the promising features and morphology dependant properties of SnO_2 and ZnO as well as their composites (TZO), it may be expected that their nanostructure forms represent building blocks for special nanodevices and may offer exciting opportunities for fundamental research and technological applications. The present investigation is an effort in this direction to measure the structural and optical properties of such nanoparticles.

2. Materials and methods

Appropriate amount of $SnCl_2{\bullet}2H_2O$ and Zn $(O_2CCH_3)_2$ (AR grade, Merck) were dissolved in distilled water to make the solution having pH 1.8 to 2.0. Required amount of Sodium Oxalate (AR grade, Merck) was dissolved in distilled water to make the pH of the solution to 8.0. Both these solutions were added drop wise from two burettes simultaneously, which took 15 minutes to make a solution of 300ml. The mixture was stirred with the help of magnetic stirrer and saddle for 75 minutes when it becomes the clear solution. To this clear solution, liquid ammonia solution having strength of 25 wt % of NH_3 was added drop by drop. In this way 30 drops were added when white precipitates were seen. Different such solutions having their final pH of 10.30 were subjected to precipitation, filtering and continuous washing with distilled water the final powders were obtained after drying these powders in the oven at 120°C for 2 hrs. Several samples were prepared with varying composition of Sn and Zn at different pH conditions. However, the present investigation deals with three samples which were prepared at pH =10.30 at varying Sn/Zn ratios in atomic % (S_1: 0.87, S_2: 1.75 and S_3: 3.75).

3. Result and Discussion

Phase analysis of the composite SnO_2-ZnO powders were made by using X-ray diffraction technique. The diffractometer used for this purpose was X-Pert Pro with copper target and other accessories given in our earlier work [8]. The patterns have been displayed in Figure 1. Three samples having different Sn/Zn ratio prepared under same conditions have been shown in this figure. The average crystallite size for each sample was calculated by using the Debye-Scherer formula, $D = 0.89\lambda/(\beta\cos\theta)$ where 'λ' is the X-ray wavelength, 'θ' is the Bragg diffraction angle and 'β' is the peak width at half maximum [9]. The phase analysis was made by comparing inter-planer spacing'd' and intensity 'I' values with those of standard XRD patterns of ICDD (Ref. code: 98-000-6056). This has been given in Table-1. One can see that the crystallite size increases from (46±0.5) nm to (58±0.5) nm under Sn/Zn ratio goes from 0.87 to 1.75. But for Sn/Zn = 3.75, this size decreases to (34±0.5) nm.

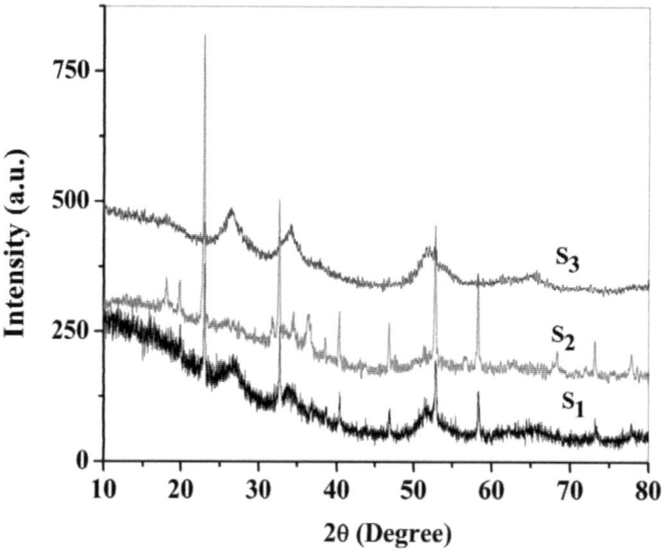

Figure 1. XRD patterns of S_1 (Sn/Zn = 0.87), S_2 (Sn/Zn = 1.75) and S_3 (Sn/Zn = 3.75)

Further, the phase analysis indicates formation of pure SnO_2 casiterite phase for Sn/Zn = 3.75 whereas two phases of SnO_2 and $ZnSn(OH)_6$ are found for Sn/Zn = 1.75 [1, 6]. It is really interesting to see that a lower Sn/Zn ratio of 0.87 results in multiphase system having SnO_2, Zn_2SnO_4 and $ZnSn(OH)_6$ phase. For the SnO_2 casiterite phase, the "d" values are 3.349, 2.643 and 1.764 with corresponding (h k l) values (1 1 0), (0 1 1) and (1 2 1) respectively (ICDD: Ref. code: 98-000-6056). For the impurity phases [Zn_2SnO_4 and $ZnSn(OH)_6$], the corresponding 2θ values are found to be around 22, 32, 53, 58 and 19, 46, 68, 73 respectively [10,11].

Table 1. Average crystallite sizes and phase identified from XRD patterns

Sample No.	Sn/Zn ratio	Average Crystallite Size (± 0.5nm)	Phases Identified
S_1	0.87	46.00	Zn_2SnO_4, SnO_2(Cassiterite), $ZnSn(OH)_6$
S_2	1.75	58.00	SnO_2 (Cassiterite), $ZnSn(OH)_6$
S_3	3.5	34.00	SnO_2(Cassiterite)

To correlate these observations, the differential thermal analysis curves along with the weight loss during heating have been shown in Figure 2.

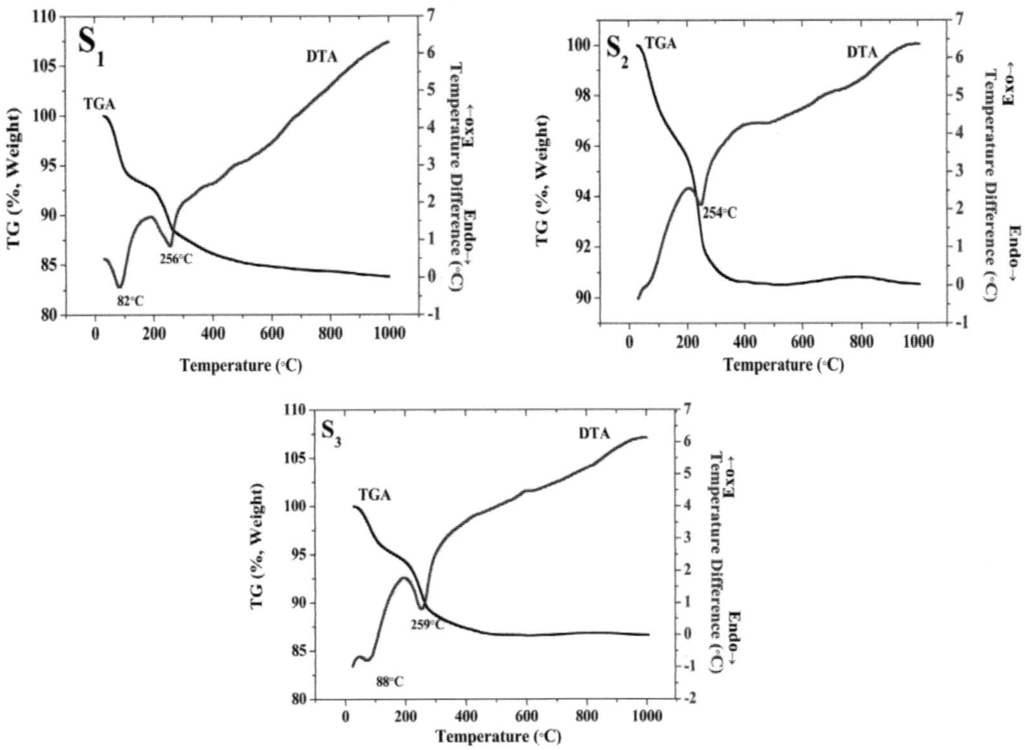

Figure 2. DSC and TGA analysis of S_1 (Sn/Zn = 0.87), S_2 (Sn/Zn = 1.75) and S_3 (Sn/Zn = 3.75)

The results show two peaks (one strong endohermic peak around 256 °C and weak endothermic peak around 82 °C) for compound having Sn/Zn = 0.87 and one (endothermic peak around 254 °C) for Sn/Zn = 1.75while one strong endothermic peak around 259°C and a weak endothermic peak around 88 °C for Sn/Zn = 3.5.This indicates the water molecules on the surface of the compounds go out while heating the sample whereas transformation of phase like $SnZn(OH)_6$ taken place leading to SnO_2 or Zn_2SnO_4 phases. The peak temperature shifts from 254°C to 259°C when complete SnO_2 is formed. This is in accordance with the XRD results presented in table 1 and figure 1. The optical absorption spectra of the samples taken in the ethanol medium have been shown in the Figure 3. The three samples taken for observations show interesting structures in UV region. In Sn/Zn = 0.87 the band gap calculated from absorption spectra corresponding to maximum absorption peak at wavelength (λ_m= 276 nm) was found to be 4.49 eV whereas for Sn/Zn = 1.75 the band gap was found to be 4.90 eV (λ_m = 253 nm). In the third sample where Sn/Zn = 3.75, the absorption shows a flat structure with minimum absorption and the transmission in the UV region was found to be change thus to a constant value in the visible range of spectrum studied. However, the inset shows the magnified image showing the λ_m = 275 nm giving the band gap to be 4.50 eV. Thus it is obvious that pure SnO_2 phase is responsible for such behavior.

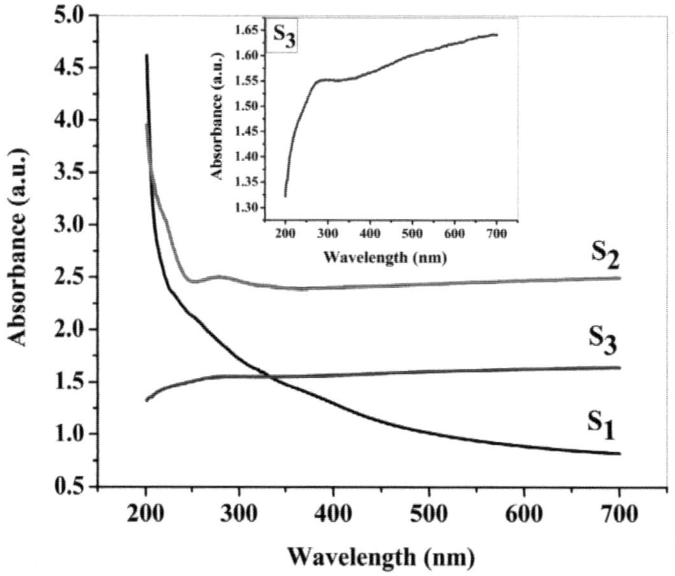

Figure 3. UV – visible absorption spectra of S_1 (Sn/Zn = 0.87), S_2 (Sn/Zn = 1.75) and S_3 (Sn/Zn = 3.75). In-set shows the magnified image of absorption spectra of S_1 (Sn/Zn = 0.87).

The photoluminescence spectra of these samples were taken under different excitation energies (figures not shown) have been given in Table-II. The excitation spectra were taken in UV range i.e., 240 nm to 400 nm and the emission spectra were recorded in the visible range of 400 nm to 800 nm. For brevity typical emission and excitation spectra for sample having Sn/Zn = 1.75 have been shown in Figure 4(b) and Figure 4(b).

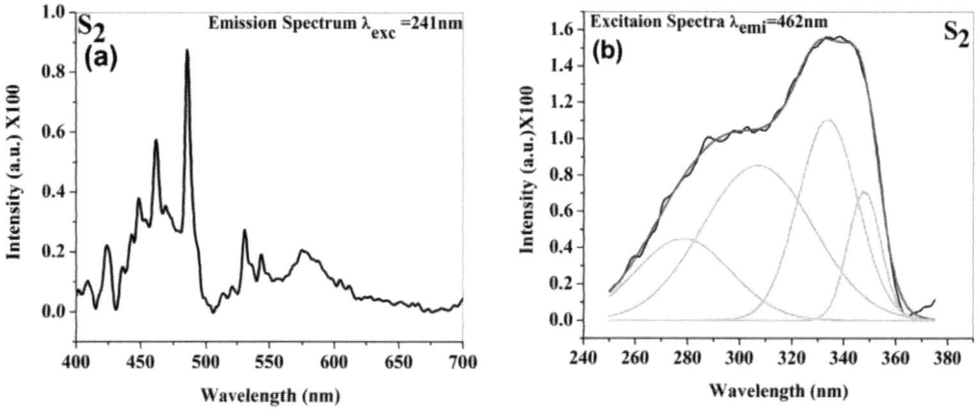

Figure 4. Typical photoluminescence (a) emission spectra and (b) excitation spectra for sample having Sn/Zn = 1.75

There is several emission bands found in the photoluminescence spectrum. But excitation spectra have been recorded by placing the monochromator at 448, 462, 485, and 531 emission bands respectively.

Table 2. Details of PL (excitation and emission) spectra obtained for S_1 (Sn/Zn = 0.87), S_2 (Sn/Zn = 1.75) and S_3 (Sn/Zn = 3.75)

Sample	Emission Spectra		Excitation Spectra	
	λ_{exc} (nm)	Peak position (nm)	λ_{emi} (nm)	Peak position (nm)
S_1	241	405.64, 423.85, 441.42, 447.36, 460.45, 485.16, 531.12, 544.18, 574.31	448	238.76, 290.03, 325.32, 373.25
	287	422.41, 435.94, 448.42, 461.95, 484.86, 530.66, 543.16	462	274.31, 353.81, 370.66
	351	424.12, 462.21, 486.23, 530.18, 543.47	485	284.64, 344.46, 390.22
			531	206.64, 297.12, 320.14, 364.61,395.95
S_2	241	408.46, 424.21, 448.74, 461.12, 485.56, 530.16, 543.54, 573.68	448	238.74, 289.80, 317.33, 332.71
	287	422.41, 448.43, 461.95, 485.12, 530.83, 543.32	462	278.36, 306.92, 333.75, 347.88
	351	423.28, 462.19, 485.42, 509.34, 530.83, 543.64	485	284.00, 342.06, 368.47
			531	283.01,341.14,367.08
S_3	241	408.76, 424.38, 448.91, 461.18, 484.61, 530.34, 543.71, 5556.14	448	238.87, 286.34, 317.06, 333.10
	287	423.04, 448.02, 461.56, 474.05, 485.51, 531.31, 543.82	462	283.68, 303.24, 335.55, 347.60
	351	424.85, 435.31, 442.58, 447.78, 462.35, 485.26, 531.06, 543,.55	485	283.14, 345.54, 368.71
			531	283.61, 311.09, 367.67

The presences of several common bands were then taken for excitation purpose. In this way, the excitation wavelengths 241 nm, 287 nm, and 351 nm were chosen. The PL emission spectral bands have been given in the table after curve fitting of the complex spectra [12].

Scanning electron microscopic observations of the surfaces of nanopowders have been shown in Figure 6. The powders were found to be nanocrystalline in nature. The nanocrystals aggregate to form bigger agglomerates and these are homogenously distributed. One can find different shapes such as spherical, hexagonal and pyramidal etc which aggregate to give larger grains. This has been well defined in sample S_1 with Sn/Zn = 0.87.

Figure 5. FESEM micrographs of S_1 (Sn/Zn = 0.87), S_2 (Sn/Zn = 1.75) and S_3 (Sn/Zn = 3.75)

When this ratio is increased to 3.75 well defined cubes are seen [10]. The size of the cubes varies from 2µm to 1µm. Also seen are rhombohedral shaped crystallites. On the rhombohedral shaped surface smaller crystallites are seen to be adherent [7, 12-14].

The transmission electron microscope photographs given in Figure 7 for sample S_2 and S_3 show the segregation of nanocrystallites leading to well defined grain boundaries and lattice fringes.

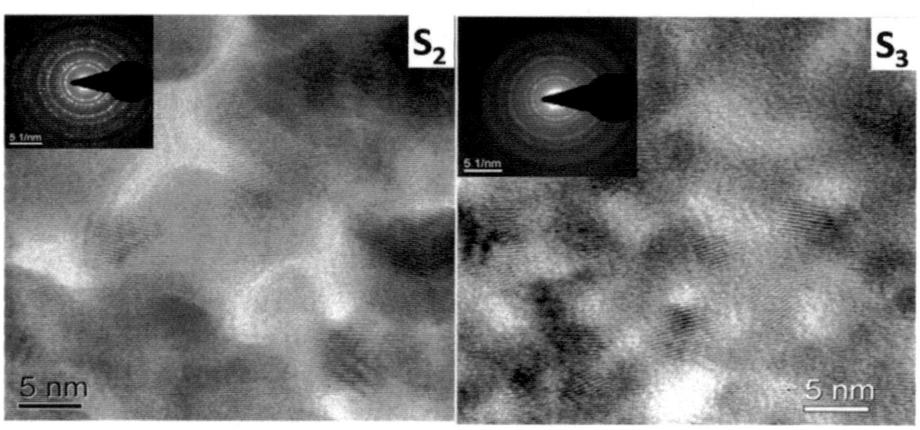

Figure 6. The transmission electron microscope photographs sample S_2 (Sn/Zn = 1.75) and S_3 (Sn/Zn = 3.75). The insets show selected area diffraction pattern

The selected area diffraction pattern given in the inset for a particular grain shows spotty nature while these nonocrystallites give well defined diffraction pattern. From this and the lattice fringes, the lattice constant calculated for c-axis oriented grains to be 0.299 nm. When well defined nanocubes or rhombohedral shape crystallites were seen in SEM, their electron microscopic photographs and SAED patterns show the characteristic feature of crystallinity in the nano range.

4. Conclusion

From these above observations, one can conclude that ,

a) composition plays an important role in nanostructure formation in SnO2-ZnO composite powders,
b) with a pH of 10.3 and Sn/Zn = 3.75, the powder shows cassiterite structure both through X-ray and electron diffraction,
c) the photoluminescence studies also support these results

More comprehensive studies with pH variation and composition change will be reported in detail in future publication.

References

[1] Wang J, Fan H, Yu H 2016 *Optik* **127** 580
[2] Wang Z L 2004 *J. Phys.: Condens. Matter* **16** R829
[3] Peng H, Ndione P F, Ginley D S, Zakutayev A, Lany S 2015 *Phys. Rev.* X **5** 021016
[4] Fouad O A, Glaspell G, El-Shall M S 2008 *Top Catal* **47** 84
[5] Fouad O A, Glaspell G, El-Shall M S *Nano* 2010 **05** 185
[6] Guan Y, Wang D, Zhou X, Sun P, Wang H, Ma J, Lu G 2014 *Sens. Actuators B* **191** 45
[7] Wang Q, Zhang L S, Wu J F, Wang W D, Song W G, Wang W 2010 *J. Phys. Chem. A* **114**
[8] Panda N R, Sahu D, Mohanty S, Acharya B S 2013 *J. Nanosci. Nanotechnol* **13** 1
[9] Klug H P, Alexander L E 1974 X-Ray Diffraction Procedures: For Polycrystalline and Amorphous Materials (New York : Wiley)
[10] Fu X, Wang X, Ding Z, Leung D Y C, Zhang Z, Long J, Zhang W, Li Z, Fu X 2009 *Appl. Catal., B* **91** 67
[11] Joseph L A, Jeronsia J E, Jaculine M M, Das S J 2016 Phys.res.int. **1801795** 1
[12] Md Sin N D, Mamat M H, Malek M F, Rusop M 2014 *Appl Nanosci* **4** 829
[13] Hemmati S, Firooz AA, Khodadadi AA, Mortazavi Y 2011*Sens Actuators B Chem* **160(1)** 1298
[14] Chang S P, Chang S J, Lu C Y, Li M J, Hsu C L, Chiou Y Z, Hsueh T J, Chen I 2010 *Superlattices Microstruct* **47(6)** 772

IOP Conf. Series: Materials Science and Engineering **178** (2017) 012002 doi:10.1088/1757-899X/178/1/012002

Designing and synthesis of a polymer matrix piezoelectric composite for energy harvesting

Asutya Kumar Biswal[1], Satyabati Das[2], Amritendu Roy[3]
School of Mineral Metallurgical and Materials Engineering (SMMME)
Indian Institute of Technology, Bhubaneswar, Odisha, India, 752050

Email: akb11@iitbbs.ac.in

Abstract. Now a day, a large variety of electronic and network devices require small yet steady power supply for operation. Traditionally, these devices are battery operated and the batteries are periodically charged for continuous operation. Often, the devices are so located that supply of power to recharge the batteries becomes challenging. Electrical energy harvesting by means of principle of piezoelectricity could be a viable solution to the above problem by means of providing a permanent power source. In this regard, piezoelectric lead zirconium titanate (PZT) was found to be a potential material. However, poor mechanical properties (brittleness) of bulk ceramic materials have been a concern for energy harvesting by means of mechanical motion (footsteps). In the present work, $Pb(Zr_{0.52}Ti_{0.48})_{1-x}Nb_xO_3$ at x=0.05 was prepared by conventional solid state synthesis route. XRD and SEM analyses were performed for structural characterization. PZT powders were found to be in single phase with tetragonal symmetry without any trace of a second phase. To render the required mechanical properties (flexibility), in the present work, we designed a polymer matrix ceramic composite without much compromising the piezoelectric properties. We prepared composite thick films of lead zirconium titanate (PZT) ceramic in poly vinylidene fluoride (PVDF) polymer matrix with varied composition of PZT from 10-50 vol %. The study of surface morphology by scanning electron microscope (SEM) shows good degree of dispersion of PZT in PVDF matrix. Ferroelectric characteristics of the composite films were studied by measuring the polarization-electric field hysteresis loops. Generated output voltage and current from the composite films are found to be approximately 0.35 volt and 4 nA, respectively.

1. Introduction

Over the past decades, piezoelectric materials have been considered due to their favorable physical properties such as high dielectric constant, low dielectric loss and high energy storage density which have been considered for possible energy harvesting applications such as transducers, actuators, capacitors, energy storage devices.[1] Nevertheless, the ceramic piezoelectric materials are brittle, often with low dielectric breakdown strength and thus poses difficulties in practical applications though they are associated with high dielectric constant which can provide good energy conversion efficiency. .[2] On the other hand, polymer piezoelectrics owns high breakdown strength, good flexibility and mechanical strength as well as they are easy to process at low temperature, albeit having very low dielectric constant.[3]

Thus, to get optimum properties, a blend of properties of both piezoelectric ceramic and the polymer is needed to make a composite where polymer act as the matrix and ceramic particle as the reinforcement. Currently, polymer matrix ceramic composites are being used in a number of applications such as embedded capacitors, sensors and gate dielectrics.[1, 2]

A Piezoelectric lead zirconium titanate (PZT) was found to be a potential material owing to its impressive piezo as well as ferroelectric properties.[4] Recently, piezoelectric polymer, polyvinylidene fluoride (PVDF) has also been paid increased attention due to its high breakdown strength, easy processing, good thermal stability and most importantly nontoxic in nature.[3] PVDF is a semi-crystalline polymer having different phases (α,β,γ,δ) depending on the process parameter such as solvent, melt temperature, annealing temperature, method of casting, starching of the film and pressure on film .[5] Among the four phases β phase is known for its ferroelectric nature.[6-8] Table 1 depicts the important physical properties of PVDF.

Table 1 The physical Properties of PVDF [reference]

PVDF (Sigma Aldrich)	Density (gm/ml)	Melting point (°C)	Crystallizing (°C)	Glass transition (°C)	Thermal stability (°C)
CAS Number 24937-79-9	1.74	171	138	-38	>400

2. Experimental Details

2.1 Synthesis of PZT Powder

The starting materials for making lead zirconium titanate (PZT) with composition $Pb(Zr0.52Ti0.48)1-xNbxO3$ where x=0.05 are PbO (sigma Aldrich, >99.0% purity, granular form), ZrO_2 (sigma Aldrich,99% purity,5μm powered), TiO_2 (Himedia,99% purity, powered) and Nb_2O_5 (sigma Aldrich, >99.9% purity, -325 mesh powdered). This composition is near to the morphotropic phase boundary (MPB) to achieve better electromechanical properties due to coexistence of ferroelectric tetragonal $PbTiO_3$ (T_C=490°C) and ferroelectric rhombohedral $PbZrO_3$ (T_C=230°C) at room temperature [9, 10] precursors with stoichiometric ratio were taken to achieve the aimed composition with 2% extra PbO to compensate for vaporized lead during calcinations and sintering at high temperature due to its high partial vapor pressure.[10] Precursors were mixed in a ball mill with isopropyl alcohol with zircon balls (ZrO_2) as grinding media for 24 Hrs. to achieve homogenization throughout the bulk sample and after that dried in an oven until all alcohol is removed from the prepared sample at a temperature around 80°C.[9] Calcination was done at 900°C for 2 Hrs. The conditions of temperature rate and soaking time are 5°c/min and 2 Hrs. respectively. Calcined powder was subsequently ground for 48 Hrs. in isopropyl alcohol with zirconia ball (ZrO_2) as grinding media to break down the agglomerated particles. The dried powdered were mixed with polyvinyl alcohol as binder solution for making green pellet of 10 mm diameter in a hydraulic pressure about 12 kg/cm^2. Afterwards sintering was done in a closed alumina crucible with PZT powered bed to compensate PbO vapor pressure at high temperature around 1250°C for 4 Hrs.[11-13]

2.2 Synthesis of PZT-PVDF Composite

For making thin film composite, the materials are Poly vinylidene fluoride (PVDF) (Sigma Aldrich) and sintered lead zirconium titanate (PZT). Initially PZT powder of 600 nm size was stirred around 30 minutes at 80°C in water and ethanol in order to achieve OH order on the surface of PZT powder.[2] Surfactant, Span 80 (CDH, chemical name Sorbitane monooleate) is added to the OH order PZT powder and stirred for 10Hrs at room temperature. After that, solvent acetone was removed from it PVDF and DMF (dimethyl formamide) solution was added. PVDF was mixed with dimethyl formamide (DMF) and stirred until a transparent solution is achieved. When the transparent solution is achieved that shows all PVDF is well mixed with the solvent, then add the pre-prepared PZT powered in it. Mixed and stirred at around 70°C for 30 minutes. After that it is probe-sonicated for 1 minute. The sonicated solution contains air bubble, which was removed in an evacuated desiccator for 10 minutes. Next, the solution was poured it into a petri dish and heated to around 110°C for 30 min to remove all residual solvent present. The process is called solvent casting method.[2]. Different volume fraction from 10 to 50% PZT were made successfully by the above described process.

2.3 Characterization

Room temperature XRD was performed on both bulk ceramic and PZT-PVDF composites of different composition by a Bruker D8 Advanced A25 X-Ray diffractometer using Cu Kα radiation of wavelength λ = 1.5418Å. This was done over a wide range of 2θ ($20^o \leq 2\theta \leq 80^o$). The morphology study were examined with an environmental scanning electron microscope (EVSEM). Silver electode was applied on both side of the sample to study the V-I characterstics and ferroelectric behaviour by using Kiethly 6517B high resistance electrometer and Radiant ferroelectric tester.

3. Results and discussion

3.1 XRD and SEM Study of Both PZT and PZT-PVDF Composite

Powder XRD patterns of PZT powder is depicted in Fig. 1 which was matched with JCPDS card no 00-050-0346. It was observed that pure tetragonal phase of PZT without any trace of any secondary phase was formed upon sintering.

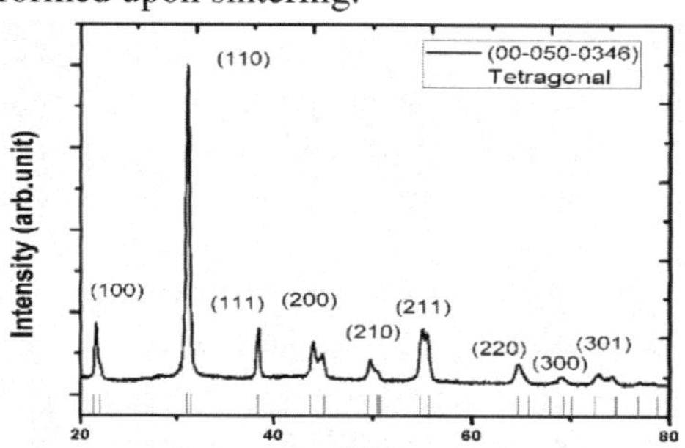

Fig. 1 XRD patterns of PZT powder

Powder diffraction patterns of PZT-PVDF composite films having different PZT contents of 10-50 vol. % were studied as shown in Fig. 2. Phase evolution of PVDF depends on processing conditions. The first peak at 2θ corresponding to 18.5 was allotted to α-phase, at 2θ = 20 corresponding to β-phase and broad peak corresponding to γ-phase.[2] With the increase in the PZT content into the PVDF matrix, the diffraction peaks of PVDF gradually reduced in intensity and PZT peaks became sharper. No other peaks are visible other than the matrix and reinforcement suggesting lack of significant phase evolution due to PZT-PVDF interaction. After 40 volume fraction of PZT, PVDF peaks are complitely vanished because of the strong and intense diffraction peaks of the PZT ceramic particle overshadowing the PVDF peaks.

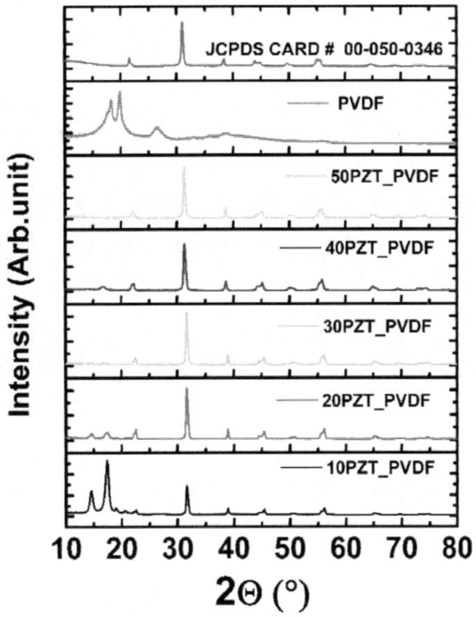

Fig. 2 XRD pattern of PZT-PVDF composite thin film

Fig. 3(a) shows the SEM images of pure PVDFwhile Fig. 3(b) and 3(c) show SEM images PZT-PVDF composites with two different volume fractions of PZT. Fig. 3 (d) shows a photograph of as processed PZT-PVDF composite. Fig. 3(b) and (c) clearly indicate that PZT powders are well compatible in the matrix phase. When volume fraction increases particles are being denser in the matrix. Thus, from the above morphology study it could be concluded thatPZT particles are uniformly dispersed throughout the matrix phase. Fig. 3 (d) depicts a 50% PZT composite film with good flexibility.

3.2 Electrical measurements

Fig. 4 depicts the polarization-electric field (P-E) hystersis loops for the 30% and 050% PZT-PVDF composites. While a saturated hysteresis loop of the composite could not be found in our study, the shape of the loops suggests ferroelectric nature whereinapplied electric field induces rearanged domains towards the direction of electric field. The non-linear behaviour clearly indiates that different rates at which domain walls and electric field changes which causes a phase difference between electric field and polarisation.[2] Table 2 shows a summary of P-E study for different composition of PZT-PVDF composite.

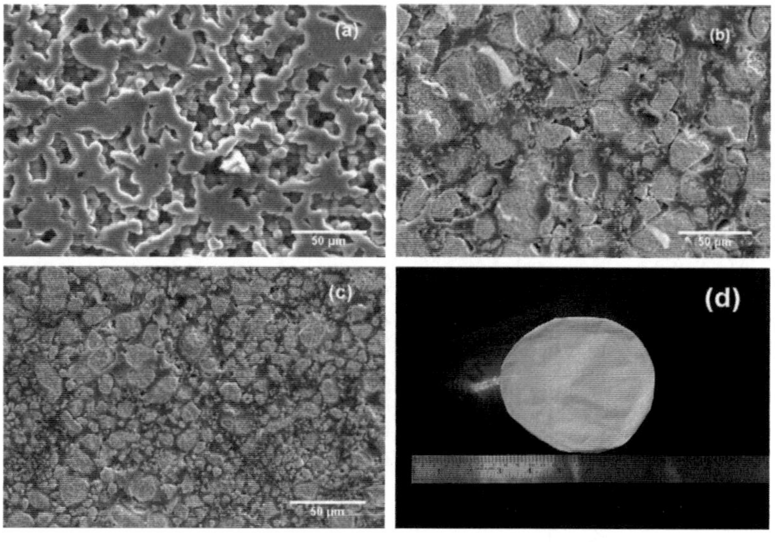

Fig. 3 (a) SEM images of pure PVDF, (b) 30% PZT composite, (c) 50% PZT composite and (d) composite film.

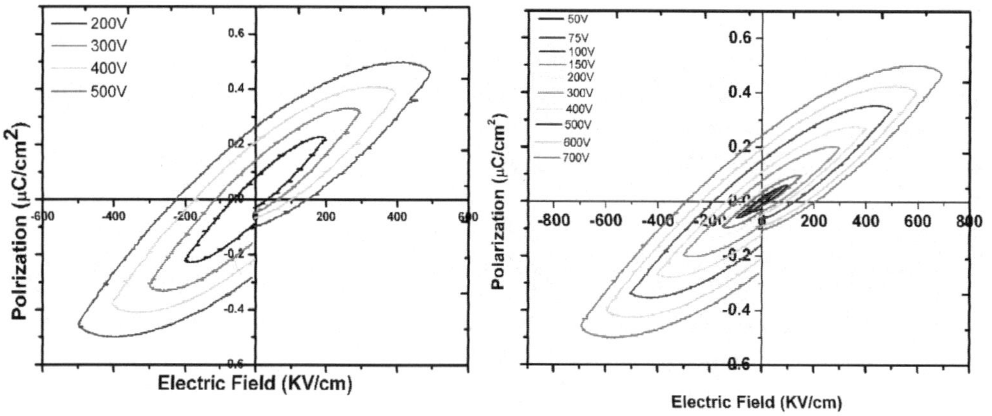

Fig. 4 Polarization Electric field hysteresis loop (PE Loop) of 30 vol. % and 50 vol. %.

Table 2 Remnant polarization and Coercive Field of different volume fraction composite.

Vol. Fraction	Remnant Polarization (Pr) [μC/cm2]	Coercive Field (Ec) [KV/cm]
0.1	0.01	0.3
0.2	0.02	0.4
0.3	0.26	0.5
0.4	0.03	0.4
0.5	0.24	0.7

V-I characteristics depicts that maximum voltage and current output generated by the 0.5 PZT volume fraction. Maximum voltage was d 0.35 V and current of ~4 nA. Fig. 5 shows the data for a 50 vol. % composites, wherein output current profile is shown on left panel and right panel depicts bout the output voltage.

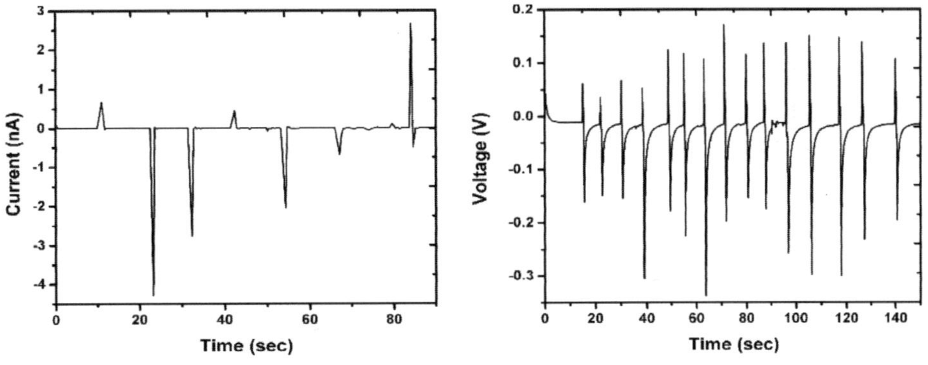

Fig. 5 depicts the V-I characteristics of 0.5 PZT volume fraction composite.

4. Conclusion

Solid state synthesis process was used successfully to produce PZT powder. Solvent casting method produced well dispersed PZT-PVDF thick film composite. XRD and SEM analysis showed pure single phase of PZT was synthesized in our study which was uniformly dispersed in PVDF matrix. P-E loops indicate the signature of ferroelectric character and remnant polarization of about 0.24 $\mu C/cm^2$ for 50% PZT- PVDF film. Maximum output voltage of 0.35V and current of 4nA were obtained.

5. References

[1] Arbatti M, Shan X, and Cheng Z. Y 2007 Ceramic–polymer composites with high dielectric constant *Advanced Materials* **19** pp. 1369-1372.

[2] Luo B, Wang X, Wang Y, and Li L 2014 Fabrication, characterization, properties and theoretical analysis of ceramic/PVDF composite flexible films with high dielectric constant and low dielectric loss *Journal of Materials Chemistry A* **2** pp. 510-519.

[3] Chu B, Zhou X, Ren K, Neese B, Lin M, Wang Q, Bauer F, and Zhang Q 2006 A dielectric polymer with high electric energy density and fast discharge speed *Science* **313** pp. 334-336.

[4] Babu I and de G 2014 With Highly flexible piezoelectric 0–3 PZT–PDMS composites with high filler content *Composites Science and Technology* **91** pp. 91-97.

[5] Thomas P, Satapathy S, Dwarakanath K, and Varma K 2013 Dielectric properties of poly (vinylidene fluoride)/CaCu3Ti4O12 nanocrystal composite thick films *arXiv preprint arXiv:1301.4056*.

[6] Hasegawa R, Takahashi Y, Chatani Y, and Tadokoro H 1972 Crystal structures of three crystalline forms of poly (vinylidene fluoride) *Polymer Journal* **3** pp. 600-610.

[7] Martins P, Lopes A. C, and Lanceros-Mendez S 2014 Electroactive phases of poly(vinylidene fluoride): Determination, processing and applications *Progress in Polymer Science* **39** pp. 683-706.

[8] Jain A, Siva Subba Rao P, and Jayanth Kumar S 2013 PVDF-PZT Composite Films for Strain Sensing Applications.

[9] Necira Z, Boutarfaia A, Abba M, and Abdessalem N 2012 Synthesis of PZT powder by conventional method at various conditions in *EPJ Web of Conferences* p. 00038.

[10] Amer A. M, Ibrahim S. A, Ramadan R. M, and Ahmed M. S 2005 Reactive Calcination Derived PZT Ceramics *Journal of Electroceramics* **14** pp. 273-281.

[11] Hammer M and Hoffmann M. J 1998 Sintering Model for Mixed-Oxide-Derived Lead Zirconate Titanate Ceramics *Journal of the American Ceramic Society* **81** pp. 3277-3284.

[12] Kingon A. I and Clark J. B 1983 Sintering of PZT Ceramics: I, Atmosphere Control," *Journal of the American Ceramic Society* **66** pp. 253-256.

[13] Kingon A. I and Clark J. B 1983 Sintering of PZT Ceramics: II, Effect of PbO Content on Densification Kinetics *Journal of the American Ceramic Society* **66** pp. 256-260.

IOP Conf. Series: Materials Science and Engineering **178** (2017) 012003 doi:10.1088/1757-899X/178/1/012003

Characterization of iron ore pellets with dextrin added organic binders under different time and temperature conditions

Anand babu kotta*[1], Swapan kumar karak[1] and M. Kumar[1]

[1]Department of Metallurgical and Materials Engineering, National Institute of Technology, Rourkela, Orissa, India

Corresponding author E-mail: anand7850@gmail.com

Abstract: In the present work, Dextrin is used as a binder for iron ore pelletization, as it's free from Silica and Alumina. Green pellets were prepared by mixing of iron ore particles (-75 μm) with varying binder percentage (1 wt. % bentonite, 0.5 and 1 wt. % of dextrin) separately. The green pellets were first dried in air for 24 h and then in an electric oven at 383 K for 4 h. The dried pellets were fired at varying temperatures (1173, 1273, 1373, and 1473 K) and indurated for 1 and 2 h. The samples were characterized for physical (pellet size) and mechanical (compressive strengths, porosity and hardness) properties. Good quality pellets were prepared with organic binder, and which increases the compressive strength of dry and fired pellets. Strength results are matching with the bentonite binder pellets and it is well above the industrially acceptable limit (250 kg/pellet). Porosity of the fired pellets decreases with increasing temperature and induration time. The hardness of the pellet varies from surface to core of the pellet. Moreover, the influence of wear mechanism (based on collisions) on pellets and its characteristics of generated particle size distribution (PSD) have been investigated. It was noticed that the material loss during wear test decreases with increasing strength of the pellet. From PSD analysis, the coarse particles were revealed that the collisions are dominating during wear.

1. Introduction

Iron-ore is a common element in the earth crust (5-6%) and it is most used metal, accounting for 93% of global metallic and alloy products [1]. During mining process, huge amount of fines is generated which cannot be used directly in the furnaces, these fines are need to be recycled and re-utilized. The extraction of Fe content from low grade iron ores are not suitable for iron making process due to its production cost and they are crushed in to fines, these fines can be enhanced through mineral-agglomeration processes [2]. C.V Firth 1944. Published the first systematic research works on the pelletization of magnetite and hematite iron ores. He showed that they have roughly similar pelletization characteristics [3]. In Popular pelletization process moisture, flux and binders are added to the iron ore powder and make pellets subsequently thermally treated to attain the required strength (ISO). The cold strength of the pellet was investigated using a compressive strength and drop number [4]. The use of bentonite is favorable concerning physical, mechanical and metallurgical properties. However, because of its acid constituents (SiO_2 and Al_2O_3), it is considered as a chemical impurity, especially when concentrate with high SiO_2 content. These have adverse effects on the process. De Souza, Kater and Steeghs (1984) reported that with the addition of 1 % bentonite, containing 85 % SiO_2+Al_2O_3 decreases the iron content (0.6-0.7 wt %) in the pellets [5]. The increase in silica content can lead to increases in the steel production cost as reported by Chizhikova et al., 2003, Schmitt, 2005. Different researchers have tried to find alternative to bentonite to overcome problems as mentioned above. In recent times they have focused on organic binders due to its outstanding binding properties and free from acid constituents. Mostly, Carboxymethyl cellulose (CMC), Peridur C-10 and

Peridur CX3, organic polymers, lime and Pridor, domestic CMC, Na lignosulphonate (NLS), starch and glue.

A.I. Arol et al. (1989) comparatively studied bentonite and certain organic binders namely Carboxymethyl cellulose (CMC), Peridur C-10 and Peridur CX3 with magnetite concentrates from Divrii, Turkey. They observed that the organic binders have achieved the wet and dry (673 K) pellet properties. However, the fired compressive strength of pellets produced with organic binders at 1473 K for 120 minutes was found insufficient as these organic binders do not contribute to the compressive strength of fired pellets [6]. Osman Sivrikaya et al. (2012) have focused on the effect of organic binders, (calcined colemanite ($Ca_2B_6O_{11}$. $5(H_2O)$), DPEP06-0007 Polymer and Superfloc A150-LMW) individually and along with combinations on wet, dry, preheated and fired (thermally-treated) pellet strengths. It was concluded that the organic binders alone have satisfy the wet and dry pellet properties, but preheated and fired pellet have resulted in insufficient strengths. Combination of binders (Technical CMC 0.10%+Calcined colemanite 0.66%, DPEP06-0007 Polymer 0.10% + Calcined colemanite 0.66% and Superfloc A150-LMW 0.10% + Calcined colemanite 0.66%) have both sufficient wet and dry pellet qualities and improved preheated and fired pellet strengths [7]. Guanzhou Qiua et al. (2004) have studied the effect of organic polymer (Funa) as binders for agglomeration. It behaves like surfactants and colloids. Organic polymer sharply diminishes the contact angle of the iron ore from 46° to 0° and expands on the surface of the concentrate, this improves the hydrophilic ability due to this, they becomes more negative with addition of Funa indicates that the iron ore particles will not adhere together through electrostatic attraction when Funa is used as binder[8] . Ammasi et.al. (2015) used Na lignosulphonate (NLS) and copper smelting slag (Cu-SS) as a binder with varying compositions. The Pellets prepared with NLS as a binder, it shows a good wet and dry compressive strength and increasing with increasing NLS percentage. Even at higher temperatures the compressive strength has become constant even with an increasing NLS percentage. When The Pellets made with 1.0 wt. of Cu-SS along with 0.5 wt. % NLS, it was observed that the compressive strength was increased with increasing temperature and required compressive strength was attained at 1523 K [9]. J. Kamalabadi et al. (2013) have studied the effect of lime and Pridor binder and compared with the bentonite binder in dried conditions. They were observed that the dry strength of the samples prepared through bentonite binder was higher than that of the samples prepared by the lime binder, strength of the pellet has increased with small amount of addition of peridor binder [10].

Dextrin is an organic binder that has been successfully used in various applications such as a thickening and binding agent in pharmaceutical products and paper coatings. In the present study dextrin is used to study the effect on agglomeration of iron ore at various firing temperatures and induration times. This present study was focused on the effect of firing temperature and induration time on dust generation during mechanical wear of iron ore pellets. The influence of the dextrin binder and its composition on the physical, mechanical and wear rate of the pellets has been investigated and compared with the standard binder bentonite. Furthermore, the particle size distributions in the dust generated during the collision of pellets were investigated. Establish a relationship among the critical diameters of dust particles, and flow behavior of off-gases from the blast furnace.

2. Materials and methods

2.1. Raw Materials

2.1.1. Iron ore concentrate. In the present study, Hematite Iron ore concentrate was collected from Orissa Mining Corporation (OMC). The as-received ore was dried at 378 K overnight. The ore was divided into representative samples of 1.5 kg by sampling methods and sealed in plastic bags, and its nominal chemical composition has been outlined in Table 1. From the chemical composition, Hematite concentrate has a relatively high amount of Fe_2O_3 and aluminum, and silicon oxides are major gauge constituents. Particle size measurements of the iron ore al fines were performed by using

IOP Conf. Series: Materials Science and Engineering **178** (2017) 012003 doi:10.1088/1757-899X/178/1/012003

the Vibratory Laboratory Sieve Shaker (Fritsch, Germany) ''Analysette3''. The required particle sizes that is -75 μm and it was separated from raw materials.

Table 1. Nominal Chemical composition and loss on ignition of raw materials

Iron Ore Source	Chemical Composition (weight percent on dry basis)						
	Fe (Total)	Fe_2O_3	Al_2O_3	SiO_2	TiO_2	MnO	Loss on Ignition
Orissa Mining Corporation (OMC)	62.63	89.56	4.09	3.80	0.11	0.01	2.43
Bentonite binder	-	3.96	58.36	21.14	-	-	6.75

2.1.2. Binder. Two types of binders used for this study: bentonite and dextrin. Tests conducted with (0.5 and 1 w.t % of dextrin and I w.t % bentonite) and without binder added to the iron ore and pellets were prepared.

2.1.3. Preparation of Iron Ore Pellets. The green pellets were prepared by mixing of dry concentrate of iron ore particles, required amount of moisture (Water ≈10%) and pre-defined binder percentage, for 5-10 min. Then the mixture is added to the disc pelletizer, and it operated at a speed of 25 rpm to make pellets. The pellets were pulling out from the pelletizer and screened to required size ratio which is between 9-16 mm diameters. The pellets were exposed to the atmosphere for 24 h and then dried in an electric oven at 383 K for four hours. The dried pellets fired in the muffle furnace from room temperature to predefined temperatures ranging from 1173 to 1473 K and indurated at varying time periods (1 and 2 h) to attain the workable strength (ISO 4700 standards), pellets to allow for furnace cooling. Then pellets were allowed to characterization.

2.2. Pellet quality testing

Pellet quality were determined by testing the pellets at different stages of the process. Tests were conducted for the pellets: 1) drop test for dry pellets, 2) compressive strength and porosity for dry and heated pellets, and 3) Hardness for heated pellets 4) wear rate 5) Particle size distribution analysis.

2.2.1. Drop number of dry pellets. 10 dried pellets were utilized to focus the dry pellet drop number by dropping a solitary fired pellet over and over from a stature of 46 cm (18 inch) onto a steel plate. The drop test for a solitary fired pellet was preceded until a split happened on the pellet and the last esteem recorded. A pellet should withstand at least 4 drops [11].

2.2.2. Compressive Strength. Compressive strengths of individual pellets depend on the chemical composition and physical properties of the concentrate, additives used in the pelletizing method, particle size, firing technique, and temperature. A dried pellet is crushed and the maximum load is recorded. It measures the ability of dried pellets to survive handling during the firing process. It should be a mean value at least 2.24 kg/pellet. A fired pellet is crushed and maximum load is recorded. It must satisfy the ISO 4700 standards [12].

2.2.3 Porosity. Total porosity of balls is expressed as a portion of pore volume out of the total volume of the input raw material, the apparent porosity values of iron pellets were determined in accordance Archimedes principle. In the porosity measurements kerosene was used as a medium. It is calculated accordance with the following formula.

$$\text{Apparent porosity} = \frac{W - D}{W - (S - s)} \quad \text{........ (1)}$$

Where, '*D*' is the weight of dried piece;
'*W*' is the weight of oil saturated piece;
'*S*' is the weight of the piece + thread while immerse in oil, and
'*s*' is the weight of thread only while immerse in oil

Lower porosity leads to a better particle–particle contact but obstruct the diffusion of oxygen throughout the pellet during reduction process. The porosity of typically fired pellets is in the range of 18 - 32% [13].

2.2.4. Hardness. The hardness of the fired iron ore pellets was determined by using a Vickers hardness tester to measure the microhardness (HV) according to the ISO6507-1 standard. The hardness measurement performed on a cross section of a polished surface by moving from the surface to focal point of the pellets [14].

2.2.5. Mechanical Wear of Pellets. In order to evaluate the mechanical wear and fines generation rate from fired pellets, the method of Muhammad Nabeel (2016) was directly followed. Predefined pellets (60 g) were placed in a metallic jar and rotated with a 300 rpm up to 30 minutes. Effect of wear rate was calculated in two methods as shown in Fig .1. In method 1 the generated fines were kept in the jar and weight of the pellets was measured at a regular time interval of 5 minutes, so that the mass of the jar was constant throughout the experiment. In method 2, take out formed fines from the pellets at regular intervals. The fines generation was calculated from the mass loss of the pellets after rotation and wear rate was calculated equation (2) [15] accordingly.

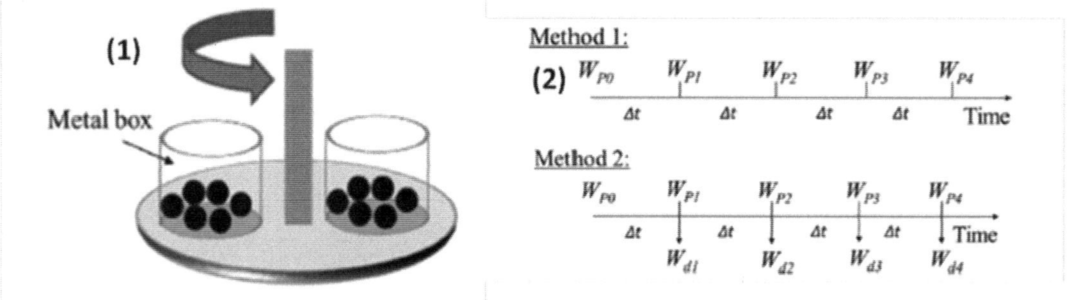

Figure 1. (1) Experimental equipment (2) Main operations during different experimental trials

$$\text{Wear rate (\%)} = \frac{W_t - W_{t+\Delta t}}{W_t \Delta t} X100 \quad \text{.......... (2)}$$

W_t = weight of the pellet before rotation

$W_{t+\Delta t}$ = weight of the pellet after rotation

2.2.6. Particle Size Distribution. Particles generated from the wear tests have collected from the jar and the Particle Size Distribution of the fines was measured by using scanning electron microscope (SEM) at different magnifications.

3. Results and Discussion

A set of experiment was performed on pellets to investigate the quality of pellets which are made-up of organic binders, concerning drop number, porosity and compressive strength of dry pellet and, porosity and compressive strength of fired pellet.

3.1. Dry pellet drop-number (DPDN)

The DPDN of the iron ore pellets was obtained by using different types of binders, and the results were compared with pellets made using bentonite as binder and pellets made without any binder. Reference bentonite binder was found to be 4.82 as shown in Figure 2, which are slightly over the limit so they are sufficient. However, DPDN values for pellets made without binder was 2.8 and with dextrin (0.5 and 1 Wt. %) addition were determined to be 3.6 and 4.7, without binder and 0.5 Wt. % of dextrin is a little lower than the required value.

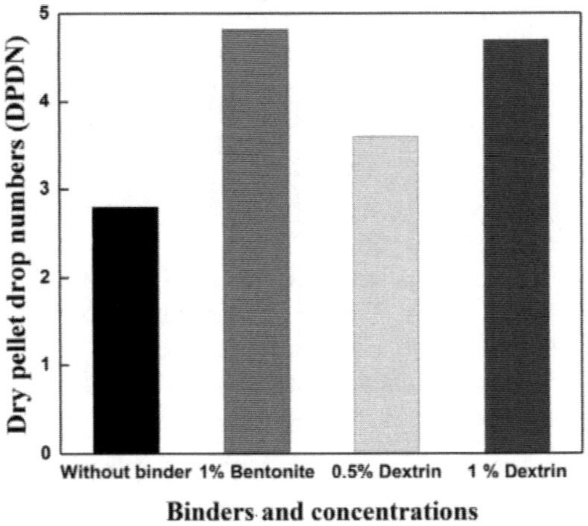

Figure 2. Effect of binders on Dry Pellet Drop number

3.2. Dry pellet compressive strengths (DPCS)

The DPCS of the iron ore pellets was determined by using different types of binders. Bentonite samples were considered as the reference point for strength calculation. The DPCS of the iron ore pellets obtained by bentonite binder was 4.6 kg/pellet which are over the limit, so they are sufficient. However, DPCS values for pellets made without binder and with dextrin addition were determined to be 1.34, 3.9 and 4.3kg/pellet respectively as shown in Figure 3. The pellets made without binder is a lower than the required value. Therefore, they can be considered insufficient, and the production of pellets without binder is not suitable. The DPCS produced with organic binders were found to be much greater than required value 2.24 kg/pellet dextrin was suitable for the pelletization process. On the other hand, the DPCSs of pellets by organic binders were found to be lower than those obtained by bentonite, but they are still greater than the minimum required limit.

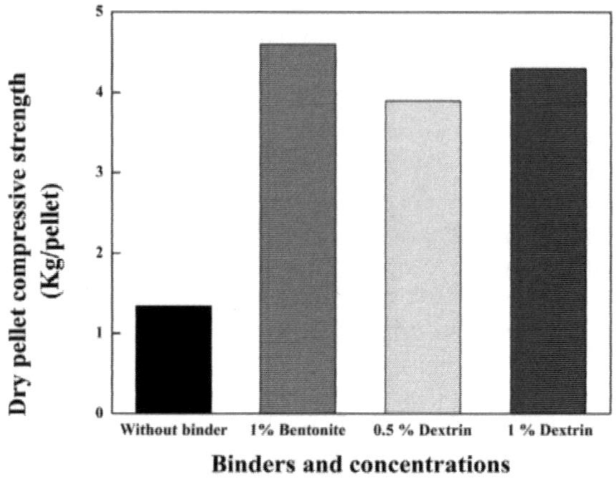

Figure 3. Effect of binders on Dry pellet compressive strengths

3.3. Porosity of Dried pellets

Porosities of pellets produced with different binders were determined after drying. The porosity of the iron ore pellets of bentonite binder has shown 32.8 % which are lower when compared to ISO standards. However, porosity values for pellets made with no binder and with dextrin addition were determined to be 35.6, 33.5 and 33.2 % in all the conditions they have satisfied the required value as shown in figure 4. From the results it is confirm that these pellets can be used for pelletization Therefore, they can be considered sufficient to the production of pellets without binder. The porosity of the dried pellets produced with organic binder was found close to the bentonite samples.

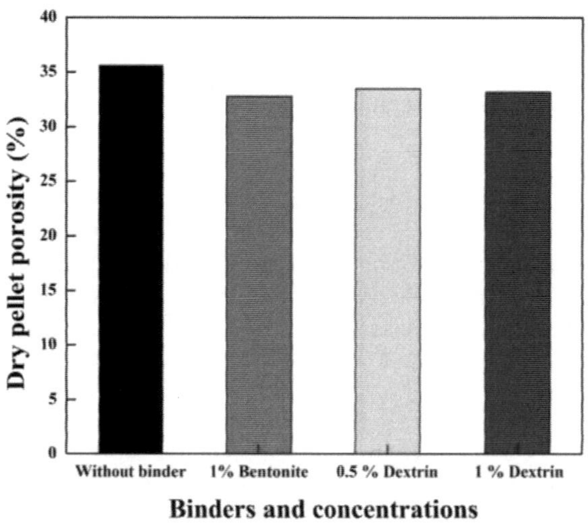

Figure 4. Effect of binders on Porosity

3.4. Characterization of fired pellets

In thermal heat treatment process, pre-heating strength is the most important factor because the pellets are moving from traveling grate to the rotary kiln. If the pre-heating strength is insufficient, fragments and dust will form from the pellets due to this both pellet quality and operation efficiency will be affected adversely. The properties of the pellets are depending upon the thermochemical conditions like firing temperature and induration time. Due to the thermo-chemical reaction, recrystallization and slag formation takes place. Slag phase formed around the pellet and it moves towards the center of the pellet. The slag phase is densely sintered and it contributes to the pellet strengthening

3.5. Effect of binder, firing temperature and induration time on compressive strength

The compressive test of pellets having different binders was determined, which are produced with without binder, 1.0 % bentonite, 0.5 and 1 % of dextrin addition. The compressive strength of the pellets has increased with increasing firing temperature and induration time. It was observed that the pellets were heated at 1173 K for 1 hr. is 31.32, 98.32, 86.3 and 94.3 kg/ pellet respectively. Further, it was increased to 35.6, 102.36, 98.36 and 110.5 kg/pellet with increasing the induration time to 2 hr. A similar trend was observed at 1273, 1373 and 1473 K as shown in Figure 5, it is due to increasing induration time, the binder melts and fuse the particles, increase the compressive strengths of fired pellets. Pellets fired At 1473 K compressive strengths of without binder, 1% bentonite and 0.5 and 1 % of dextrin were, 145.8, 3.2.32, 238.5 and 275.4 Kg/pellet respectively, these values are increased with increasing induration time up to 2 h 154.23, 350.5, 280.32 and 295.6 kg/pellet, which were the highest values of all results. Reference binder (1% bentonite) gives the best results than the organic binder but at 1% dextrin at 1473 K for 1 and 2 h induration time, and 0.5 % of dextrin at 1473 K for 2 h induration time was satisfy the industrial requirement.

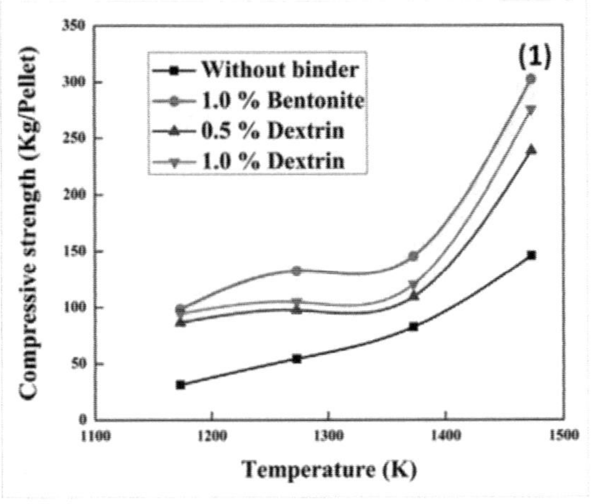

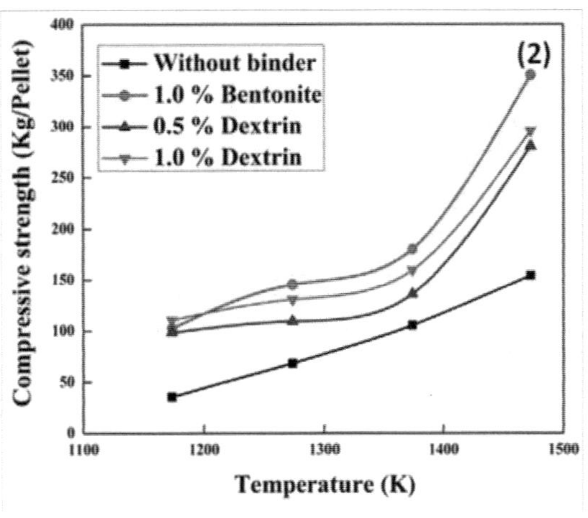

Figure 5. Effect of firing temperature and induration time on Compressive strength of pellets with different binders for (1) 1 hr. and (2) 2hrs

3.6. Porosity of fired pellets

Porosity of pellets made with different binders was determined to determine the effect of different firing temperatures and induration times. With increasing temperature, the porosity decreases for all pellets as shown in Figure 6. The porosity differences between pellets contain dextrin and bentonite was only about 2-3%. The pellets produced without binder and fired at 1173 K have the highest porosity about 31.8 and 31.03 % at 1 and 2hrs induration time respectively, it was observed for pellets produced with 1% bentonite, 0.5 and 1% Dextrin are 27.5, 28.3 and 27.8 % at 1hr and 26.3 28.3 and 27.4 % 2 h induration time respectively at the same temperature. The porosity is decreased with increasing temperature up to 1473 K they are 25.2, 20.3, 21.9 and 21.8 % at 1 h and 24.3 18.5, 21.5 and 21.3 %at 2 h induration time for without binder, 1% bentonite and 0.5 and 1% dextrin respectively. In literature, typical porosities for dry and fired pellets from high-grade iron oxide are mentioned in the range 31-36%, whilst a porosity of 22-30% for fired pellets is associated with good reducibility. The porosities for pellets produced with different binder addition were found in the industrial pellet porosity percentage interval.

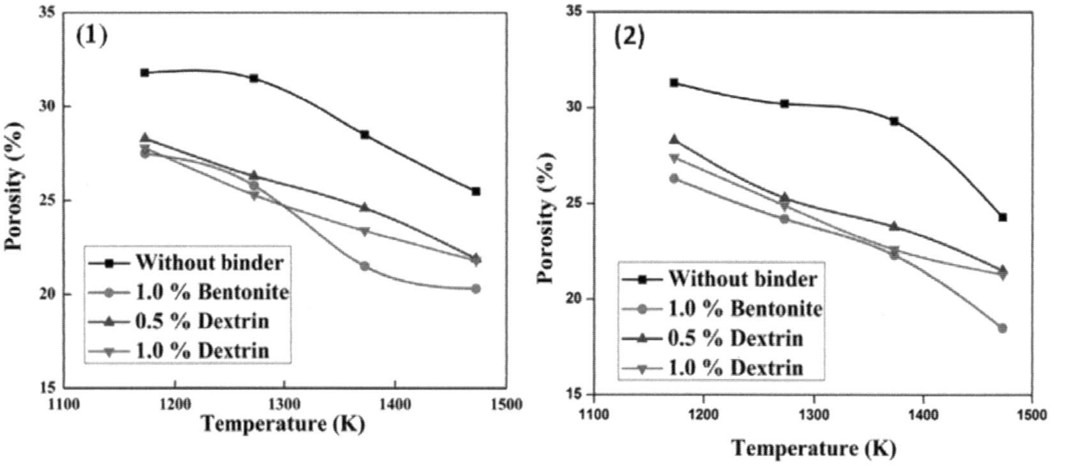

Figure 6. Effect of firing temperature and induration time on Compressive strength of pellets with different binders for (1) 1 hr. and (2) 2hrs

3.7. Hardness of fired

The hardness of the fired pellets made with binders (different) and without binder, were determined, and studied the effect of firing temperature and time on hardness. The increasing temperature and induration time shown a pronounced effect on hardness of the pellets. However, it was observed that the hardness values are decreasing from surface to the center/core of the pellet in all conditions. The hardness of fired pellets made with organic binders has shown improvement when compared to without binder and less than the bentonite binder. Figure 7 shows the effect of 1 h induration time on the hardness of the surface at 1173 K is 128, 225, 185 and 218 HV increased to 251, 340, 315 and 332 HV with increasing temperature up to 1473 K for without binder, 1% bentonite and 0.5 and 1% Dextrin respectively and the same trend was observed for 2 h induration time as shown in Figure8. However, it was found that the hardness values of the surface of all pellets are 3 to 4 times higher than the values observed in the center of the pellet. It is due to the presence of a hematite and slag content on the surface of the sintered pellets.

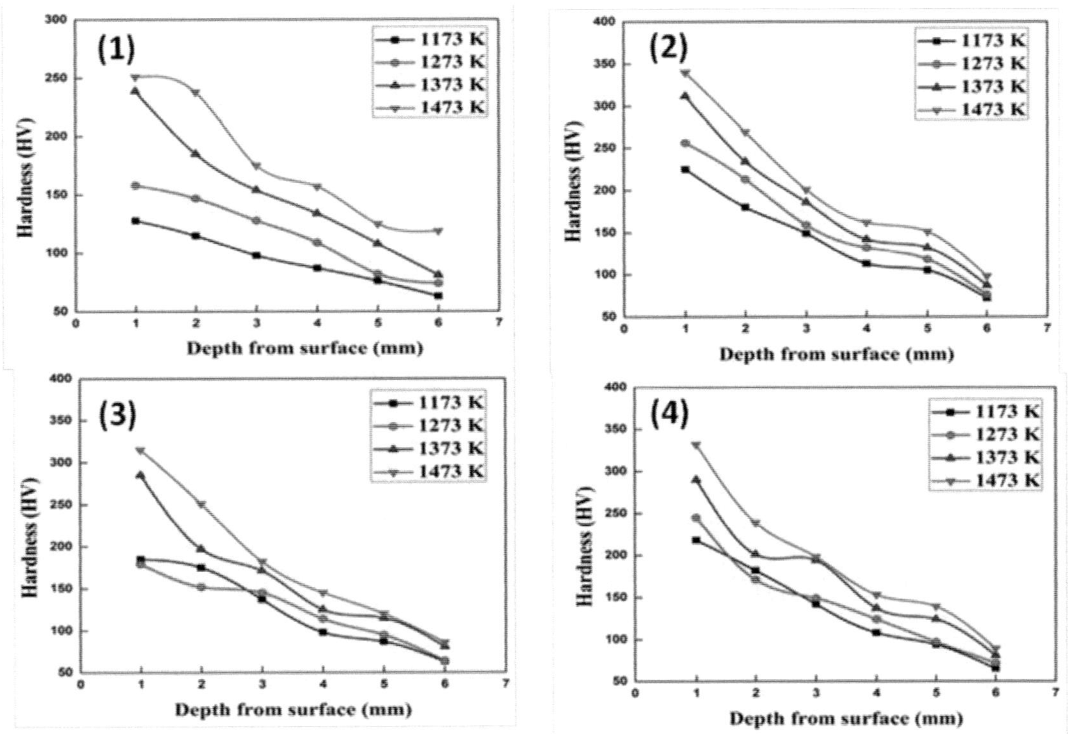

Figure 7. Effect of firing temperature and 1 hr induration time on hardness of pellets with different binders 1) Without Binder 2) 1% Bentonite 3) 0.5% Dextrin and 4) 1% Dextrin

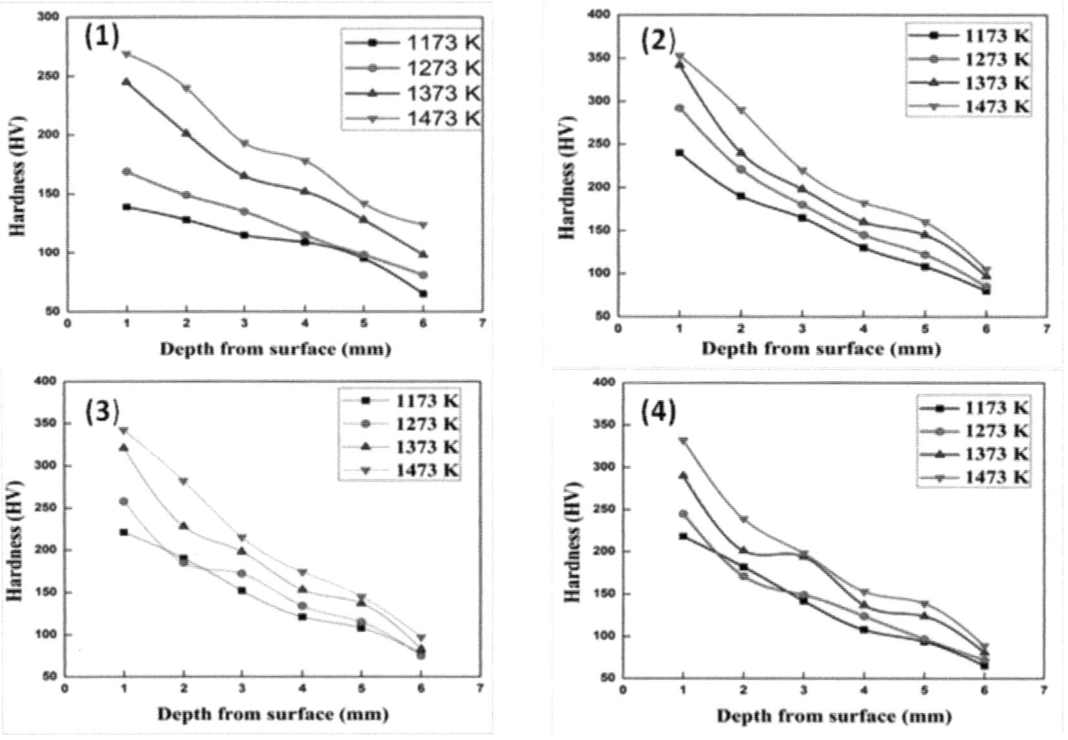

Figure 8. Effect of firing temperature and 2 hr induration time on hardness of pellets with different binders 1) Without Binder 2) 1% Bentonite 3) 0.5% Dextrin and 4) 1% Dextrin

3.8. Mechanical Wear of Pellets

To evaluate the effect of strength on mechanical wear and dust generation were investigated by using collision test. The results are shown in Figure 9 it can be seen that there was a significant difference between fines generation rates for pellets with good strengths. From the compression test results. 1% of bentonite (302 and 350 Kg/pellet) and 1% dextrin (275.6 and 295 Kg/pellet) at 1473 K for 1 and 2 h induration time, and 0.5 % of dextrin (280 kg/ pellet) at 1473 K for 2 h induration time were satisfying the industrial requirement and rest of compositions were not attaining the required strength so, the pellets were not suitable for DRI process. Wear test was conducted on the above-mentioned compositions. It observed that the fines generated from the pellets decreased with increasing strength which in turn lead to less material loss from the pellets. In method 1, the weight of the pellets maintains a constant throughout the experiment, it can be observed that the wear rate decreases (~2 times) with an increased rotational time of the test experiments as shown in Figure 9 (1). In method 2, removing of formed fines from the jar at regular intervals, the wear rate was increased with increasing the rotational time as shown in Figure 9 (2). The difference in the wear rate between the two methods was affected by the cushioning effect of dust, that means the generated dust in the jar was reduce the wear rate and it reported by several authors. The increase of the wear rate values during the wear tests can have revealed by that a hardness gradient exists in all pellets. It is clear that the harder materials have a higher have a higher resistance to wear.

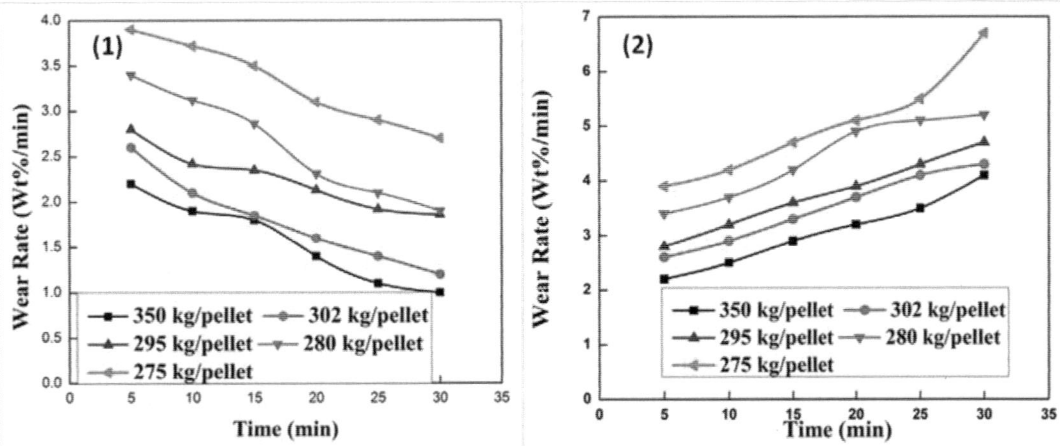

Figure 9. Changes of wear rates for different strength of pellets depending on the time of wear test 1) Method 1 and 2) Method 2

3.9. Size Distributions of Particulates Generated by wear test

Dust particles were collected from the wear test to study the particle size distribution by using the SEM. 1 gram of each dust sample was taken from the jar to determine the effect of strength on powder fineness of each pellet. From the investigation, the particles were divided into three categories, i.e. fine particles (0–10 μm), medium size particles (10–20 μm) and coarse particles (>20 μm). The size distributions of the dust particle are shown in Figure. From the SEM analysis, high amount of coarse particles was present in the method 2 for all type of pellets as shown in Figure10, which is significantly larger weight fraction than the method 1. Evidence of Coarse particles, collisions is a dominating mechanism during the wear.

Figure 10. SEM images for the particle size distributions for Wear particles

4. Conclusions

In the present study, the addition of dextrin binder into pellet mixture was proposed to increase the low pellet compressive strength encountered with the use of organic binders in iron ore pelletizing. The performances of organic binders on physical-mechanical pellet qualities were comparatively tested against bentonite binder performance.

IOP Conf. Series: Materials Science and Engineering **178** (2017) 012003 doi:10.1088/1757-899X/178/1/012003

- The DPDN produced without binder is a little lower than the required value. Therefore, they can be considered insufficient, The DPDN values of 1 Wt. % of bentonite and dextrin has shown marginally higher value than required value that is 4 and is suitable for pelletization.

- The DPCS of pellets by organic binders were found to be lower than those obtained by bentonite pellet but they are still greater than the minimum required limit, due to the presence of Dextrin between the particle to particle.

- It was observed that firing temperature, time, compositions were affecting the properties of iron ore pellets. The required compressive strength 250 Kg/Pellet was obtained at 1% of bentonite (302 and 350 Kg/pellet) and 1% dextrin (275.6 and 295 Kg/pellet) at 1473 K for 1 and 2hrs induration time, and 0.5 % of dextrin (280kg/ pellet) at 1473 K for 2 h induration time. From this we can conclude that with increasing the induration time bonding between the particle increases as well as improves the strength

- The porosity of the fired pellets decreases with increasing temperature and induration time. The porosity differences between pellets contain dextrin and bentonite were only about 2-3%.

- The hardness of the pellet varies from surface to the center of the pellet. The surface has shown 3-4 times high hardness compared to the center of the pellet.

- The strength of pellets under the current experimental conditions can influence the wear rate of pellets. The wear rate of method 1 decreases (~2 times) and increases in method 2 with an increased rotational time of the test experiments. The difference in the wear rate between the two methods was affected by the cushioning effect.

- After wear test the generated particles were characterized and which are marginally coarser in case of method 2 than method 1 from the aforementioned we could say that wear collisions are dominating during wear, were as fine dust particles were observed in case method 1.

References

[1] Taylor S. R 1964 *Geochim. Cosmochim. Acta*, **28** 1273
[2] Pal J, Ghorai S, Goswami M. C, Prakash, S, Venugopalan T 2014 *ISIJ International*, **54(3)** 620
[3] Abouzeid A. Z, Seddik A. A. 1981 *Powder Technol.* **29** 233
[4] Forsmo S. P. E, Apelqvist A. J, Björkman B. M. T, Samskog P. O 2006 *Powder Technol.* **169** 147
[5] Srivastava U, Kawatra S.K, Eisele T.C 2013. *Metallurgical and Materials Transactions B* **44(4)** 1000
[6] Haas L. A, Zahl. R. K, Aldinger J.A 1989 *Rep. Investig. / United States. Bur. Mines* **9230** 21
[7] Sivrikaya O, Arol A.I, Eisele T, Kawatra S. K 2013 *Miner. Process. Extr. Metall. Rev.* **34** 210
[8] Qiu G, Jiang T, Fa K, Zhu D, Wang D 2004 *Powder Technol.* **139(1)** 1
[9] Ammasi A, Pal J 2015 *Ironmak. Steelmak.* **43(3)** 203
[10] Kamalabadi khorasani J, Saeidi A, Shafie A 2013 *Int. J. ISSI.* **10** 34
[11] Zhucheng H, Yi L, Jiang T 2012 *Powder Technol.* **221** 284
[12] Kisan M, Sangathan S, Nehru J, Pitroda S. G 2002 *Iso 6507-1:1*
[13] Amal T, Diana C, Véronique S, Philippe M, Véronique F 2012 *Powder Technol.* **230** 86
[14] Kawatra S.K, Ripke S 2002 *J Int. J. Miner. Process.* **65** 165
[15] Nabeel M, Karasev A , Jönsson 2016 P. G. *ISIJ Int.* **56** 960

Dry Sliding Wear Behaviour of Micro oven Treated Fly Ash Reinforced Epoxy Composite using ExtendedTaguchi Approach Optimization

A Pattanaik[a], S K Bhyan[b], S C Mishra[b], S. S. Patro[c]andAjitBehera[b]
[a]Department of Mechanical Engineering, C V Raman College of Engineering, Bhubaneswar
[b]Department of Metallurgical & Materials Engineering, National Institute of Technology, Rourkela
[c]Department of Mechanical Engineering, CVRCE, Bhubaneswar
Corresponding author Email: ashungr@gmail.com

Abstract
Fly ash epoxy composite has been developed in ultrasonic mixing method with the aim of reducing the impact of hazardous waste like fly ash on the environment. Post curing has been done in micro oven to modify the surface so as to have better wear resistance properties. An orthogonal array, L_{32} has been designed with the influencing parameters like % fly ash, applied normal load, speed, track diameter and time of operation. A multi objective parametric condition, TOPSIS is used to convert it into a single objective optimization technique and ANOVA helps in getting the percentage contribution of each parameter for the wear of the material. In general, this article emphasizes the importance of micro oven treatment on the surface modification of polymeric material which retards the wear.

Keywords: Fly ash; Epoxy resin;micro oven; TOPSIS; Taguchi; ANOVA

1. Introduction

High strength at low weight makes the polymer composites usable in automobile, aerospace industries and various household appliances. Particulate reinforced polymer composites are advantageous over fibre reinforced polymer composites due to low fabrication cost and less raw material cost. The choice of material for a particular application depends on the variables like material cost, thickness, quality and working conditions. Particulate reinforced polymer composites can be used for low energy transfer where wear plays an important role in determining the life of the material. The possible reason in sudden increase in wear is due to frictional heat generated at the contact surface and loss of material [1]. So, wear is an important field which has to be analysed in perspective of increasing the life span of the material. Various researchers have found the effect of various parameters like sliding distance, sliding speed, time and applied load on the wear of the material [2].Some researchers have also concluded that the particulate filled polymer composites are mostly encouraging building materials because of their reasonable determination of lattice andfortifying strong molecule stage. It prompts a composite with a blend of quality modulus better than those of traditional metallic materials [3].Some polymer composites find application in high wear applications like pump wear rings, bushings, line shaft bearing and pressure reducing bushings. Ceramic filler like fly ash has been introduced to epoxy resin with the aim of increasing the mechanical, electrical and thermal properties simultaneously. Fly ash is a major hazardous waste generated in million tons from the thermal power plants has silica and alumina as the major constituent. Earlier researchers have done some work on the mechanical properties of the fly ash epoxy composite but very little work has been done on tribology of the material [4-9]. Previously, effort has been made to show the effect of parameters without any curing process. A slight but effective modification has been done in the form of micro oven curing to increase the wear resistance of the material. And it has been found that wear resistivity increases after micro oven curing. It may be due to the total modification of the polymer composite.

2. Experimental details

2.1 Fabrication of composite

Class C fly ash used in this experimental work is collected from CPP-2 of Rourkela Steel Plant, Odisha, India. SEM micrograph shown in figure 1 reveals that the particles are round in shape but having different diameters. Particle size analysis done in MALVERN MASTERIZER and a graph has been plotted between particle size and volume percentage of particle analysed as shown infigure 2. The mean diameter is found to be 27.26μm. Detail chemical compositional analysis shown in table 1 reveals that silicon and aluminium only constitute more than 70% of the total volume which make it a good ceramic material. Epoxy resin (LY556), chemically named Bisphenol-A-Diglycidyl-ether and hardener (HY951) supplied by ATUL industries limited, Kolkata is used as a matrix material due to its easy availability and low cost. It is a high dense polymer having density 1129gm/cm^3. The mixing ratio between epoxy resin and hardener is fixed as 10:1 by weight percentage. Fly ash is mixed with epoxy resin at four different weight percentages like 10, 20, 30 and 40. It is put under an ultrasonic sonicator with the horn dipped inside for 30 minutes. A pulse time of 5 sec is given during the mixing time. After the mixture is ready, pre weighed hardener is mixed gently by hand stirring method to avoid trapping of bubbles. Once the mixture is ready, it is poured gently to plastic moulds having diameter 10mm and allowed to harden in atmospheric condition for 24 hours. Once it is hardened, the plastic pipe is cut and solid cylindrical composites are removed and kept carefully for further use. The composites are further micro oven treated at 120 °C for 30 minutes. The micro oven treated condition is determined by trail basis.

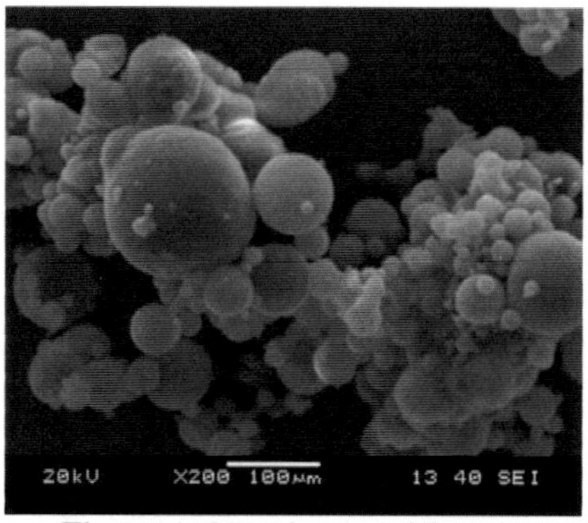

Figure 1: SEM image of fly ash.

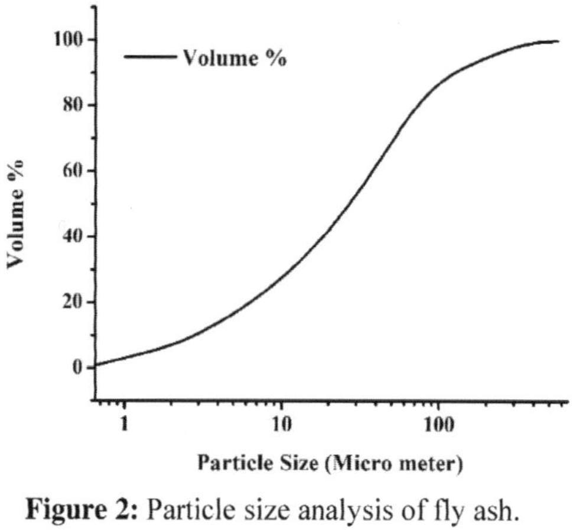

Figure 2: Particle size analysis of fly ash.

Table 1: Chemical composition of fly ash

Constituents	Percentage (%)
Fe_2O_3	8.1
MgO	1.14
Al_2O_3	24.98
SiO_2	55.85
P_2O_5	0.15
SO_3	1.16
K_2O	0.85
CaO	2.54
Na_2O	0.2
TiO_2	1.75
CO_3	1.56
LOI	1.72

Dry sliding wear test has carried out as per ASTM G99 standard. A schematic diagram in figure 3 show various parts of a dry sliding wear test rig. Surface morphology has been carried out using JEOL 6480LV scanning electron microscope. Thermal analysis has been carried out using Mettler Toledo-822 low temperature DSC (Differential Scanning Calorimetry) with temperature -100 °C to +400 °C. Fourier transform infrared spectroscopy (FTIR) applied to measure radioactive thermal properties of thin polymer films for application in an intermediate temperature range. A "Shimadzu", IR presige-21, automatic infrared microscope with a wavelength range 350 cm^{-1} to 7800 cm^{-1} is used for structural characterization of the composite.

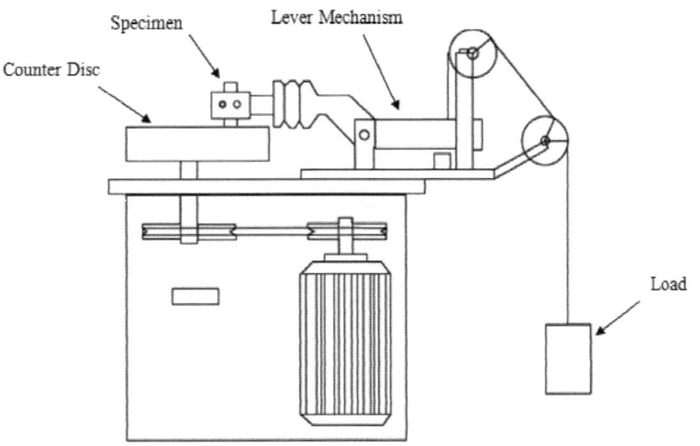

Figure 3: Schematic diagram of sliding wear test rig.

2.2 Taguchi design

Wear plays an important role in characterization of any material as it is directly concerned with the weight loss of material. To reduce the number of experiments Taguchi L32 orthogonal array is applied and results have been obtained. Each wear testgives three outputs like wear in micrometre, frictional force and coefficient of friction. As it is giving three different outputs, so initial steps are taken to make it a single parametric optimization technique. To make it a single objective optimization, TOPSIS method has been applied. After making it a single objective optimization, Taguchi optimization technique has been applied to get the suitable parametric combination and ANOVA gives the percentage contribution in each situation. Confirmatory test has been done to find the percentage of error in the experiment. Percentage of reinforcement of fly ash, duration of sliding, speed, track diameter and load are considered as five control variables and altered at different levels as demonstrated in table 2 in this experimental work.

Table 2: Levels of variables used in the experiment

Control Factors	Unit	Levels			
		1	2	3	4
Time (A)	min	5	10		
% of FA (B)		10	20	30	40
Speed (C)	RPM	200	400	600	800
Track Diameter (D)	mm	40	50	60	70
Load (E)	N	10	20	30	40

Taguchi method is used to find out the influence of each parameter. In this experiment no noise factor or interactions between parameters is considered. Table 3 shows the L32 orthogonal array with three desired outputs in the form of wear, frictional force and coefficient of friction. The three output parameters have been converted into a single relative closeness coefficient which is also mentioned in table 3 and it has been found out by using TOPSIS method. Signal to noise ratio is calculated from the output results. As we require minimum value of these results, so it is a lower is better type experiment and the result having lower value of S/N ratio is treated as the most suitable result.

The S/N ratio is calculated as, $\eta = -10\log_{10}[1/n\sum y_i^2]$ where $1 \leq i \leq n$ (1)

Whereη is the number of experiments carried out. Therefore, the ideal level of the input factors is the combination of individual parameters with levels having the maximum signal to noise ratio.Despite the performance characteristics, a noteworthy signal to noise ratio resembles to better performance with least variation. Therefore, the ideal level of the input factors is the combination of individual parameters with levels having the maximum signal to noise ratio.

2.3 Basics of TOPSIS method

The traditional TOPSIS strategy is taking into account the thought that the best option ought to have the most limited separation from the positive perfect arrangement and the best separation from the negative perfect arrangement. TOPSIS model shows the alternative ways of normalizing the data and measuring the distances from the mean position. TOPSIS method has superior advantages with regard to the adaptability of its evaluation method and the accuracy of evaluation result. The TOPSIS method consists of the following steps:

- Normalizing the decision matrix

$$r_{ij} = \frac{X_{ij}}{\sqrt{\sum_{k=1}^{m} X^2{}_{kj}}}, i = 1,...m; \ j = 1,...n \tag{2}$$

Multiplying the columns of normalized matrix with associated weights. Here similar weights have been assigned to each parameter i.e.0.33.

$$Vij = W_j \times r_{ij}, \tag{3}$$

- Determine the positive ideal and negative ideal solutions respectively,

$$A^{+} = \{v_1^{+}, v_2^{+}, \ldots v_n^{+}\} = \{(\max v_{ij} | j \in K_b)(\min v_{ij} | j \in K_c)\} \quad (4)$$

$$A^{-} = \{v_1^{-}, v_2^{-}, \ldots v_n^{-}\} = \{(\min v_{ij} | j \in K_b)(\max v_{ij} | j \in K_c)\} \quad (5)$$

Where, K_b= Set of benefit criteria and K_c= Set of cost criteria
Obtaining the distances of the existing alternatives from the positive ideal and negative ideal solutions, two distances for each alternative are, respectively, calculated as follows:

$$S_i^{+} = \sqrt{\sum_{j=1}^{n} (v_{ij} - v_j^{+})^2}, i = 1,2 \ldots m \quad (6)$$

$$S_i^{-} = \sqrt{\sum_{j=1}^{n} (v_{ij} - v_j^{-})^2}, i = 1,2 \ldots m \quad (7)$$

- Calculate the relative closeness to the ideal alternatives;

$$RC_i = \frac{S_i^{-}}{S_i^{+} + S_i^{-}} \quad i=1, 2\ldots m \quad 0 \le RC_i \le 1 \quad (8)$$

- Rank the alternatives according to their relative closeness to the ideal alternatives, the bigger RC_i, the better alternative A_i.

3. Results and discussion

3.1 Statistical Analysis

For each treatment condition 32 numbers of tests have been carried out and results in the form of wear, frictional force and coefficient of friction are obtained with accordance to the change in parameters as demonstrated in table 3. In all the three cases, these are multi objective parametric conditions are converted to single objective condition by applying TOPSIS. A relative closeness value is obtained for each condition which is the combined output of all the outputs. The optimal parametric combination is then assessed which would come about most noteworthy relative closeness value.

Table 3: L32 Orthogonal array with results for micro oven curing condition

Sl no	% FA	Time (Min)	Speed (RPM)	Load (N)	Track diameter (cm)	Wear (µm)	Frictional force (N)	COF (µ)	Relative closeness coefficient (P)
1	10	5	200	10	40	80.32	4.41	0.2	0.052048
2	10	5	400	20	50	126.66	11.16	0.3	0.080975
3	10	5	600	30	60	103.55	13.47	0.2	0.115109
4	10	5	800	40	70	110.72	20.83	0.4	0.158343
5	10	10	200	40	40	98.13	17.01	0.3	0.147733
6	10	10	400	30	50	70.91	12.32	0.3	0.148024
7	10	10	600	20	60	280.53	7.07	0.3	0.024583
8	10	10	800	10	70	41.8	6.82	0.6	0.140271
9	20	5	200	10	50	102.74	1.98	0.17	0.018908
10	20	5	400	20	40	39.79	3.99	0.23	0.091138
11	20	5	600	30	70	198.63	13.93	0.5	0.065534
12	20	5	800	40	60	143.25	18.63	0.72	0.115085
13	20	10	200	40	50	147.05	21.27	0.3	0.126366

14	20	10	400	30	40	82.43	16.91	0.22	0.170223
15	20	10	600	20	70	67.92	9.03	0.24	0.117349
16	20	10	800	10	60	58.89	5.02	0.39	0.078548
17	30	5	200	20	60	50.68	11.92	0.35	0.190415
18	30	5	400	10	70	23.77	5.88	0.38	0.198314
19	30	5	600	40	40	54.41	17.73	0.21	0.245772
20	30	5	800	30	50	72.43	12.37	0.18	0.145873
21	30	10	200	30	60	181.22	15.06	0.17	0.076727
22	30	10	400	40	70	221.49	17.92	0.11	0.074851
23	30	10	600	10	40	183.17	5053	0.8	0.965018
24	30	10	800	20	50	163.77	3.46	0.6	0.02069
25	40	5	200	20	70	84.55	5.22	0.2	0.058149
26	40	5	400	10	60	41.57	4.04	0.32	0.088577
27	40	5	600	40	50	138.49	11.36	0.4	0.075809
28	40	5	800	30	40	23.98	10.11	0.1	0.296568
29	40	10	200	30	70	58.63	10.85	0.23	0.15616
30	40	10	400	40	60	84.72	14.1	0.28	0.142684
31	40	10	600	10	50	35.8	4.46	0.34	0.11078
32	40	10	800	20	40	36.87	10.15	0.41	0.215866

The average wear values for diverse levels of selected parameters are indicated in tables 4. From the table4 below, the average value of time at level 2 is higher than the average value of time at level 1 which indicates that level 2 is better than level 1. Similarly, better choices have been found out in other parameters.

Table 4: Mean response table for relative closeness coefficient for micro oven curing

Level	TIME	FLY ASH	RPM	LOAD	TRACK DIA
1	-1.770	-2.058	-1.562	-3.967	-3.485
2	-3.088	-2.052	-1.693	-2.039	-1.965
3		-4.020	-4.127	-1.526	-2.189
4		-1.586	-2.334	-2.184	-2.078
Delta	1.317	2.434	2.565	2.441	1.520
Rank	5	3	1	2	4

Figure 4 shows the main effects plots for mean values which is the output result of table 4. Taguchi suggested a robust method having a minimum sensitive to all noise factors. In this, loss function is used to find out the deviation between the experimental value and desired value. Moreover, this loss function can be converted from signal to noise ratio. Generally, the loss functions are categorized into lower-the-better, higher-the-better and nominal-the-better [10,11].

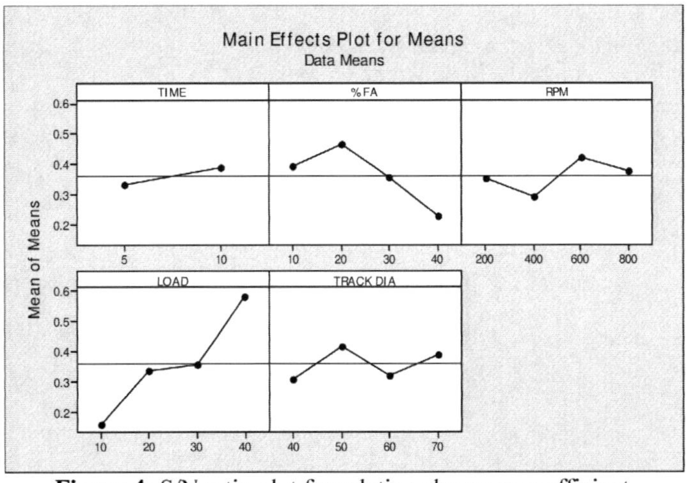

Figure 4: S/N ratio plot for relative closeness coefficient.

The above main effect plots ultimately show the optimum parametric condition to get minimum wear, frictional force and coefficient of friction.

In order to find out the contribution of each parameters like time, % of reinforcement, speed, load and track diameter on wear rate ANOVA has been carried out and results have been shown in table 5. The procedure is to a great degree accommodating to reveal the level of essentialness of effect of factor(s) on a specific reaction. It includes the total variability of the reaction (total of squared deviations about the mean) into commitments rendered by each of the parameter and the blunder [12]. The analysis has been carried out with at 95% level of confidence. The last column in table below show the significance of each parameter having very lower value. At the same time it also shows which parameter has no direct impact on the outputs.

Table 5: ANOVA table with adjusted sum of square for tests (micro oven treatment)

Source	DF	Seq SS	Adj SS	Adj MS	F	P
Time	1	0.039258	0.039258	0.039258	214.81	0.000
Fly ash	3	0.044797	0.044797	0.014932	81.71	0.000
RPM	3	0.023814	0.023814	0.007938	43.44	0.000
Load	3	0.031945	0.031945	0.010648	58.27	0.000
Track diameter	3	0.004484	0.004484	0.001495	8.18	0.001
Error	18	0.003290	0.003290	0.000183		
Total	31	0.147587				

From Table 5 only track diameter has the least significance while other parameters have influence on the output.

3.2 Confirmatory Experiment for Minimum Wear Rate

The confirmatory experiment helps in determining and analyzing the actual experimental value with the predicted value. In the wake of assessing the optimal parameter settings, the following step is to anticipate and check the enhancement of quality characteristics utilizing the optimal parametric combination. The estimated relative closeness coefficient γ using the optimal level of the design parameters can be calculated as:

$$\hat{\gamma} = \gamma_m + \sum_{i=1}^{n} \left(\bar{\gamma} - \gamma_m\right) \tag{9}$$

Where, γ_m = total is mean grey relational grade

$\bar{\gamma}$ = mean grey relational grade at the optimal level

n =number of the main design parameters

Table 6 compares the value between predicted value of input parameters and actual value using optimal parametric condition. Error is calculated by the difference in actual value and the predicted value. In all the three conditions, a minimum percentage of error is obtained which confirms that the results are least varied.

Table 6: Confirmatory test results

	Initial parameter setting	Optimal parameter condition	
		Predicted value	**Experimented value**
Level of factors	A2B3C1E3D2	A1B4C1D3E2	A1B4C1D3E2
Time	5		5
% FA	20		40
Speed	200		200
Load	10		30
Track diameter	50		50
Relative coefficient value	0.870022		
S/N ratio of relative coefficient value		1.30663	1.24129

Improvement in relative coefficient value = 0.95%

The main difference in atmospheric curing and micro oven curing is the treatment condition. In micro oven curing, the temperature generated inside the micro oven is transferred to the material. The heat moves through the material uniformly. Due to this inward movement, there may be modification in the internal structure of the material. As the surface is modified its hardness value increases as compared to normal atmospheric treated sample.

3.3 Morphology of wear surfaces

Morphology of worn out surfaces reveals the actual cause of wear during the process of sliding. Figure 5 show the various wear tracks and plastic flow of material. In the dry sliding wear test, frictional force plays an important role for the wear of the material. This frictional force is responsible for the loss of material and generation of heat at the contact surface. In case of polymeric material the heat generated helps in plastic flow of material which lairizes the material removed at the initial rotations of the disc. These regions will be called the primary energy dissipation zones. It is noted that the wear increases with increase in normal applied load and sliding distance [13,14]. Rate of wear is maximum in case of abrasion wear and the mechanism of wear is completely localised and the plastic deformation plays an important role in it [15]. Each experiment is carried out thrice and mean values have been taken into consideration for the calculation of relative closeness coefficient.White marks in the figure5 indicate the plastic flow of the material. The plastic flow of material is due to the unidirectional rotation of the sample and also may be due to heat generated at the sliding surface. Although there is no micro-crack was found around any void, continuity of the specimen was deteriorated so as to decrease the wear resistance [16].

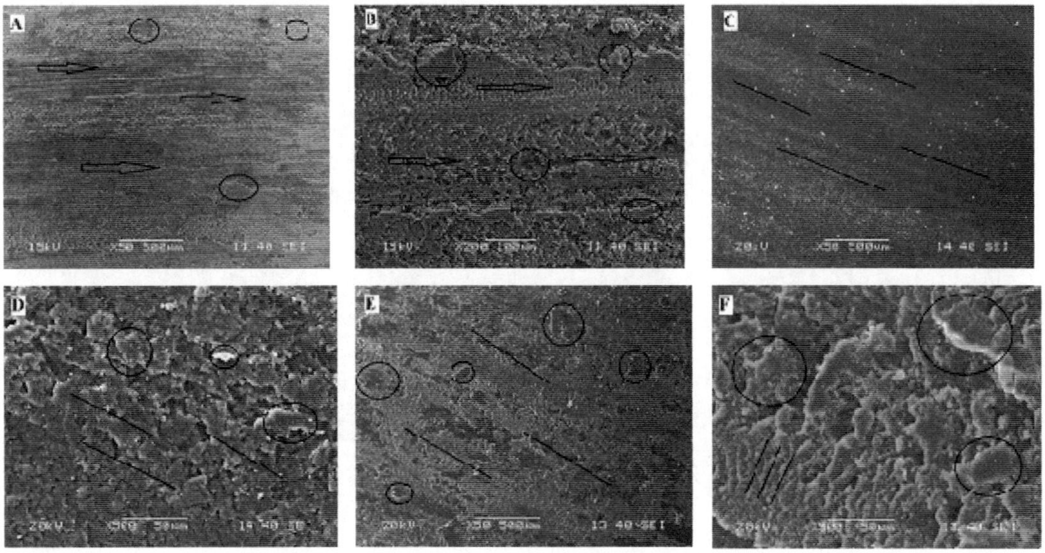

Figure 5: SEM micrographs of wear surfaces.

The straight lines in the above figures show the direction of sliding. The circular marks indicated the protruded surfaces where only matrix epoxy material is accumulated due to overheating. Figure 5 C seems to be smoothest of all as it is micro oven treated. In figure5 B and D crack propagation and erosion of reinforcement material is clearly visible. Track broadening has taken place due to the secondary erosion that is due to the entrapment of eroded fly ash particle at the initial stages of wear. Figure 5 F which is at 500X magnification clearly shows that matrix material is highly responsible for the erosion of the material. River pattern which is clearly visible in figure 5 F is due to the overheating and overlapping of material. Regardless of material, temperature and load, more or less tribo-layers existed on the worn surface [17].

3.4 DSC Analysis

Figure 6a and figure 6 b show the Tg curve of the polymeric material before and after micro oven curing. The graphs have been plotted between temperature and amount of heat absorbed. The protruded portion represents the enthalpy of that material. From the figure it is clear that, the enthalpy of the material decreases with curing. It indicates that, the material property has been modified and strength has been enhanced. The hardness of a polymer increases with crystallinity so as the mechanical strength.

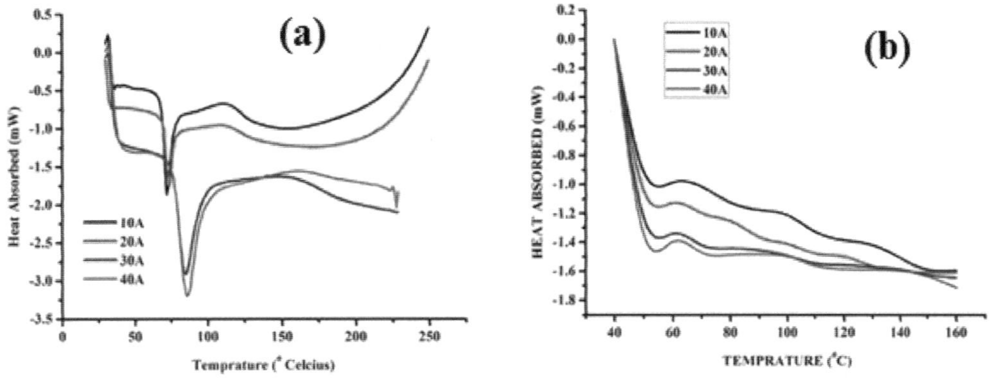

Figure 6: Comparison of DSC curves (a) for untreated and (b) micro oven treated samples.

3.5 FTIR Analysis

Figure 7 show the comparison of graphs plotted between wave number vs. % transmission for atmospheric treated and micro oven cured samples.

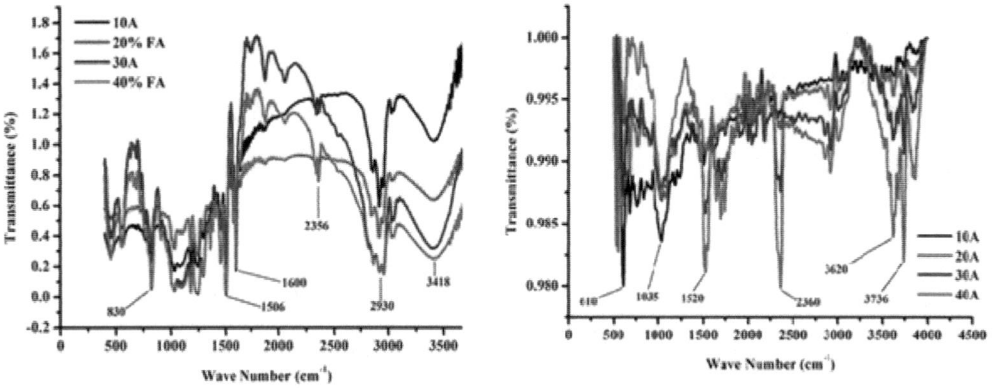

UNTREATED **MICRO OVEN TREATED**

Figure 7: Comparison of FTIR spectroscopy for untreated and micro oven treated samples

From the above comparison, it is clear that more number of sharp peaks are available.inmicro oven cured samples. It may be due to the strengthening of bonds after curing. There is change in wave numbers also. Table show the most possible bonding that may be available before and after curing.

Table 7: Wave numbers showing possible bonding in curing samples

Sl No	Wave number	Probable chemical bonding	Existence in 10%reinforced FA composite	Existence in 40% reinforced FA composite
i.	610	C-Cl / C-Br	Absent	Present
ii.	830	C-Cl / C-Br	Present	Absent
iii.	1035	=C-O-C=	Absent	Present
iv.	1506/1520	C-N=O	Present	Present
v.	1600	$-NH_2$	Present	Absent
vi.	2356/2360	$=NH_2^+$	Present	Present
vii.	2930	$-CH_3$	Present	Absent
viii.	3418/3736	$=CONH_2$	Present	Absent

Due to these variations of bonding, there is structural strengthening of the material. This may be the possible reason of increase in wear resistivity of the material.

4. Conclusion

In wear, some part of work is transmitted to heat energy and some part is dissipated as frictional loss. As it is a composite material, both matrix and the reinforcement material share the loss due to the wearing of the material. As the fly ash is ceramic filler, it is easier to erode out the matrix material. Below is the list of findings after the experiment is carried out-

a. Fly ash is a hazardous industrial waste which can be used as a polymeric composite which is novel method of utilizing the waste material.
b. TOPSIS method can be utilized for converting the multi objective parameters into a single objective parameter and Taguchi optimization technique can be used to find out the best possible with less number of experiments.
c. ANOVA helps in determining the influencing parameter. The sequence of influencing parameters is different for different treatment condition. It is in order of RPM, load, %fly ash, track diameter and time for the micro oven treatment condition.
d. From the above conclusion, time is found to be the most negligible parameter which affects the wear. It may be due to the initial over heating which cause the plastic deformation and layer formation of the material.
e. Taguchi optimization technique is helpful as there is minimum percentage of error.

f. Micro oven treatment condition give wear resistant property as the treatment is related to structural modification.
g. FTIR analysis suggests the change in bonding of the material, so this may be the possible reason of change in strength and wear resistivity.

References

[1] Mao K, Langlois P, Hu Z, Alharbi K, Xu X, Milson M 2015 *Wear***333** 822-826.
[2] Chih-Chung T Y, Wen-Cheng J. W 2000*Wear***242** 97-104.
[3] Deuis R L, Subramanian C, Yellup M 1996 *wear***201** 132-144.
[4] Biswas S, Satapathy A 2010 *Tribol. Trans.***53** 520-532.
[5] Xu W, Ma X, Tang N, Zhu L, Li W, Ding Y 2015 *Manuf. Rev.***2** 25.
[6] Kiran T S, Kumar M P, Basavarajappa S, Viswanatha B. M 2014 *Mater. Des.***63** 294-304.
[7] Pattanaik A, Mohanty M K, Sathpathy M. P 2015 *J. Mater. Metall. Eng.***5** 11-17.
[8] Renukappa N M, Suresha B, Devarajaiah R M, Shivakumar K N 2011 *Mater. Des.***32** 4528-4536.
[9] Satapathy A, Patnaik A, Pradhan M. K 2009 *Mater. Des.***30** 2359-2371.
[10] Sivapragash M, Kumaradhas P, Retnam B S J, Joseph X F, Pillai U T. S 2016 *Mater. Des.***90** 713-722.
[11] Mukhopadhyay A, Duari S, Barman T K, Sahoo P 2015 *J. Inst. Eng. Ser. D***97** 215-231.
[12] Biswas S, Satapathy A 2009 *Mater. Des.***30** 2841-2853.
[13] Maru M M, Tanaka D K 2006 *J. Brazil Soc. Mech. Sci. Eng.***XXVIII** 278-285.
[14] Alidokht S A, Abdollah-zadeh A, Assadi H 2013 *Wear***305** 291-298.
[15] Bertagnoli D, Borrero-Lopez O, Rodríguez-Rojas F, Guiberteau F, Ortiz A. L 2015 *Ceram. Int.***41** 15278-15282.
[16] Gao X, Jiang Z, Wei D, Kosasih B 2015 *Tribol. Int.***92** 117-125.
[17] Zhang Q Y, Zhou Y, Liu J Q, Chen K M, Mo J G, Cui X H 2015 *Wear* **344-345** 22-31.

Simulation Study of Stress and Deformation Behaviour of Debonded Laminated Structure

C K Hirwani[1], H Mittal[2] S K Panda[3] S S Mahapatra[4] S K Mandal[5] and A K De[6]

[1,2,3,4]Department of Mechanical Engineering, National Institute of Technology Rourkela Odisha, India.

[5,6]DRDO Integration Centre, Panagarh, West Bengal, India

Email[1]: chetanhirwani111@gmail.com

Abstract. The bending strength and deformation characteristics of the debonded laminated plate under the uniformly distributed loading (UDL) have been investigated in this research article. For the simulation study, an internally damaged laminated plate structure model has been developed in ANSYS based on the first-order shear deformable kinematic theory via ANSYS parametric design language (APDL) code. The internal debonding within the laminated structure is incorporated using two sub-laminate approach. Further, the convergence (different mesh densities), as well as the validity (comparing the responses with published results) of the present simulation model, have been performed by solving the deflection responses under the influence of transversely loaded layered structure. Also, to show the coherence of the simulation analysis the results are compared with the experimental bending results of the homemade Glass/Epoxy composite with artificial delamination. For the experimental analysis, Glass/Epoxy laminated composite seeded with delamination at the central mid-plane of the laminate is fabricated using an open mould hand lay-up composites fabrication technique. For the computational purpose, the necessary material properties of fabricated composite plate evaluated experimentally via uniaxial tensile test (Universal Testing Machine INSTRON-1195). Further, the bending (three-point bend test) test is conducted with the help of Universal Testing Machine INSTRON-5967. Finally, the effect different geometrical and material parameters (thickness ratio, modular ratio, constraint conditions) and magnitude of the loading on the static deflection and stress behaviour of the delaminated composite plate are investigated thoroughly by solving different kinds of numerical illustrations and discussed in detail.

1. Introduction

Since last few decades, the application of the layered composite material for the development of structure and the structural element has been increased in many industries such as aerospace, automobile, marine, aircraft and civil construction. The composite material is known for their specific strength and the stiffness and tailor-made property as per application. Instead of having all the all the favorable property the layered structure is always subjected to an unavoidable defect called delamination. The delamination is the detachment of the consecutive layer of the laminate. The presence of the debonding in the layered structure affect the structural performance largely. Therefore, it is important to study the effect of damage or delamination on the structural behavior of the layered structure. In this regard, many works have already been presented in the past. In order to have a progress, as well as a present status review of few important of them, are discussed further.

IOP Conf. Series: Materials Science and Engineering **178** (2017) 012005 doi:10.1088/1757-899X/178/1/012005

Alfano and Crisfield [1] presented an FE interface model of the delaminated composite using interface element and interface damage law. Zak et al. [2] model the delamination using the additional constraints condition at the delaminated front and consider the contact forces with the help of node to node contact. Parhi et al. [3] established the FE model to analyse the flexural characteristic of the first ply failure of the composites plates with multiple internal damages under low-velocity impact loading. Yam et al. [4] model the arbitrary delamination using FE approaches and obtained the structural behaviour. Bruno et al. [5] model the debonding using the concept of fracture and contact mechanics in addition with the FSDT kinematics to investigate the mixed-mode delamination behaviour. Cetkovic and Vuksanovic [6] investigated the bending, buckling and free vibration characteristics of a laminated composite plate using generalized laminated plate theory. The flexural response of the layered composites beam is analysed by Chen et al. [7] by developing a mathematical model based on modified couple stress theory based on FSDT. The crack closure integral technique along with finite element method (FEM) has been used by Sathish et al. [8] to investigate the debonding propagation within the woven fabric composite plate. Das et al. [9] investigated the stress and the energy release rate of the edge delaminated composites structure by developing a simulation model in ANSYS. Thai and Choi [10] investigated the bending and free vibration characteristic of a functionally graded plate using an FSDT kinematic based eloped mathematical model. Aslan and Daricik [11] examined the effect of multiple damages on the tensile, compressive, flexural and buckling characteristic of Glass/Epoxy layered composites structure. Transverse crack growth analysis of the edge delaminated composites plate and the shell is studied by Li [12] using the extended layer-wise method. Michalcova and Kadlec [13] studied the crack growth behaviour of the edge delaminated Carbon/Epoxy composite under various unlike environment condition, experimentally.

From the extensive literature review, it has been clear that work on modeling and analysis of delaminated layered composite structure has been reported by many authors in the past. Most of the work are related to the edge delamination compare to the internal debonding, but both have the equal importance. Therefore, **the aim of the present work is to develop a simulation model of the layered structure with internal debonding to analyse the bending and strength characteristics of the same**. In order to archive the objective a simulation model in ANSYS environment has been developed using APDL code. Further, the stability and the accuracy of the present model has been checked by computing the deflection responses for different mesh refinement and compare them with existing published literature. In addition, the present responses are compared with experimentally evaluated bending responses of Glass/Epoxy delaminated composite. Finally, the effect of design parameter on flexural behaviour of the delaminated composites plate have been reported and discussed in detail.

2. Theoretical Formulations

2.1. Mathematical model for without delamination

The mathematical model of the laminated composite plate as shown in figure 1 is developed in the present study. The geometrical parameters of the laminate are considered as length *a,* width *b* and thickness *h*. The laminate is composed of N number of orthotropic layers. The displacement field of the laminate is considered based on the FSDT and presents as:

$$\begin{aligned} u(x,y,z) &= u_0(x,y) + z\theta_x(x,y) \\ v(x,y,z) &= v_0(x,y) + z\theta_y(x,y) \\ w(x,y,z) &= w_0(x,y) + z\theta_z(x,y) \end{aligned} \tag{1}$$

where, u, v and w represents the displacements of any point along the x, y and z coordinate axes respectively. u_0, v_0 and w_0 are corresponding displacements of a point on the midplane. θ_x and θ_y are the rotations of normal to the mid-surface, i.e., z=0 about y and x-axes, respectively.

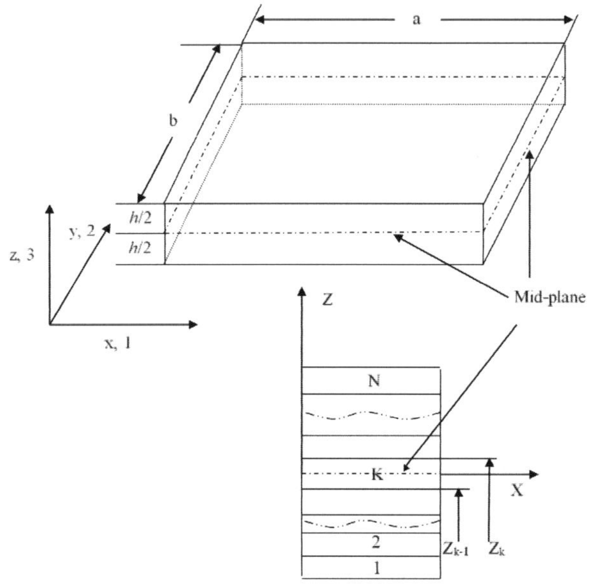

Figure 1. Geometry and stacking sequence of laminated composite plate.

2.2. Laminate constitutive relation
The laminate constitutive relation i.e. stress-strain relations for k^{th} lamina oriented at an arbitrary angle ϕ about any arbitrary axes is written as:

$$\{\sigma\} = \left[\overline{Q}_{ij}\right]\{\varepsilon\} \tag{2}$$

where, $\{\sigma\}$, $\left[\overline{Q}_{ij}\right]$ and $\{\varepsilon\}$ are the stress tensor, transformed reduced stiffness matrix and strain tensor, respectively.
2.3. Finite element formulation
The model is discretized using four-noded shell 181 elements, chosen from the ANSYS element library. The displacement vector 'δ' at any point within the mid-plane can be presented as:

$$\delta = \sum_{i=1}^{4} N_i(x, y)\delta_i \tag{3}$$

where, $\{\delta_i\} = \{u_{0_i}\ v_{0_i}\ w_{0_i}\ \theta_{x_i}\ \theta_{y_i}\ \theta_{zi}\}^T$ is the nodal displacement vector and N_i is the corresponding shape function for the 'i^{th}' node.
The strain vector can be expressed in the matrix form after introducing the FEM steps and conceded as:

$$\{\varepsilon\} = [H]\{\overline{\varepsilon}\} \tag{4}$$

where, $[H]$ is the thickness coordinate matrix and $\{\overline{\varepsilon}\}$ is the midplane strain vector which can be further expressed as:

$$\{\overline{\varepsilon}\} = [B_L]\{\delta\} \tag{5}$$

where, $[B_L]$ is a general strain displacement relation matrix.
The total strain energy of the laminate can be expressed as:

$$U = \frac{1}{2}\iint\left[\int_{-h/2}^{+h/2}\{\varepsilon\}^T\{\sigma\}dz\right]dxdy \tag{6}$$

The total work done due to the externally applied mechanical load (F) is expressed as:

$$W = \int_A \{\delta\}^T \{F\} dA \tag{7}$$

The stiffness matrix ([*K*]) and the mass matrix ([*M*]) can be written as:

$$[K] = \int_A (\sum_{k=1}^{n} \int_{z_{k-1}}^{z_k} [B_L]^T [D][B_L] dz) dA \tag{10}$$

The modeling of the delamination segment is as same as the modeling of the laminated segment. The modeling approach and the intermittent continuity condition can be seen in hirwani et al. [14]

2.4. Governing equations

The final form of governing equation for bending analysis of laminated plate is obtained using variational principle:

$$\partial\Pi = \partial(U - W) = 0 \tag{11}$$

where, Π is the total potential energy.

Further, equation (11) is modified by substituting the equations (6) and (7) into equation (11) and expressed as:

$$[K]\{\delta\} = \{F\} \tag{12}$$

The equation (12) is solved by using the following sets of support conditions to avoid any rigid body motion as well as reduce the number of unknowns.

Simply supported

$$v_0 = w_0 = \theta_y = \theta_z = 0 \text{ at } x=0 \text{ and } a;$$

$$u_0 = w_0 = \theta_x = \theta_z = 0 \text{ at } y=0 \text{ and } b; \tag{13}$$

Clamped

$$u_0 = v_0 = w_0 = \theta_x = \theta_y = \theta_z = 0 \text{ at } x=0 \text{ and } a; y=0 \text{ and } b; \tag{14}$$

Free

$$u_0 \neq v_0 \neq w_0 \neq \theta_x \neq \theta_y \neq \theta_z \neq 0 \text{ at } x=0 \text{ and } a; y=0 \text{ and } b; \tag{15}$$

3. Result and discussion

The central deflection and the stress of the delaminated composite structure have been examined in the present investigation. For the computation purpose, a simulation model of the internally debonded plate structure in ANSYS environment using APDL code has been established. The model is discretised using four noded shell 181 elements chosen from the ANSYS element library. Further, the computed responses of the layered composite structure are compared with that of the numerical responses of the earlier published literature. In addition, the present responses are also compared with that of the experimentally computed central deflection responses of the Glass/Epoxy delaminated composite. Finally, the effect of various design parameter central deflection parameter and strength behaviour of delaminated composites plate has been examined via different numerical example and discussed in detail. **The procedure used to model the delaminated composite in ANSYS APDL has been discussed in the following line.**

- **First of all, the geometry of the plate is defined as length, *a*; width *b*; and thickness *h*.**
- **The orientation of each layer is provided and the material property is assigned to the corresponding layers.**
- **The delamination is model using two sub-laminate approach and the intermittent continuity condition has been given using the coincident node under coupling command.**
- **The model is discretised with the help of Shell 181 element, chosen from the ANSYS element library.**
- **Now the end/boundary condition, loading as well as analysis type has been provided.**
- **Finally, the bending as well as stress responses is computed via choosing suitable syntax from the post processing.**

3.1 Convergence and comparison study

In this section, the stability of the present model has been checked by computing the responses for different mesh division. In general, the structural domain is discretised via an equal number of elements in both the direction to maintain the symmetry of meshing. With increasing the number of the element, the responses changes and the computational cost increases simultaneously but after attaining certain mesh sizes the responses repeated and the percentage difference between the subsequent steps also decreases. In order to maintain the accuracy of the desired responses with minimum computational cost, the particular mesh sizes have selected. In this present example, the nondimensional central deflection of the square (a=b=0.2 m) simply supported three-layer cross-ply layered composites plate subjected to uniformly distributed load (q=100KN/m^2) have been examined for different mesh refinement. The responses are evaluated via present simulation model using the material properties from the material data-1 and presented in figure 2. The figure shows that the responses are converging well for various mesh refinement. Based on that the (13×13) mesh size is used for further investigation. The responses are nondimensionalized using the formulae: $\bar{w}=100E_y h^3 w / qa^4$. Now, a square laminated composite plate with simply supported boundaries have been analysed for the bending responses, compared with Xiao et al. [15] and Reddy [16] and presented in Table 2. The material and the geometrical parameter are taken as same as the reference.

Table 1. Material properties used in present analysis

Material Properties	Material data-1	Material data-2	Material data-3 (Experimental)
E_x	250 GPa	172.5GPa	8.739 GPa
E_y	10 GPa	6.9GPa	7.926 GPa
E_z	10 GPa	6.9GPa	E_2
G_{xy}	5 GPa	3.45GPa	3.75 GPa
G_{yz}	2 GPa	1.38GPa	1.875 GPa
G_{xz}	5 GPa	3.45GPa	3.75 GPa
υ_{xy}	0.25	0.25	0.17
υ_{yz}	0.25	0.25	0.17
υ_{xz}	0.25	0.25	0.17

Table 2. Comparison study for square laminated composite plate

h/a	Present	Xiao et al.[15]	Reddy [16]
0.05	0.7951	0.8029	0.7694
0.1	1.138	1.1401	1.0250
0.2	2.312	2.2383	-

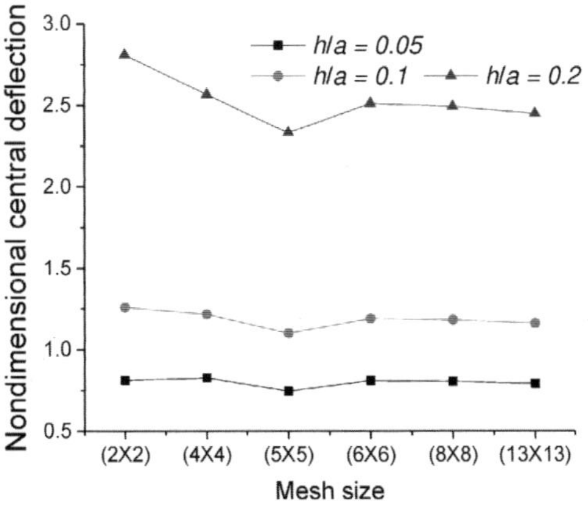

Figure 2. Convergence test central deflection for various mesh density

3.2 Preparation of the woven Glass/Epoxy delaminated composite plate

For the fabrication of the Glass/Epoxy laminated composites plate with debonding, a well-known composite fabrication technique hand layup is utilized. The steps used to fabricate the laminate are as follows:

- First of all, a square wooden mold is prepared with the help of flat plywood and wooden strip.
- The mylar sheet is placed at the bottom of the mold.
- Now, the mold release agent is applied to the Mylar sheet so that there will not be any difficulty at releasing the fabricated composite.
- The epoxy polymer (Lapox L-12) and the corresponding hardener (K-6) are mixed together in 10:1 proportion.
- The first thin layer of the mixture is applied to the mylar sheet.
- Further, the woven glass fiber sheet of the required dimension is placed in the mold.
- A small pressure is applied with the help of iron roller to ensure the proper binding of the mixture and the glass fiber.
- The above three steps are repeated till the plane of delamination reached.
- Now, the artificial delamination of the required size is created using the thin Teflon tape.
- Again the thin layer of mixture, another woven fabric and the small pressure are applied.
- The process is repeated until the required number of layers of the laminate is completed.
- The another mylar sheet, treated with mold release agent is placed at the top of the laminate.
- Now, the stacked laminate is pressed via hot-press for the proper stacking as well as pre-curing for half an hour at 60°C.
- After pre-curing the composite is kept at the environment condition for post curing for 72 hours.
- Finally, the mylar sheet is released and the laminate is kept at the air tied container.

In the present work, eight layers of Glass/Epoxy laminated composite with 6.25% of the total area of debonding, located at the center mid-plane of the laminate has been fabricated.

3.3 Material property evaluation

For evaluating the material property of the fabricated laminate, sample specimen for the tensile test has been prepared according to American Society for Testing and Materials (ASTM) standard ASTM D 3039/D3039M. The laminated composite is cut into three different directions along length, along with and 45° to the direction of the length to compute the young's modulus in x (E_x), young's modulus in y (E_y) and young's modulus in 45° to the length direction (E_{45}°), respectively. The tensile test is conducted in Universal Testing Machine (UTM) INSTRON-1195 at the parent Institute National Institute of Technology Rourkela, Rourkela, Odisha. The loading rate of the test is kept fix through the test as 1 mm/minute. Further, all the evaluated young's modulus is used are used to compute the shear modulus using the formulae as presented in equation (16). The poisons ratio of the GFRP is taken from the Crawley et al. [17] as 0.17. The experimentally evaluated material properties are shown in Table 1.

$$G_{xy} = \frac{1}{\dfrac{4}{E_{45^\circ}} - \dfrac{1}{E_x} - \dfrac{1}{E_y} - \dfrac{2\upsilon_{xy}}{E_x}} \tag{16}$$

3.4 *Experimental bending test*

The experimental three-point bending test has been conducted using UTM INSTRON-5967 at NITR Rourkela, Odisha. The loading rate for the experiment has been fixed as 2 mm/minute throughout the test. The sample specimen has been prepared according to the ASTM standard ASTM-D790. The size of the sample specimen is (48mm×16mm×3mm). Three sample of each type of the composite is tested and the average of them is taken as a final result. The experimentally evaluated material property is used to compute the bending responses using ANSYS and compared with those of the experimental responses and presented in Table 3.

Table 3. Comparison study central deflection of delaminated Glass/Epoxy composite

Delamination size (%)	Load (N)	Experiment (mm)	Current Simulation (mm)
6.25%	10	0.0840	0.0772
	20	0.1651	0.154
	30	0.2474	0.232
	40	0.3290	0.309

3.5 *Numerical illustrations*

The comparison study clearly indicates that the simulation model is capable enough to evaluate the bending response of the laminated and delaminated composite with adequate accuracy. Further, the applicability of the simulation model is expanded by solving the various illustration and highlighted the influence of the geometrical and material parameter. The examples are computed for the twenty-layer angle-ply laminate with 56% of debonding located at the center mid-plane of the laminate. The material data-2 is used for further study, else stated otherwise.

3.5.1 *Influence of the modular ratios on bending behavior of the delaminated composite plate*

In the present example, the influence of the modular ratios on bending behavior of the delaminated composite plate has been examined. For evaluating the responses, a square (a=b=0.5m) clamped twenty-layer angle-ply $(\pm60^\circ)_{10}$ laminated composite (a/h=100) with 56.25% of centrally located debonding at the mid-plane of the laminate under UDL loading have been utilized. The responses are evaluated using simulation model and presented in figure 3. The figure clearly indicated that the central deflection parameter decreases with the increase of modular ratio. The depicted trend on the result is expected because with increasing the modular ratio the longitudinal modulus increase, which increase the global stiffness resulting in the deflection decreases.

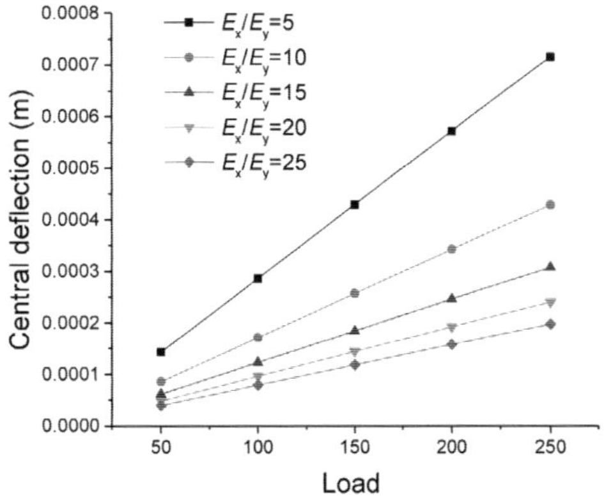

Figure 3. Effect of modular ratio on bending behavior of delaminated composite plate

3.5.2 Influence of the thickness ratios on bending behavior of the delaminated composite plate
The central displacement of square (a=b=0.5m) clamped twenty-layer ($\pm60°$)$_{10}$ 56% delaminated composite plate subjected to UDL loading have been computed for the different thickness ratio (a/h=10, 20, 50, 80,100) using the presently developed simulation model. The computed responses are shown in figure 4. From the figure it depicts that as the thickness ratio increases the central deflection responses increases irrespective of the magnitude of the loading. This is because of the decreasing thickness as well the global stiffness of the layered structure with increasing the thickness ratio.

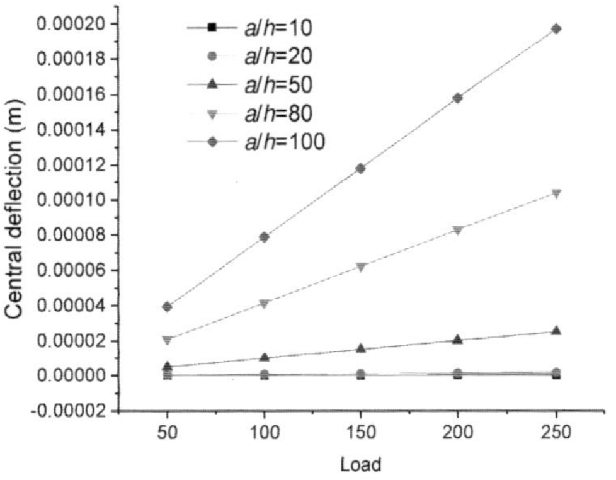

Figure 4. Effect of thickness ratio on bending behavior of delaminated composite plate

3.5.3 Influence of the boundary condition on bending behavior of the delaminated composite plate
In this illustration, bending behavior of the square (a=b=0.5m) angle-ply ($\pm60°$)$_{10}$ laminated composites plate with 56.25% of delaminated composite plate is examined for various boundary condition such as all side clamped (CCCC), three sides clamped and one side free (CCCF), two sides clamped and two sides free (CFCF). The responses ate evaluated via present simulation model and presented in figure 5. The figure indicated that the response follows the increasing trend with the CCCC, CCCF and CFCF independent of the magnitude of the loading. This is because the constraint degree of freedom decreases with CCCC, CCCF and CFCF.

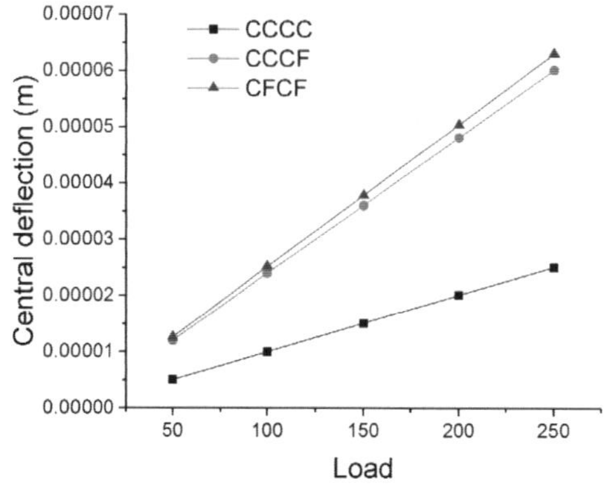

Figure 5. Effect of boundary condition on bending behavior of delaminated composite plate

3.5.4 In-plane normal and shear stress of the delaminated composite plate

The in-plane normal as well as the shear stress of the square (a=b=0.5m) clamped angle-ply $(\pm 60°)_{10}$ 56% delaminated layered structure for two different thickness ratios (a/h=50 and a/h=100) have been evaluated in the present illustration. The stress responses are computed with the help of the current ANSYS simulation model and depicted in table 4. The table indicates that the stress value increases with increasing the magnitude of the load as well as the thickness ratio. The in-plane stress contour plot for the load magnitude 40 and thickness ratio 50 is presented in figure 6(a)-(c).

Table 4. In-plane normal as well as the shear stress

Load	a/h=50			a/h=100		
	σ_x (N/mm^2)	σ_y (N/mm^2)	τ_{xy} (N/mm^2)	σ_x (N/mm^2)	σ_y (N/mm^2)	τ_{xy} (N/mm^2)
40	59690.2	67382.6	48969.5	237768	270550	195806
80	119380	134765	97939	475536	541101	391612
120	179071	202148	146909	713304	811651	587417
160	238761	269530	195978	951071	1080000	783223
200	298451	336913	244848	1190000	1350000	979029

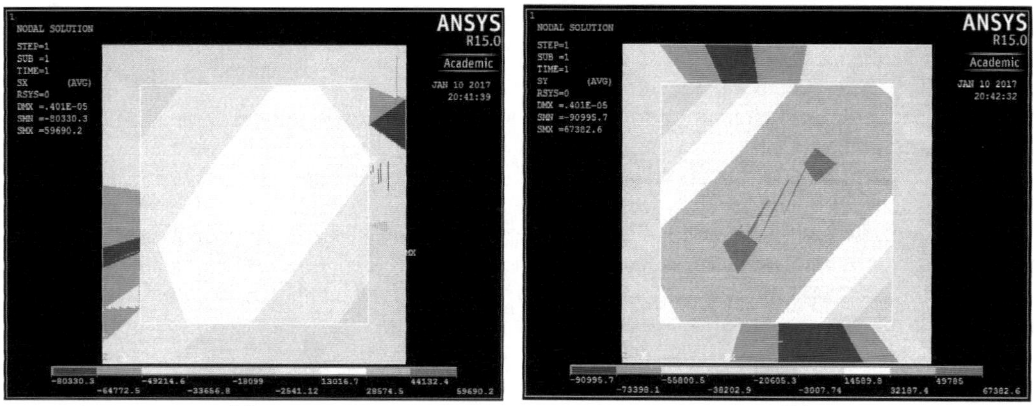

(a) (b)

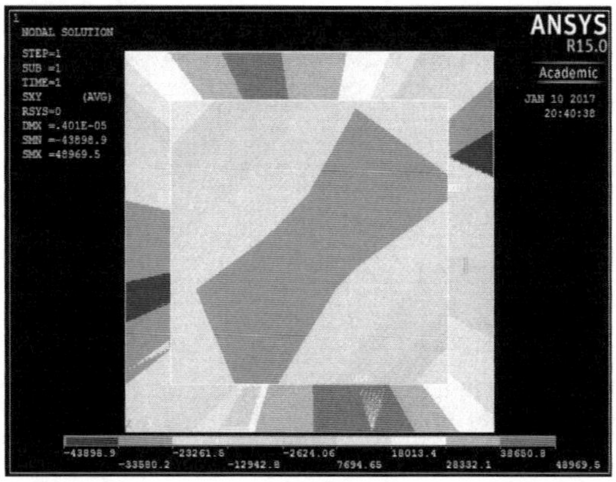

(*c*)

Figure 6. In-plane stress (*a*) Normal stress (σ_x); (*b*) Normal stress (σ_y); (*c*) Shear stress (τ_{xy})

4. Conclusion

The static and the strength behaviour of the delaminated composites plate has been analysed in the present work. A simulation model of the delaminated plate has been developed using the two sub-laminate method in commercial simulation software ANSYS with the help of ANSYS parametric design language (APDL) code. Now, the stability of the proposed model is checked by computing the responses for the different mesh refinement. The computed responses of the simulation model are compared with those of the numerically calculated responses of published literature. In addition, the simulation responses are also compared with that of the experimentally calculated responses. The woven glass fiber reinforced polymer (GFRP) composite is fabricated and the material property is evaluated using hand layup method and tensile testing, respectively. Finally, the influence of the design parameter is obtained and discussed in detail. Based on the outcome of the different illustration some important conclusions have been outlined and presented in the next lines.

- The validation study indicates that the model is capable enough to solve the bending problem of the damaged laminated composite.
- In addition, the validation with experimental study shows that the model is also stable with the experimental responses.
- The central displacement increases with the increasing the thickness ratio.
- Further, the deflection responses decrease with the increase in the modular ratio.
- The in-plane normal and the shear stress increase with the increase in the loading.

Acknowledgements
This work is under the project sanctioned by the Defence Research and Development organization (DRDO) through the grant DIC/MMG/LP/14-15/01/05 Dated: 15/09/2014. Authors are thankful to DRDO, Govt. of India for their consistent support.

References

[1] Alfano G and Crisfield M A 2001 Finite element interface models for the delamination analysis of laminated composites: mechanical and computational issues *International Journal for Numerical Methods in Engineering* **50** 1701-1736
[2] Zak A, Krawczuk M and Ostachowicz W 2001 Vibration of a laminated composite plate with closing delamination *Journal of Intelligent Material Systems and Structures* **12** 545-551
[3] Parhi PK, Bhattacharyya SK and Sinha PK 2001 Failure analysis of multiple delaminated composite plates due to bending and impact *Bulletin of Materials Science* **24** 143–149
[4] Yam LH, Wei Z, Cheng L and Wong WO 2004 Numerical analysis of multi-layer composite plates with internal delamination *Computers and Structures* **82** 627–637

[5] Bruno D, Greco F, and Lonetti P 2005 A 3D delamination modelling technique based on plate and interface theories for laminated structures *European Journal of Mechanics A/Solids* **24** 127–149
[6] Cetkovic M and Vuksanovic 2009 Bending, free vibrations and buckling of laminated composite and sandwich plates using a layerwise displacement model *Composite Structures* **88** 219-227
[7] Chen W, Li L and Xu M 2011 A modified couple stress model for bending analysis of composite laminated beams with first order shear deformation *Composite Structures* **93** 2723-2732
[8] Sathish S, Prasath J, Reddy KS and Naik NK 2012 Damage evolution in delaminated woven fabric E-glass/epoxy composite plates under transverse static patch loading *International Journal of Damage Mechanics* **22** 982–1005
[9] Das S, Choudhury P, Halder S and Sriram P 2013 Stress and free edge delamination analyses of delaminated composite structure using ANSYS *Procedia Engineering* **64** 1364-1373
[10] Thai H T and Choi D 2013 A simple first-order shear deformation theory for the bending and free vibration analysis of functionally graded plates *Composite Structures* **101** 332-340
[11] Aslan Z and Daricik F 2016 Effects of multiple delaminations on the compressive, tensile, flexural, and buckling behaviour of e-glass/epoxy composites *Composites Part-B* **100** 186-196
[12] Li D H 2016 Delamination and transverse crack growth prediction for laminated composite plates and shells *Computers and Structures* **177** 39-55
[13] Michalcova L and Kadlec M 2016 Carbon/epoxy composite delamination analysis by acoustic emission method under various environmental conditions *Engineering Failure Analysis* **69** 88-96
[14] Hirwani C K, Sahoo S S and Panda S K 2015 Effect of delamination on vibration behaviour of woven Glass/Epoxy composite plate-An experimental study, *IOP Conf. Series: Materials Science and Engineering* **115**
[15] Xiao J R, Gilhooley D F, Batra R C, Gillespie Jr. J W and McCarthy M A 2008 Analysis of thick composite laminates using a higher-order shear and normal deformable plate theory (HOSNDPT) and a meshless method *Composites Part-B* **39** 414-427
[16] Reddy J N Mechanics of laminated composite plates CRC press 1996
[17] Crawley E F 1979 The natural modes of Graphite/Epoxy cantilever plates and shells *Journal of Composite Materials* **13** 195-205

Tensile behavior of MWCNT enhanced glass fiber reinforced polymeric composites at various crosshead speeds

K K Mahato*, D K Rathore, R K Prusty, K Dutta, B C Ray

Composite Materials Group, Metallurgical and Materials Engineering Department, National Institute of Technology, Rourkela-769008, India

*Email: kishorepce@gmail.com

Abstract. Fiber reinforced polymeric (FRP) composite materials are subjected to different range of crosshead speeds during their in-service life. The work has been focused to investigate the effect of carbon nanotube (CNT) addition in glass fiber reinforced polymer (GFRP) composite on tensile behavior. The Control GFRP composites and CNT modified composites were tested at different crosshead speeds viz. 1, 10, 100 mm/min. CNT modified matrix was processed with epoxy as a matrix materials and multi-walled carbon nanotube (MWCNT) as a filler with different MWCNT content (i.e. 0.1, 0.3 and 0.5 wt. %). Increase in the CNT content upto 0.3% the tensile strength increasing for all the crosshead speeds as compared to the control GFRP composite. The tensile strength are dependent on the CNT content in GFRP composite. It has been observed that addition of 0.1% CNT and 0.3% CNT enhanced the tensile strength by 6.11% and 9.28% respectively than control GFRP composite. The tensile modulus is found to be mostly unaffected on an optimum CNT content in the GFRP composite. The tensile strength of control GFRP and all CNT modified GFRP composites were found to be crosshead speed sensitive and increased with increasing crosshead speeds in the aforesaid loadings. However, slight decrease in tensile modulus was observed with addition of CNT due to agglomeration of the CNT in the polymer matrix composites. The DSC analysis was also carried out to understand the effect of the CNT content on the glass transition temperature (Tg) of GFRP composites. Different failure patterns of GFRP composite tested at 1, 10, and 100 mm/min crosshead speeds were identified.

1. Introduction

In this present era, the demand and usage of FRP composites was increasing in rapid manner. The applications of these composites are not limited to a field, but it is widely spread over to various arena. The field of applications includes automotive, sporting goods, aerospace, structural, cryogenic vessels, solar panels and oil and gas pipelines. Incorporation of nanofillers into the polymer matrix has been an active area of research around the world. Reproduction of the distinct behaviors of nano-fillers in the polymer matrix is one of the crucial objective to achieve better mechanical, electrical and/or thermal properties. The usage of CNT as nanofillers is due to its strength (about 22 GPa [1]) and modulus (about 1 TPa [2]). Various researchers have studied on the mechanical performance of GFRP composites and incorporation of different nanofillers [3–5] at various environments and loading conditions [6–9]. There is a dearth in the literature of tensile behavior of CNT embedded into polymer matrix with crosshead speed variations.

Property	Epoxy	Woven Fabric E- Glass Fibers
Density (g/cm)	1.162	2.58
Tensile Strength (GPa)	0.11	3.4
Tensile modulus (GPa)	4.1	72.3
Strain at failure (%)	4.6	4.8
Areal weight of fabric (g/m^2)	-	360

Table 1. Mechanical Properties of E-glass fibers and epoxy

The current experimental study deals with the incorporation of multi-walled carbon nanotubes (CNTs) in epoxy resin that caused improvement in tensile strength over conventional GFRP. In the present study, epoxy of GFRP composite was modified with 0.1 wt. %, 0.3 wt. % and 0.5 wt. % of CNT and laminates were fabricated. Also, the variation in the loading rate with CNT content is evaluated.

2. Materials and Methods

2.1 Testing Materials

The current experimental investigation consists of fabrication of GFRP composite and Diglycidyl Ether of Bisphenol A (DGEBA) based epoxy resin was used as matrix and triethylene tetra amine (TETA) was used as hardener. The woven fabric E-glass fibers used in the experimental program was supplied by Saint Gobin industries. Epoxy resin and hardener were obtained from Atul Industries Ltd, Gujarat having the trade name of Lapox L-12 and K-6 respectively. The epoxy to hardener weight ratio was taken as 10:1 as per the supplier's recommended standard.

2.2 Preparation of GFRP and CNT enhanced GFRP laminates

The GFRP composite laminated was fabricated with 10 layers of woven fabric E-glass fibers having 60:40 weight fraction of glass fibers and epoxy respectively. The fabrication was done using hand layup method. To ensure appropriate distribution of CNTs in the polymer matrix is the crucial issue to get better mechanical properties. Therefore, the CNTs and polymer suspension were thoroughly mixed and stirred with the help of magnetic stirrer and ultra-sonicator. The control GFRP and CNT-GFRP laminates were cured in hot press compression moulding machine at 60°C temperature for 20 minutes at a pressure of 5 kgf/cm^2. As per the ASTM D3039 standard the samples were cut using diamond cutter. After that the samples were post cured in an oven at 140°C for 6 hours [10].

2.3 Testing Methods of GFRP composite

2.3.1 Tensile Test

The test was performed with tension fixture in Instron 8862 Universal testing machine (UTM) shown in figure 1(b) in accordance to the ASTM D3039 standard samples to evaluate the tensile properties. The samples were tested at room temperature with different loading rates viz. 1, 10, 100, 500 and 1000 mm/min.

2.3.2 *Scanning Electron Microscopy (SEM) Analysis.*

The fractured surfaces of the tested specimens were further analysed using scanning electron microscope (SEM) for determining the various dominating modes failure mechanisms using JEOL-JSM 6480 LVSEM at 20KV.

3. Results and discussion

3.1 Effect of tensile behavior of MWCNT enhanced glass fiber reinforced polymeric composite at various crosshead speeds

The tensile stress-strain curves for control GFRP, 0.1wt% CNT, 0.3wt% CNT and 0.5wt% CNT at different loading rates is shown in figure1. It is observed that with increase in the CNT content upto 0.3% the tensile strength increasing (figure1) for all the loading rates as compared to the control GFRP composite. Further, addition of CNTs causes decrease in the strength value and it may be attributed to the agglomeration of CNTs in the polymer matrix, where due to poor stress transfer across the fiber/matrix interface.

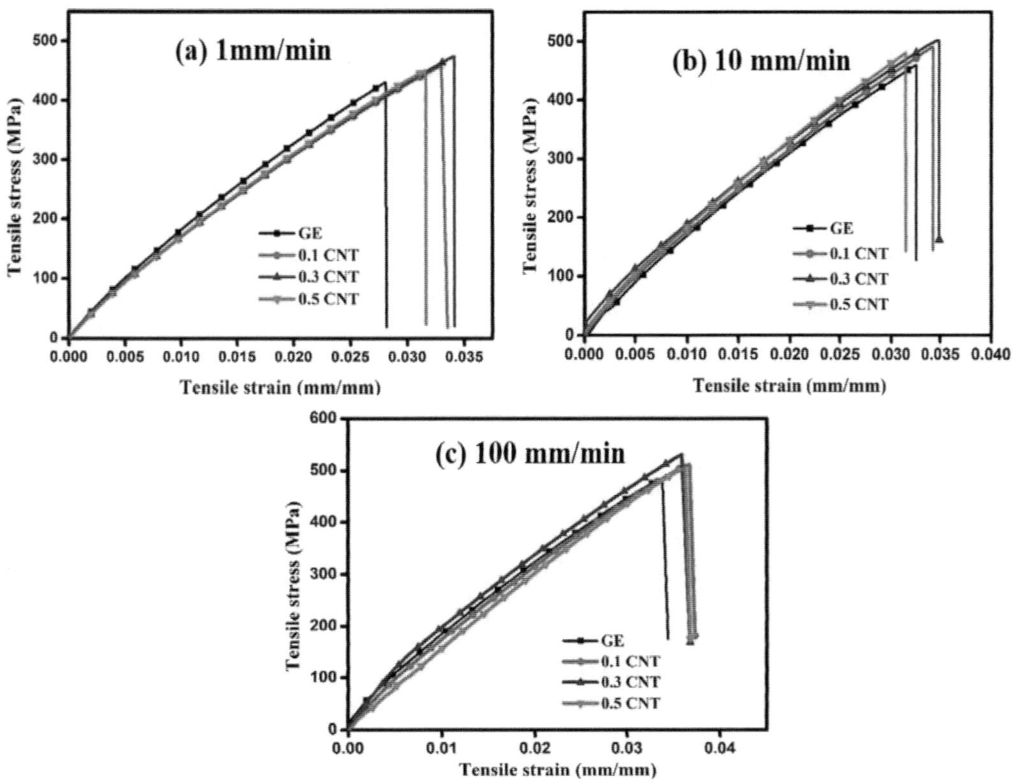

Figure 1. Stress–strain curves for GFRP composites at various MWCNT contents at room temperature with different loading rate (a) 1 mm/min, (b) 10 mm/min and (c) 100 mm/min.

The tensile properties i.e. tensile strength and tensile modulus obtained from figure1 were plotted against various CNT content for GFRP composite in figure 2 as revealed below. It is evident from figure 2 that the tensile strength and tensile modulus are strongly dependent on the CNT content for GFRP composite. Incorporation of 0.1% CNT and 0.3% CNT resulted in enhancement in tensile properties at room temperature than control GFRP composite as revealed from figure 2(a). It has been observed that addition of 0.1% CNT and 0.3% CNT enhanced the tensile strength by 6.11% and

9.28% respectively than control GFRP composite. The reason can be attributed to very high specific surface area of CNTs, which in-turn exhibits very high interfacial area in the final composite. Hence the stress transfer ability across the interface is easily facilitated, as a result higher strength in case of CNT/GFRP composite is observed than control GFRP composite at room temperature environment.

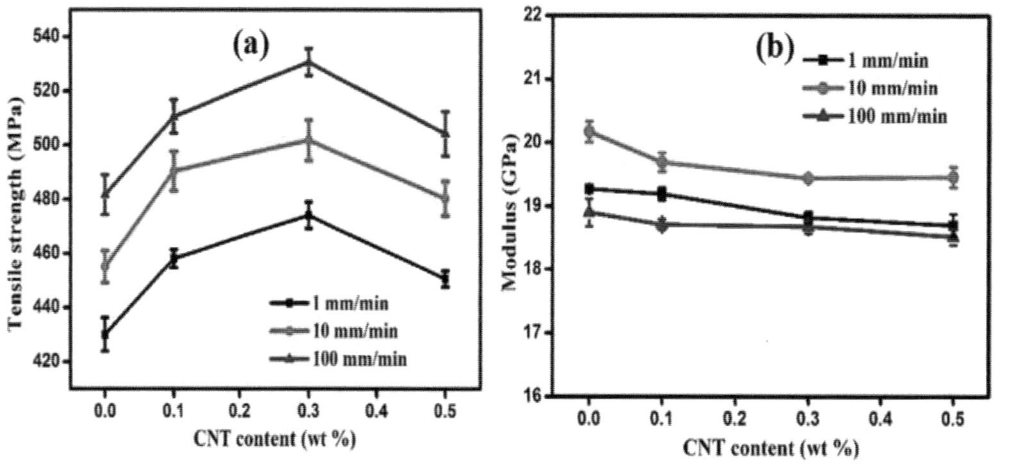

Figure 2. Variation in (a) Tensile strength and (b) Tensile modulus with CNT content in GFRP composite at various crosshead speeds.

But in the meantime drop in tensile modulus is observed with addition of CNT and the reason may be attributed to agglomeration of the CNT in the polymer matrix. Agglomeration causes poor wettability in the fiber/matrix interface resulting in weaker adhesion in the interfacial bonding between the CNT enhanced fiber/matrix composite. It is observed from figure 2 (a) that with increase in CNT content upto 0.3 wt% with loading rate, the tensile strength of the GFRP and CNT/GFRP composite is increasing. Generally, thermal stresses induces micro-cracks in the polymer matrix and/or, at the fiber/matrix interface may possibly grow without blunting at a stable state. Some of these microcracks turn upto potential cracks at low loading rates and causes substantial reduction in tensile stress of the composite system. But, at the higher loading rates, the time availability to propagate the microcracks is very less. The modulus value represented in figure 2 (b) shows no significant variation in the tensile modulus for control GFRP and CNT-GFRP composites at all the crossheads speeds. In the meantime slight drop in tensile modulus is observed with addition of CNT due to agglomeration of the CNT in the polymer matrix. But, with respect to loading rate increase in modulus value was observed in case of 10 mm/min crosshead speeds at all the CNT content. This may be attributed to proper adhesion of fiber/matrix/CNT in the composite and propagation of cracks at 10mm/min is neither very fast nor very slow constitutes proper stress transfer from matrix to the fiber.

3.2 Different failure patterns of GFRP composite

Figure 3 shows the bulk pictures of GFRP composites with different failure pattern tested at (a) 1mm/min, (b) 10 mm/min (c) 100 mm/min loading rates. At lower loading rates the initiation of crack generates at the middle portion of the specimens. But at higher loading rates multiple numbers of cracks are generating throughout the samples i.e. at middle portions as well as at the tab portions.

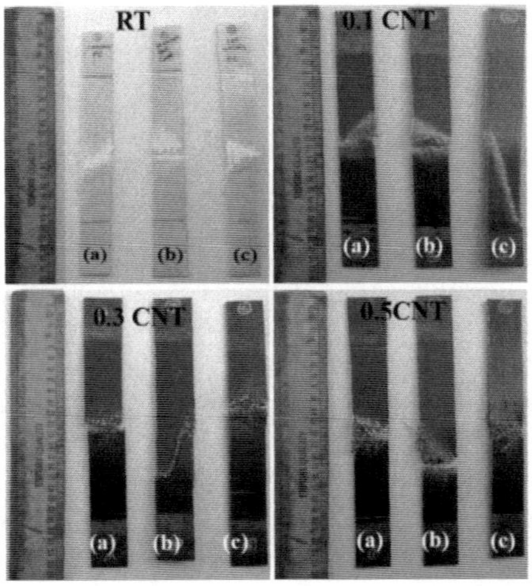

Figure 3. Figure: Different failure patterns of GFRP composite and CNT embedded GFRP composite tested at (a) 1mm/min, (b) 10 mm/min (c) 100 mm/min crosshead speeds.

3.3 Temperature modulated differential Scanning Calorimetry (TMDSC) Measurement

The glass transition temperature (T_g) of the matrix phase of the composites were analyzed using TMDSC. Figure 4 indicates the TMDSC curves for GFRP composite at RT, 0.1% CNT, 0.3% CNT and 0.5% CNT-GFRP content. The control GFRP composites indicates T_g about 95.32 °C, whereas the CNT-GFRP at 0.1, 0.3 and 0.5 wt. % of CNT exhibits 114.23 °C, 113.55 °C and 112.91 °C respectively. This indicates addition of CNT into GFRP composites significantly enhances the T_g. This may be attributed to the proper distribution of CNTs in the matrix resulting rise in the T_g.

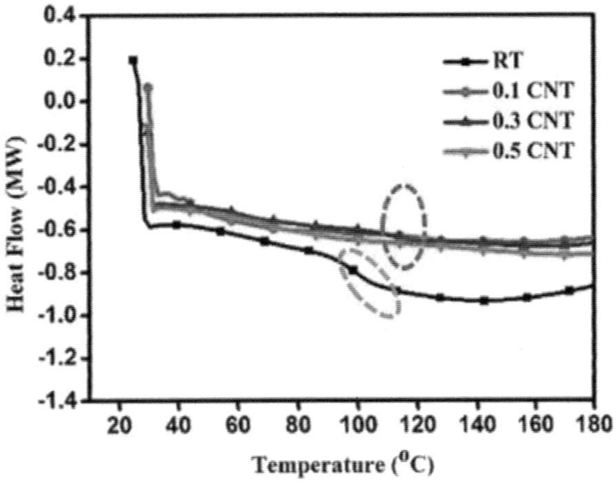

Figure 4. Temperature modulated differential scanning calorimetry (TMDSC) plot for GFRP composites at RT, 0.1% CNT, 0.3% CNT and 0.5% CNT

3.4 Fractography analysis

After, the tensile testing post failure analysis were done using nano nova FESEM (field emission scanning electron microscope) to examine the dominant mode of failures in the composites. Figure 5(a) and 5(b) indicates SEM images of 0.1% CNT/GFRP and 0.3% CNT/GFRP composites respectively. It is evident from figure 5(a) that dispersion of MWCNTs in epoxy matrix was better in case of 0.1% CNT/GFRP composite as compared to 0.5% CNT/GFRP composites [figure 5(b)]. Good dispersion of MWCNTs imparts proper stress transfer from the polymer matrix to fiber. But, when the MWCNTs particles starts to agglomerate at a particular region in the polymer matrix, the stress transfer across the composites is not uniform and hence drop in tensile modulus was observed with increasing percentage of MWCNTs in the composites.

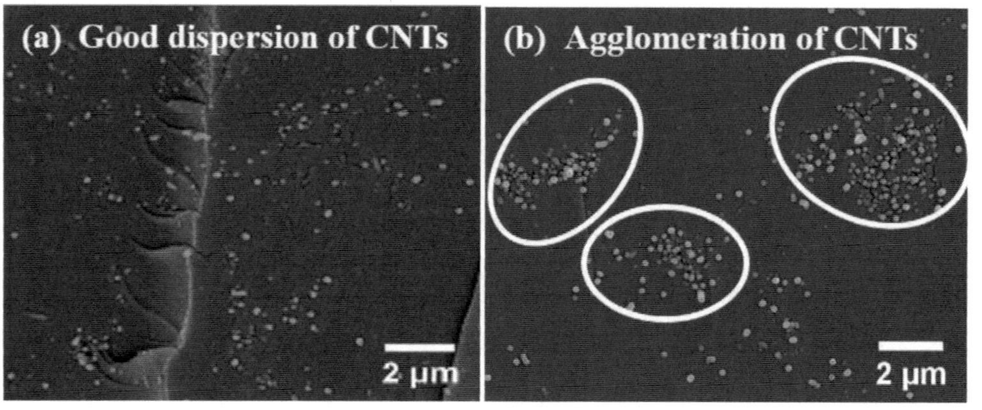

Figure 5. FESEM images of (a) 0.1% CNT/GFRP and (b) 0.5 % CNT/GFRP composites.

4. Conclusions

The present investigation comprises on the tensile response of the CNT addition with GFRP composites and further tested at various crosshead speeds includes the following conclusions

- Increase in the CNT content upto 0.3% the tensile strength increasing for all the crosshead speeds as compared to the control GFRP composite.
- Addition of 0.1% CNT and 0.3% CNT enhanced the tensile strength by 6.11% and 9.28% respectively than control GFRP composite.
- Decrease in tensile modulus was observed with addition of CNT due to agglomeration of the CNT in the polymer matrix composites.
- The glass transition temperature (Tg) of CNT embedded composites are found to be more as compared control GFRP.

Acknowledgement

The authors are heartily expresses their deep gratitude to NIT Rourkela for providing the necessary infrastructural and equipment facility.

References

[1] Li F, Cheng H M, Bai S, Su G and Dresselhaus M S 2000 Tensile strength of single-walled carbon nanotubes directly measured from their macroscopic ropes *Appl. Phys. Lett.* **77** 3161–3

[2] Salvetat J-P, Briggs G A D, Bonard J-M, Bacsa R R, Kulik A J, Stöckli T, Burnham N A and Forró L 1999 Elastic and Shear Moduli of Single-Walled Carbon Nanotube Ropes *Phys. Rev. Lett.* **82** 944–7

[3] Nayak R K, Mahato K K and Ray B C 2016 Water absorption behavior, mechanical and thermal properties of nano TiO2 enhanced glass fiber reinforced polymer composites *Compos. Part Appl. Sci. Manuf.* **90** 736–47

[4] Nayak R K, Mahato K K, Routara B C and Ray B C 2016 Evaluation of mechanical properties of Al2O3 and TiO2 nano filled enhanced glass fiber reinforced polymer composites *J. Appl. Polym. Sci.* **133**n/a-n/a

[5] Kathi J, Rhee K-Y and Lee J H 2009 Effect of chemical functionalization of multi-walled carbon nanotubes with 3-aminopropyltriethoxysilane on mechanical and morphological properties of epoxy nanocomposites *Compos. Part Appl. Sci. Manuf.* **40** 800–9

[6] Mahato K K, Biswal M, Rathore D K, Prusty R K, Dutta K and Ray B C 2016 Effect of loading rate on tensile properties and failure behavior of glass fiber/epoxy composite *IOP Conf. Ser. Mater. Sci. Eng.* **115** 12017

[7] Mahato K K, Rathore D K, Dutta K and Ray B C 2017 Effect of loading rates of severely thermal-shocked glass fiber/epoxy composites *Compos. Commun.* **3** 7–10

[8] Ray B C 2006 Loading Rate Sensitivity of Glass Fiber–Epoxy Composite at Ambient and Sub-ambient Temperatures *J. Reinf. Plast. Compos.* **25** 329–33

[9] Ray B C 2004 Effects of crosshead velocity and sub-zero temperature on mechanical behaviour of hygrothermally conditioned glass fiber reinforced epoxy composites *Mater. Sci. Eng. A* **379** 39–44

[10] Kumar D S, Shukla M J, Mahato K K, Rathore D K, Prusty R K and Ray B C 2015 Effect of post-curing on thermal and mechanical behavior of GFRP composites *IOP Conf. Ser. Mater. Sci. Eng.* **75** 12012

Structural and Dielectric Properties of Bismuth Doped Cobalt Nano Ferrites Prepared by Sol-Gel Auto Combustion Method

Krutika L. Routray[1*], Dhrubananda Behera[1]
[1]Department of Physics and Astronomy, National Institute of Technology, Rourkela, India
Email: rkruttika@gmail.com

Abstract

The present study is focused on the preparation of a series of Bismuth doped $CoFe_2O_4$ (CFO) nanoparticles and its structural and dielectric properties. The CFO particles having chemical composition $CoBi_xFe_{2-x}O_4$ (where x = 0.0, 0.1, 0.5) has been synthesized using sol-gel auto combustion method. Phase formation study of the prepared samples is carried out using X -ray diffraction (XRD) technique. It reveals single phase spinel structure having space group Fd3m. The two absorption bands ν_1 and ν_2 are observed in Fourier transform infrared spectroscopy (FTIR) spectra corresponding to the tetrahedral and octahedral sites, which show mark of spinel structure of the samples. Morphological study is carried out using Field Emission Scanning Electron Microscope (FESEM) and Transmission Electron Microscopy (TEM) which revealed that the particle size obtained is around 40 nm to 120 nm. In addition, phase formation has been confirmed from the analysis of Raman spectra. Different types of conduction mechanism of the charge carriers have been analysed with DC resistivity and AC impedance measurement at room temperature. Results demonstrate the presence of dielectric relaxation, which is found to be of non-Debye type.

Keywords
Cobalt ferrite ($CoFe_2O_4$), Solgel- auto combustion, Structural properties, Dielectric study

1. Introduction

Spinel-type ferrites have gained importance in recent years because of its extraordinary structural, magnetic, dielectric and electrical properties. $CoFe_2O_4$, being a spinel-type ceramic oxide with Co^{2+} ions at octahedral sites and Fe^{3+} ions equally occupying both tetrahedral and octahedral sites and is a suitable material for developing new technologies in the areas of strategic importance [1,2]. It is one of the essential nanoferrites with high coercivity, moderate magnetization and very high magneto-crystalline anisotropy [3]. Its high negative magneto-striction and capability to withstand at higher frequencies by minimizing the eddy current losses makes cobalt ferrite a suitable and economical material for developing oxide based magneto-strictive materials [4]. Cobalt Ferrite ($CoFe_2O_4$) is a spinel structured ceramic oxide and well recognised hard magnetic material with high coercivity and moderate magnetisation [5].

The doping composition plays a vital role in influencing the properties of Cobalt ferrite by altering its chemical as well as physical properties. $BiFeO_3$ (Bismuth Ferrite) is perhaps the only material that exhibits both magnetic and a strong ferroelectric behaviour at room temperature and can alter properties of cobalt ferrite due to its high electrical resistivity with low magnetic and dielectric losses. There is a coexistence of interrelated electric and magnetic dipole structures within a certain temperature range in $BiFeO_3$ with perovskite structure. Therefore, Bi_2O_3 is a potential dopant for enhancing the magnetic and electrical properties of ferrites [6]. It is reported that substituting bismuth in very small amount does not alter the spinal structure of Cobalt ferrite [7]. Bismuth doped Cobalt nanoferrites are used for magnetic recording applications such as audio, videotapes, high density data storage devices and microwave device applications [8]. Among several methods to synthesize nanoferrites, the sol-gel auto combustion method is a simple method which quickens up the synthesis of complex materials and brings crystal homogeneity [9]. Hence, in this study, we report the synthesis, its structural characterisations and dielectric properties of Bismuth doped Cobalt nanoferrites with the chemical composition $CoBixFe_{2-x}O_4$ (where x = 0.0, 0.1, 0.5).

2. Experimental details

2.1 Synthesis

A series of Bismuth substituted cobalt nano ferrites having the chemical formula $CoBi_xFe_{2-x}O_4$ (where x = 0.0, 0.1, 0.5) were prepared by sol-gel auto combustion method. The precursors used were Cobalt Nitrate, Iron Nitrate, Bismuth Nitrate and glycine. The stoichiometric amount of all metal nitrates mentioned above was dissolved in a glass beaker with minimum quantity of double distilled water. Glycine was added to the mixed metal solution with molar ratio of nitrates to glycine is 1:4 and mixed together with the help of magnetic stirrer at about 150^0 C with uniform stirring and evaporated to obtain a highly viscous gel. During this procedure, the solution became viscous and finally formed a very homogenous viscous brown gel. The auto-ignition was completed, yielding the black-coloured ashes similar to that of the structure of a branched tree. The as-prepared powders of all the samples were calcined separately at 600^0 C for three hours to get the final product. Further the prepared compositions were characterized by XRD, FESEM, TEM, RAMAN and FTIR followed by Dielectric studies.

2.2 Characterization techniques

The resultant powder was characterized by qualitative X-ray powder diffraction measurements to check the phase purity of the sample with the help of Rigaku X-ray Diffractometer (using Cu-Kα radiation, λ=1.54 Å) at room temperature. X-ray diffraction peaks may be indexed in the space group Fd3m. No trace of impurity was detected by X-ray diffraction. The XRD data for Rietveld refinements were collected over the range of 2θ = 200 – 800 with a step size of 0.020. Revised Rietveld analysis of the X-ray data was carried out by FULLPROF program. Microscopically investigation has been carried out with Nova Nano SEM/FEI field emission scanning electron microscope (FESEM) and FEI transmission electron microscopy (TEM) to determine the particle size. An STR 500 confocal micro-Raman spectrometer (532 nm DPSS laser) was used for Raman spectroscopy. FTIR spectroscopy was carried using Perkin Elmer, USA/ RX-I FTIR. Impedance measurement has been performed using H20KI LCR meter (IM3570) at the room temperature.

3. Results and discussion

3.1 Structural information

3.1.1 *XRD Analysis*

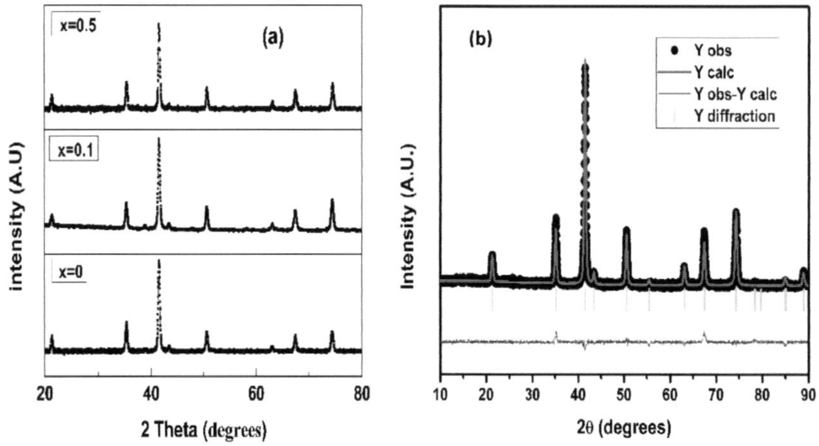

Figure 1. (a) X-ray diffraction patterns of $CoBi_xFe_{2-x}O_4$ (where x = 0.0, 0. 1, 0.5), (b) Reitveld refinement of parent cobalt ferrite ($CoFe_2O_4$).

Figure 1(a) shows X-ray diffraction patterns of $CoBi_xFe_{2-x}O_4$ (where x = 0.0, 0.1, 0.5) nano ferrite particles at room temperature. Generally for a pure cobalt ferrite, the XRD pattern exhibits eight peaks which are located in between $2\theta = 10^0$ to 80^0. The position of the peaks is assigned to cobalt ferrite in accordance with the JCPDS card no 22-1086. The phase shows the pure cubic structure of $CoFe_2O_4$ with as expected inverse spinel lattice [10]. Figure 1 (b) shows the Reitveld refinement which further confirmed the formation of the single phase of parent cobalt ferrite ($CoFe_2O_4$). The inversion parameter (χ^2) calculated from Rietveld refinement is 1.04 mentioned in the equation below:

$$\chi^2 = \left[\frac{R_{wp}}{R_{exp}}\right]^2 ; \quad (1)$$

where χ^2 is the goodness of Fit of the refinement;
$\boldsymbol{R_{wp}}$ is the weighted pattern of the refinement;
$\boldsymbol{R_{exp}}$ is the calculated pattern of the refinement.

On substitution of Bismuth in a very small amount the spinal structure of ferrite system does not get altered. From the XRD pattern it is identified for x = 0.0, 0.1, 0.5 shows peaks consistent with cubic spinel phase. From the most intense peak of the reflection (311), we have estimated the average particle size of the nanoparticle using Scherrer equation mentioned below:

$$\tau = \frac{K\lambda}{\beta cos\theta} ; \quad (2)$$

where:
τ is the mean size of the ordered (crystalline) domains, which may be smaller or equal to the grain size;
K is a dimensionless shape factor;
λ is the X-ray wavelength;
β is the line broadening at half the maximum intensity (FWHM), after subtracting the instrumental line broadening, in radians;
θ is the Bragg angle (in degrees).

3.1.2 FESEM Study

Figure 2. FESEM images of (a) $CoFe_2O_4$ nano ferrite (b) $CoBi_{0.1}Fe_{1.9}O_4$ nano ferrite (c) $CoBi_{0.5}Fe_{1.5}O_4$ nano ferrite.

The Field Emission Scanning Electron Microscope (FESEM) images of the parent and modified cobalt ferrite nano particles ($CoBi_xFe_{2-x}O_4$, x = 0 where x = 0.0, 0.1, 0.5) are shown in Figure 2(a, b, c). It is clear from the FESEM images that the particles have an almost homogeneous distribution and some particles are in agglomerated form. An apparent growth in the particle size is observed which is due to the Bi substitution. Lijun Jia et al. reported a similar behaviour in development of particle size on substitution of Bi in NiCuZn ferrite [11]. The particle size grows controllably from ~ 40 nm in parent to ~ 120 nm in Bi modified cobalt ferrite samples which may be due to the increase in lattice constant which creates internal strain. With this study it can be observed that the average particle size was in increasing trend with Bi substitution.

3.1.3 TEM Analysis

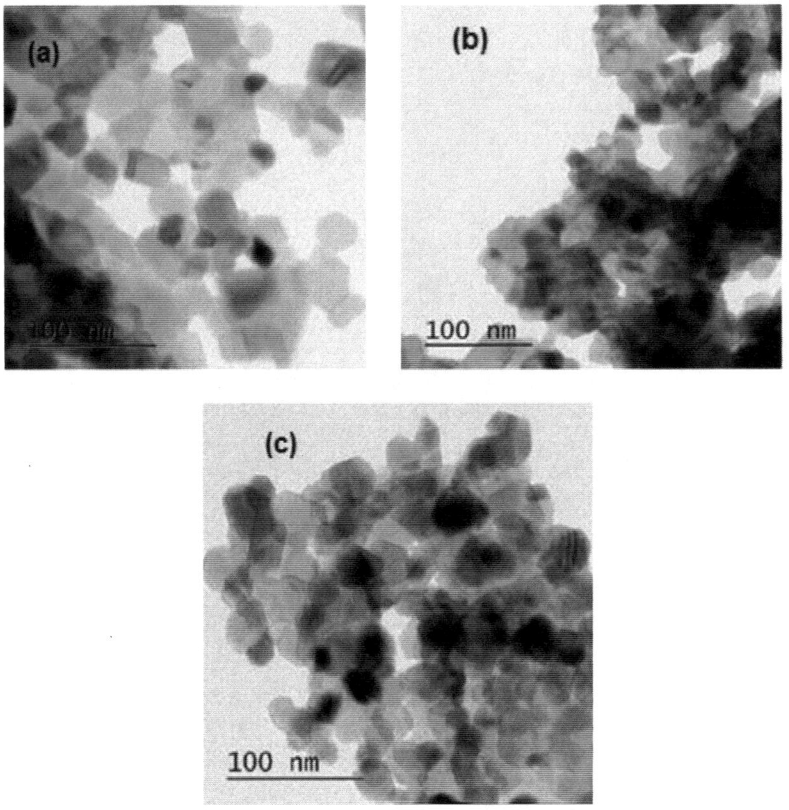

Figure 3. TEM images of (a) $CoFe_2O_4$ nano ferrite (b) $CoBi_{0.1}Fe_{1.9}O_4$ nano ferrite (c) $CoBi_{0.5}Fe_{1.5}O_4$ nano ferrite.

Figure 3 shows the transmission electron microscopy (TEM) image of $CoBi_xFe_{2-x}O_4$, (where x = 0.0, 0.1, 0.5) nanoparticles synthesized at 600^0 C. The average particle size obtained is approximately equal to 40 nm and varies to 90 nm with increase in Bismuth substitution which is in agreement with XRD pattern and FESEM images. The lattice parameter value obtained in all cases is 8.38 Å, which is also in good agreement with the standard value.

3.1.4 RAMAN Study

The samples $CoBi_xFe_{2-x}O_4$, (where x = 0.0, 0.1, 0.5) were further characterised by Raman spectroscopy for further confirmation of phase purity as shown in figure 4. Room temperature Raman spectra of the samples were recorded in the range 0-3000 cm^{-1}. As shown in the figure, strong peaks are observed in range 200 - 800 cm^{-1} and very weak peaks are observed in the range 1000 - 3000 cm^{-1}. The obtained result is in agreement with the XRD results.

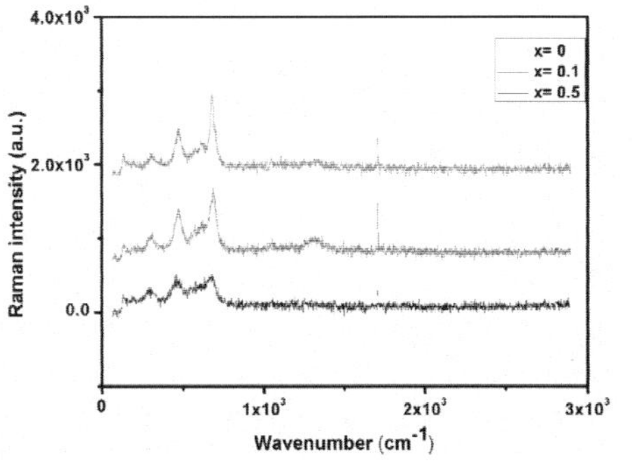

Figure 4. RAMAN spectra of $CoBi_xFe_{2-x}O_4$, (where x = 0.0, 0.1, 0.5).

The resulted peaks in the order of formation are due to the symmetric stretching of oxygen atoms along Fe-O bonds in tetrahedral coordination, symmetric bending of oxygen with respect to the metal ion, asymmetric bending of oxygen, asymmetric bending of Fe (Bi) and O, vibrational modes at octahedral position and translational movement at tetrahedron sites [12].

3.1.5 FTIR Spectra

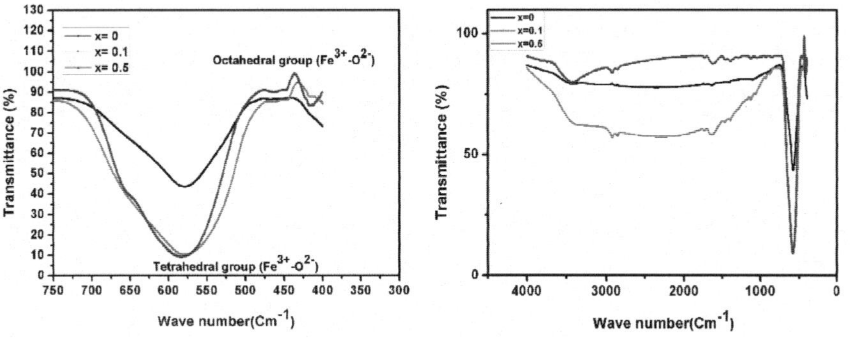

Figure 5. FT-IR spectra of $CoBi_xFe_{2-x}O_4$, (where x = 0.0, 0.1, 0.5).

The FT-IR spectra of $CoBi_xFe_{2-x}O_4$, (where x = 0.0, 0.1, 0.5) nano ferrite particles, synthesised at room temperature is represented in the figure 5. In this present work, we observe three broad features in the frequency range, 3350 - 3500 cm^{-1}, 1600-1647cm^{-1}, 440-600 cm^{-1}.The above broad frequency range corresponds to O-H stretching, H-O-H bending vibration of water molecules and spinal structure with sub-lattices respectively [13]. The presence of frequency bands in this specific range explains the normal mode of vibration of tetrahedral cluster is higher than that of octahedral cluster in the spinel ferrites. The highest one corresponds to the intrinsic stretching vibrations of the metal at the tetrahedral site, whereas the lowest band is assigned to octahedral-metal stretching. This difference between the tetrahedral and octahedral clusters of spinel ferrite is because of the variation in distance between Fe^{3+}- O^{2-} ions in tetrahedral and octahedral sites. The wave band v_1 shows an increase in the values with increase in Bi concentration which may be due to dissimilarity in the cations-oxygen bond length.

3.2 Impedance and modulus of spectroscopy study

The mechanism used to analyse the microstructure-property relationship is the Complex impedance spectroscopy (CIS) which not only helps in distinguishing between intrinsic bulk and extrinsic contributions grain boundary, surface layer, and electrode contact problem but also enables us to separate the real and imaginary component of the electrical parameters so as to get the true picture of the material properties. This technique is based on analysing the ac response of the system to sinusoidal perturbation and subsequent calculation of impedance as a function of frequency of perturbation. The frequency dependence of electrical properties of a material is represented in terms of complex impedance and Z^* electrical modulus (M^*), complex dielectric constant (ε^*) and tangent loss $(tan\delta)$ [14]. The following are the complex impedance related parameters:

Complex impedance $Z^* = Z' - jZ'' = R_s - {}^{j}/_{\omega C_s}$; (3)

Complex modulus $M^* = 1/\varepsilon^* = M' + jM'' = j\omega C_0 Z$; (4)

Loss tangent $= \varepsilon''/\varepsilon = M''/M' = -Z'/Z''$; (5)

where Z', M', ε' are the real components of impedance, modulus and permittivity and Z'', M'', ε'' are the imaginary parts of impedance, modulus and permittivity, $j = \sqrt{-1}$ is the Impedance and modulus of spectroscopy study is the imaginary factor and $\omega = 2\pi f$ is the angular frequency. Inverse spinel ferrites namely cobalt ferrite possess good dielectric properties, depending on the size of the particle, distribution of cation and the method of preparation. Dielectric properties (real and imaginary parts of impedance, relative permittivity, dielectric loss tangent) for the prepared series of $CoBi_xFe_{2-x}O_4$, (x = 0.0, 0. 1, 0.5) have been studied in the frequency range 100 Hz- 1 MHz at room temperature.

3.2.1 Impedance study

In order to analyse the electrical properties of spinel type ferrites Impedance analysis is used to determine the contribution of conductivity for both resistive (real part) and reactive (imaginary part) when AC field is applied [15]. Nyquist diagrams also known as Cole-Cole plots or Complex impedance plot gives a complete contribution of microstructure (grains and grains boundary) resistances and helps to differentiate the grain and grain boundary resistances and interfacial resistance of conducting electrodes. Figure 6 (a) and (b) shows the variations of real parts and imaginary parts of complex impedance with applied frequency (100 Hz-1 MHz). From the plots it is observed that up to a definite limiting range (~10 kHz) Z' decreases monotonically with increasing frequency thereafter they merge together and becomes nearly independent of frequency. The higher values of Z' at lower frequencies indicate that the polarization effect in the sample is larger. The values of Z" decreases identically monotonically up to a certain frequency limit (~100 kHz) beyond which they merge together at a very low value of Z" where frequency-independent nature of variation is observed till the highest frequency limit at room temperature for the varying concentration of bismuth doped cobalt nano-ferrite samples.

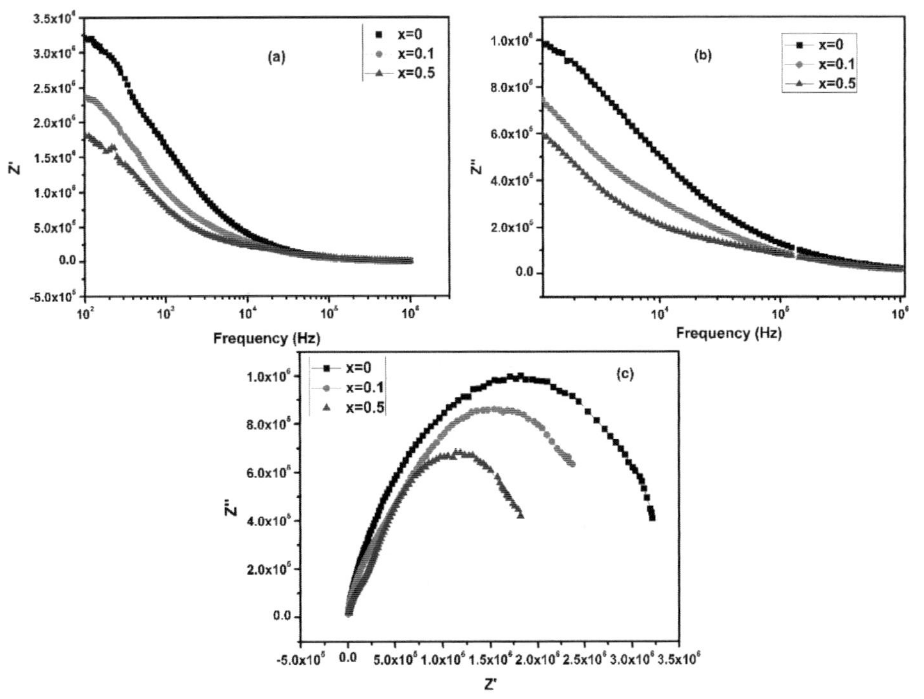

Figure 6. (a) Real part of impedance with applied frequency, (b) imaginary part of impedance with applied frequency (c) the complex impedance plane plots (Cole-Cole) for $CoBi_xFe_{2-x}O_4$, (where x = 0.0, 0.1, 0.5).

This merging of Z' and Z" at higher frequencies specifies the probable discharge of space charge accumulation at the boundaries of the homogeneous phases in the sample when on application of AC field. It is observed that both the resistance (Z') and reactance (Z") values decreases with increase in applied AC field concluding that the conduction is stimulated by application of AC field. As observed there is a monotonic decrease in the values of Z' and Z" with increasing frequencies for all the samples which indicates that relaxation behaviour in the material is dependent on the temperature and no single relaxation time is present. To further examine and differentiate the microstructure such as intrinsic (grain) and extrinsic (interfaces like grain boundary, sample-electrode effect, defects, and relaxation process etc.) Cole-Cole of impedance (Z' vs. Z") is plotted. Appearance of semicircle in the plot represents a relaxation process, whose radius measures the resistance of the sample and centre lies on the abscissa (Z' - axis) if the conduction is of Debye type. For a non-Debye type conduction process, a depressed semicircle appears in the diagram. For the appearance of three distinct semicircles in the diagram, the one appears at high frequency end is assigned to grain effect, the semicircle at intermediate frequency zone corresponds to grain boundary effect and that of at lower frequency end is ascribed to sample surface conduction effect [16]. We observe only one Cole-Cole semicircle in the plot where the radius of the semicircle decreases with increases in Bismuth concentration indicating the decrease in resistance of the material. All the curves show a tendency to bend towards the abscissa to form semicircles with their centres above the real axis, having comparatively decreased in radii. Though a single relaxation process is observed in the experimented temperature range it is tough to assign that the change in behaviour is due to intrinsic conduction effect as the resistance of the material is sufficiently low to prevent the arrival of extrinsic interfacial conduction effect.

In order to properly differentiate between the intrinsic and extrinsic consequence on conduction mechanism we illustrate the frequency dependent imaginary part of Modulus (M") in Figure 7 to verify the relaxation behaviour.

3.2.2 Complex Modulus Analysis

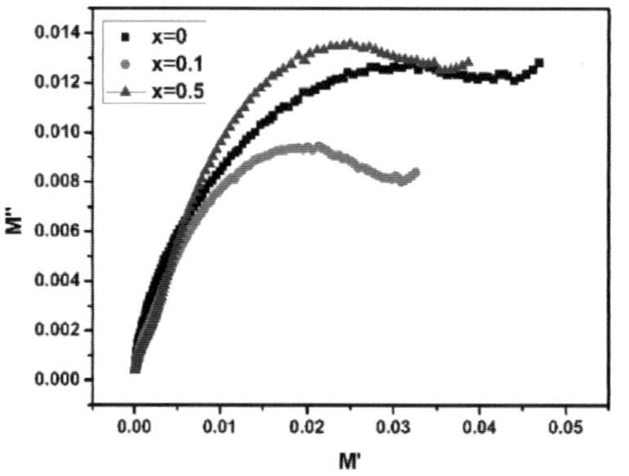

Figure 7. Complex modulus plots for $CoBi_xFe_{2-x}O_4$, (where x = 0.0, 0.1, 0.5) corresponding to the room temperature measurement.

An alternative method to explore the electrical properties of the material as a result of different relaxation time constants and expand any other properties present in the sample (which are unidentifiable or overlapping in CIS technique) complex modulus analysis is used. It is used to determine the dynamical aspects of electrical transport phenomena (carrier/ion, hopping rate, conductivity relaxation time, etc.). Electric modulus as a parameter varies inversely to the capacitance of the material and hence has the ability to suppress the highest capacitive region of sample-electrode surface conduction effect and highlight the least capacitive grain conduction effect [17]. We observed only one peak in Z' vs. Z" plots but in case of M' vs. M" plot, two peaks are observed at room temperature for all the compositions taken. Figure 7 shows the complex modulus plots for $CoBi_xFe_{2-x}O_4$, (x = 0.0, 0.1, 0.5) corresponding to the room temperature measurement. From this figure, we observe that the modulus spectrum has a typical pattern characterized by the presence of little asymmetric and depressed semi-circular arcs with its centre not lying on M' axis, thereby indicating non-Debye type of relaxation response in the investigated material system. The behaviour of the modulus spectrum on right-side semicircles represent values corresponding to higher frequencies and left semicircles corresponding to lower frequencies for the variation in concentration of Bismuth. These semicircles indicate that both grain and grain boundary capacitance play vital roles in the conduction mechanism of the material system at the measured temperature. The behaviour of electric modulus spectrum is indicative of the temperature dependent hopping type of mechanism for electric conduction (charge transport) in the investigated material system and non-Debye type dielectric relaxation.

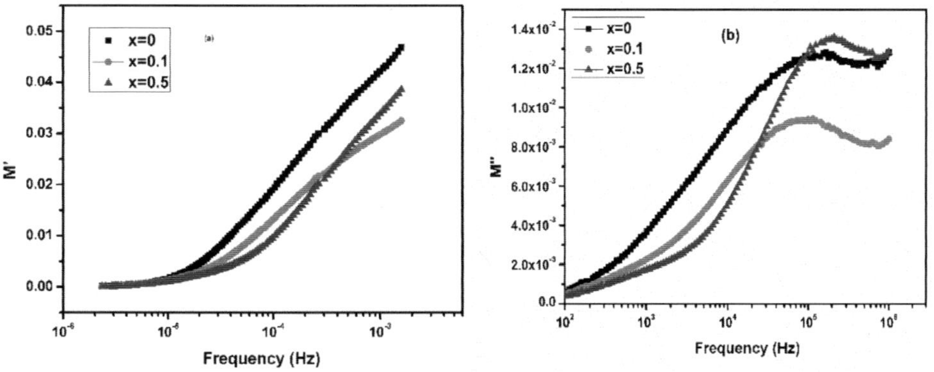

Figure 8. (a) Real part of impedance with applied frequency, (b) imaginary part of impedance with applied frequency of $CoBi_xFe_{2-x}O_4$, (where x = 0.0, 0.1, 0.5).

Figure 8(a) shows the variation of Real part (M') of modulus with frequency at room temperatures of $CoBi_xFe_{2-x}O_4$, (where x = 0.0, 0.1, 0.5). The variation of M' with frequency shows a dispersion tending towards the asymptotic plot of M' at higher frequencies which is due to the relaxation time of the sample. A monotonous dispersion occurs with increasing frequency at room temperature which is because of the short range mobility of charge carriers. Such observed behaviours may be associated with the absence of restoring force that governs the mobility of the charge carriers under the influence of an induced electric field. Figure 8(b) represents the variation of imaginary part of modulus (M") with frequency at room temperatures for variable Bismuth concentration ($CoBi_xFe_{2-x}O_4$, where x = 0.0, 0.1, 0.5). It is observed from the graph that M" indicates the presence of strong peaks and the position of the peaks (M") get shifted to higher frequencies with increase in Bismuth concentration. It is observed that the peaks are asymmetric and broader than the ideal non-Debye curve.

4. Conclusions

In this present work, structural, dielectric, and impedance / modulus conductivity studies on $CoBi_xFe_{2-x}O_4$, (x = 0.0, 0.1, 0.5) in the frequency range 100 Hz-1 MHz in the room temperature is reported. The XRD pattern confirms that sample has a cubic structure at room temperature. Dielectric properties of sample have been carried out in the frequency range 100 Hz - 1 MHz. Impedance spectroscopy concluded the presence of single semicircle arc at all reported temperatures which indicate that there is presence of grain interior (bulk) property in material. The samples showed presence of dielectric relaxation, which is found to be of non-Debye type. As has been observed, the particle size increases with Bi doping, the resistance of the material is decreased. The semicircles in the complex modulus plot indicated that both grain and grain boundary capacitance play a vital role in the conduction mechanism of the material system at the measured temperature.

References

[1] Cheng F, Liao C, Kuang J, Xu Z, Yan C, Chen L, Zhao H and Liu Z 1999 *J. Appl. Phys* **85(5)** 2782-6.
[2] Ryu J, Priya S, Uchino K and Kim HE 2002 *J. Electroceram* **8(2)** 107-19.

[3]Maaz K, Karim S, Mashiatullah A, Liu J, Hou MD, Sun YM, Duan JL, Yao HJ, Mo D and Chen YF 2009 *Physica B: Condensed Matter* **404(21)** 3947-51.
[4] Bhame SD and Joy PA 2007 *J. Phys. D: Appl. Phys* **40(11)** 3263.
[5] Chen W, Zhu W, Tan OK and Chen XF 2010 *J. Appl. Phys* **108(3)** 034101.
[6] Ghosh S, Dasgupta S, Sen A and Sekhar Maiti H 2005 *J. Am. Ceram. Soc* **88(5)** 1349-52.
[7] Kumar NS and Kumar KV 2015 WJNSE **5(04)** 140.
[8] Gupta N, Dimri MC, Kashyap SC and Dube DC 2005 *Ceram. Int* **31(1)** 171-6.
[9] Jo W, Dittmer R, Acosta M, Zang J, Groh C, Sapper E, Wang K and Rödel J 2012 J. *Electroceram* 2012 **29(1)** 71-93.
[10] JCPDS data base 22-1086.
[11] Jia L, Zhang H, Wu X, Li T, Su H. and Liu B 2012 *J. App. Phys* **111(7)** 07A326.
[12] Romþeviü M, Milutinoviü A and Romþeviü NŽ 2012 *Sci. sinter* **44** 331-9.
[13] Faheim AS, Abd El Fattah MK, Abdul Rahman AH and Badawi A 2014 *J. Nanomater. Mol. Nanotechnol* **5** 4-10.doi: http://dx. doi. org/10.4172/2324-8777.100015
[14] Pattanayak S, Priyadarshan A, Subudhi R, Nayak RK and Padhee R 2013 *J. Adv. Ceram* **2(3)** 235-41.
[15] Farea AM, Kumar S, Batoo KM, Yousef A and Lee CG 2008 *J. Alloys Compd* **464(1)** 361-9.
[16] Irvine JT, Sinclair DC and West AR 1990 *Adv. Mater* **2(3)** 132-8.
[17] Sinclair DC and West AR 1989 *J. Appl. Phys* **66(8)** 3850-6.

Investigating the Magneto Electric Coupling of [90 wt% $Na_{0.5}Bi_{0.5}TiO_3$ (NBT)-10 wt% $BaFe_{12}O_{19}$ (BaM)] novel multiferroiccomposite system by increasing of BaM grain size

Ranjit Pattanayak[1*], Subhajit Raut[1], Sourav Kuila[1], MallamChandrasekhar[1] and Simanchala Panigrahi[1]
[1] Department of Physics and Astronomy, National Institute of Technology, Rourkela 769008, India

[*]E-mail: ranjit.p20@gmail.com

Abstract. Polycrystalline three novel [90 wt% $Na_{0.5}Bi_{0.5}TiO_3$ (NBT)-10 wt% $BaFe_{12}O_{19}$ (BaM)] magnetoelctricmultiferroic composite systems were fabricated by considering the variation (increasing) of BaM grain size. The desired formation of composites was confirmed by X-ray diffraction study. The FESEM and SEM study were verified the variation of grain size and 0-3 type connectivity of composite systems. To predict the room temperature multiferroicbehaviour of theses composite systems we were taken PE and MH loop. For investigating the extrinsic and intrinsic magnetoelctric effect magneto impedance spectroscopy was considered for theses composite systems. The variation of intrinsic magnetoelctric coupling was predicted by proposing a simple mechanical model.

Keywords: Composites; Multiferroic; Magneto Impedance; Magnetoelctric Coupling (MEC)

1. Introduction

In this fastest developing technological era, high efficiency new multifunctional devices were entailing for practical applications. To fulfil the desire, recently muliferroic ceramic materials have motivated a sharply increasing number of research activities due to the presence of two or more ferroic orders (such as ferroelectric, ferromagnetic, or ferroelastic) and coupling interactions between them [1][2]. The coupling in a mutiferroic system is the control of dielectric polarization (P) by a magnetic field (H) [direct ME effect (ME_H)] and the manipulation of magnetization (M) by an electric field (E) [converse ME (ME_E)] can be most promising candidate for different applications such as: devices that can control of the magnetic (spin) state via electric fields and/or vice versa, spintronic devices, solid-state transformers, high sensitivity magnetic field sensors, and electromagnetooptic actuators, electric-field controlled ferromagnetic (FM) resonance devices, sensors, transducers, and terahertz emitters, miniature antennas, etc[3–7]. According to Landau-Devonshire theory, the free energy F of the system in terms of an applied magnetic field H and electric field E will be [1,8]:

$$-F(E,H) = P_s \cdot E_0 + \mu_0 M_s H_0 + \frac{\varepsilon_0}{2}\sum_{ij} \varepsilon_{ij} E_i E_j + \frac{1}{2}\sum_{ij} \mu_{ij} H_i H_j + \sum_{j} \alpha_{ij} E_i H_j \cdots\cdots\cdots$$

The polarization and magnetization for a single-phase multiferroic system are defined as differentiation of free energy (F);

$$P_i = -\left(\frac{\partial F}{\partial E_i}\right) = P_{si} + \varepsilon_0 \sum_{j} \varepsilon_{ij} E_j + \sum_{j} \alpha_{ij} H_j + \cdots\cdots\cdots$$

$$M_i = -\left(\frac{\partial F}{\partial H_i}\right) = M_{si} + \frac{1}{2}\sum_j \mu_{ij} H_j + \sum_j \alpha_{ij} E_j + \cdots\cdots\cdots$$

Where P_s and M_s are the electric and magnetic spontaneous polarizations; ε_{ij} and μ_{ij} are the electric and magnetic susceptibility of second-rank tensors; α is the linear magnetoelectric coupling tensor; and ε_0 is the free space permittivity. For the linear magnetoelectric effect, it will be:
$\alpha_{ij} = \left(\frac{\partial P_i}{\partial H_i}\right) = \left(\frac{\partial M_i}{\partial E_i}\right)$.
So for an isotropic single phase multiferroic material the coupling was constrained by $\alpha_{ij}^2 \leq \varepsilon_0\mu_0\varepsilon_{ii}\mu_{jj}$. It was also observed that, most of single-phase multiferroics responses either relatively weak magnetoelectric coupling even at high field or occurs at temperatures too low than room temperature [1].
In compare, magnetoelctric (ME) multiferroic composite systems both (anti) ferroelectric and (anti) ferromagnetic orders present in separate phases which can yield giant magnetoelectric coupling response above room temperature, which makes them ready for technological applications [1,8]. For these composite multiferroics, the magnetoelectric coupling term not necessarily linear, however, coupling is originating at the interface due to interaction of piezoelectric and pizzomagnetic dynamics between the two dissimilar phases. As the result through the interface symmetry breaks and allows the presence of the magnetoelectric(ME) coupling in composite multiferroics[1]. For a two phase composite system there are ten different connectivity schemes can possible (0–0, 1–0, 2–0, 3–0, 1–1, 2–1, 3–1, 2–2, 3–2, and 3–3) where their distributions as well as connectivity plays crucial role on the magnetic, electric as well as magnetoelectric coupling properties [9][10]. In these ME multiferroic composite systems, the ME coupling effect is a result of the product of the magnetostrictive effect (magnetic/mechanical effect) in the magnetic phase and the piezoelectric effect (mechanical/electrical effect) in the piezoelectric one, namely[11,12],

$$ME_H effect = \frac{Magnetic}{Mechanical} \times \frac{Mechanical}{Electric} \cdots\cdots\cdots\cdots\cdots\cdots\cdots \quad (1)$$

$$ME_E effect = \frac{Electric}{Mechanical} \times \frac{Mechanical}{Magnetic} \cdots\cdots\cdots\cdots\cdots\cdots\cdots \quad (2)$$

A series of studies on the dielectric, multiferroic and magnetoelectric properties of several composites as a function of different relative phase content and connectivity have been studied and reported such as : NiZnFeO–PbSmZrTiO[13], $BaTiO_3$–$Ni_{0.7}Zn_{0.3}Fe_2O_4$[14], $CoFe_2O_4$-$Pb_{0.7}Ca_{0.3}TiO_3$[15], laminate composite of Terfenol- D and $LiNbO_3$[16], Terfenol-D-epoxy 1–3 and 2–2 composite [17], 1-3 type composites with $CoFe_2O_4$ ferromagnetic microstrips embedded in (K,Na)NbO_3-based [18]. From the past works we were perceived, not only selection of ferroelectric and magnetic phases but also the connectivity between two phases is also very important for ME coupling. The connectivity can be tailor by varying the grain size of single or both phases for a composite system.In our present work, we have selected Barium hexaferrite (BaM) for magnetic phase (having high magnetic curie temperature T_{mc}) and Sodium Bismuth Titanate (NBT) for ferroelectric phase (having high ferroelectric curie temperature T_{fC}) to fabricate a novel composite system [19][20]. Recently, few multiferroic works upon BaM and NBT based composite systems has been reported[21–23]. But our work principally focused on fabrication of the novel composite systems [90 wt% $Na_{0.5}Bi_{0.5}TiO_3$ (NBT)-10 wt% $BaFe_{12}O_{19}$ (BaM)] and analysis of multiferroic as well as ME coupling properties with variation of BaM grain size (from nano range to micron size) keeping NBT grain size constant (in micron size).

2. Experimental details

Three polycrystalline The 90 wt% $Na_{0.5}Bi_{0.5}TiO_3$ (NBT)-10 wt% $BaFe_{12}O_{19}$ (BaM) composite systems (S1, S2 and S3) were synthesized by conventional solid state technique with variation grain size of BaM phase as keeping NBT grain size constant (micron size). Initially, polycrystalline nano $BaFe_{12}O_{19}$ (BaM- in powder format) was successfully synthesized by auto combustion technique considering stoichiometry amount of $Ba(NO_3)_2$ and $Fe(NO_3).9H_2O$ as precursors where glycine ($C_2H_5NO_2$) used as fuel. Then the combusted powder was calcinated at 900 °C for 12h for getting the pure phase of $BaFe_{12}O_{19}$ (BaM). The polycrystalline bulk $Na_{0.5}Bi_{0.5}TiO_3$ (in powder format) was synthesized by using conventional solid state technique conventional solid state technique by considering stoichiometry amount Bismuth oxide (Bi_2O_3), Sodium carbonate (Na_2CO_3) and Titanium oxide (TiO_2) as precursors with calcinations temperature as 1050 °C for 3h. Then for increasing of grain size of BaM phase the previous calcinatedBaM powder was fired at 1000 °C and 1150 °C respectively.

Then for synthesis of system S1, the calcined $Na_{0.5}Bi_{0.5}TiO_3$ (NBT) at 1050 °C and $BaFe_{12}O_{19}$ (BaM) at 900 °C powders were mixed in weight percentage 90:10 wt% thoroughly by grinding. Then polyvinyl alcohol (PVA) used as a binder in the BaM–NBT mixed powder pressed into cylindrical pellets of 10 mm diameter and 1mm thickness& sintered at 930 °C for 8h. Similarly, the systems S2 and S3 were fabricated (in pellet format) by considering the above procedure. But, for S2 and S3 systems we have taken the 1000 °C and 1150 °CcancinedBaM powders and sintered at 1000 °C and 1100 °C for 5h. The formation of the desired composite systems, was verified by XRD using DMAXB/Rigaku in a wide range of Bragg angles 20°–80° with Cu Kα radiation (λ=1.5405 Å). The surface morphology was studied by Nova Nano SEM-450 Field emission scanning electron microscopy (FESEM) and by JEOL JSM-6480LV Scanning Electron Microscopy system (SEM). The room temperature dielectric parameters, i.e. impedance, and phase angles, were measured in a frequency range of 100–1 MHz by using HIOKI IMPEDANCE ANALYZER 1352. The room temperature (RT) magneto-capacitance data were taken by HIOKI impedance analyzer coupled with a DC electromagnet at different magnetic field (0, 0.4, 0.8 and 1.2 Tesla). The RT M-H and PE loop data was taken by Microscan Vibrating sample magnetometer (VSM) and Marine India PE loop tracer.

3. Results and discussion

The figure 1 represents the X-ray diffraction pattern of all the composite systems having increasing of BaM calcined temperature (increasing of BaM grain size). From figure 1 it can be confirmed that the desired composite systems were successfully fabricated with having both ferroelectric (FE) NBT and ferrimagnetic (FM) BaM-phases without appearing of any impurity phases.

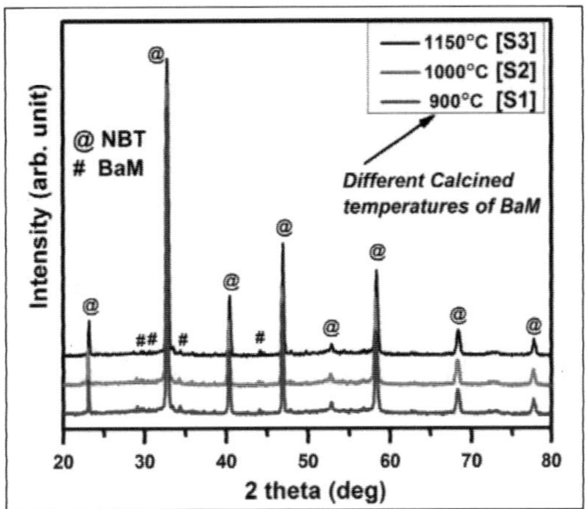

Figure 1. X-ray diffraction pattern of composite systems (S1, S2 and S3) having different calcined temperature of BaM phase.

Figure 2 represents the well dense micrograph having 0-3 type connectivity of all fabricated composite systems. From the image it can be easily indentify the two different types of grain such as: dispersed hexagonal shape grains of FM phase (BaM) and matrix of FE (NBT)phase having polygonal shape grains.

Figure 2.FESEM and SEM micrographs of composite systems having different calcined temperature of BaM phase (a) calcined at 900 °C (b) calcined at 1000 °C (c) calcined at 1150 °C.

It can also well perceive that, as calcined temperature of BaM increases the corresponding grain size increases (follow the insights-*zoom*). For calcined temperature 900 °C the BaM grain size ≈200nm (a), for calcined temperature 1000 °C the BaM grain size ≈700nm (b) , calcined temperature 1500 °C the BaM grain size ≥1µm (c). It is well-known that the distribution and connectivity of both ferrite and ferroelectric phases in a composite plays crucial role on the various properties such as: magnetic, dielectric, resistivity as well as magnetoelectric coupling [24–27].

Before we have to investigate the MEC of these composite systems, at first we have verified the presence of both ferroelectric and ferromagnetic orders (multiferroic) of these composite systems by considering the PE loop and M-H loop study at room temperature (Figure 3).

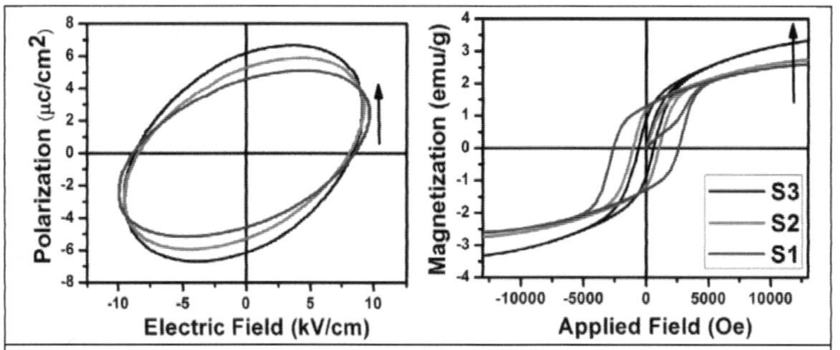

Figure 3.P-E loop and M-H loop of composite systems.

Table 1. Polarization and magnetization parameters of composite systems.

Systems	P_s (µc/cm^2)	P_r(µc/cm^2)	Systems	M_s (emu/g)	H_c (Oe)
S1	5.1	4.6	S1	2.5	2768
S2	5.9	5.2	S2	2.7	1159
S3	6.6	6.1	S3	3.2	616

From figure 3 it was observed that, all the systems shown both ferroelectric and ferromagnetic hysteresis loops which is the evidence of mutiferroicbehaviour of these composite systems at room temperature (RT). From the variation of PE loop and M-H loop, we can observe that, as BaM grain size increases the saturation polarization (P_S) and saturation magnetization (M_S) increases (the values are given in table 1). These variations for the composite systems are due to variation of BaM grain size but the results are purely intrinsic (not from interfaces). The increasing of (M_S) with increasing of BaM grain size may be due to strengthening in the Fe^{3+}–O–Fe^{3+}super exchange interaction giving higher net magnetization[28]. The reduction of coercive field for composite systems is due to pinning of magnetization by the presence of ofBaM–NBT interfaces as the result freeness of domain walls made possible [29]. To extract the intrinsic and extrinsic magnetoelectric effect for these composite systems we were analyzed the MC capacitance data at different magnetic fields. As per the *Catalan[30],* due to presence of Maxwell winger space charge polarization it was difficult to segregate the intrinsic and extrinsic MC effect for a MF system. But *Schmidt et al. and Kotnala et al.*[31,32]estimated the intrinsic magnetoelctric coupling (MEC) from magneto impedance (MI) data. Here we can consider the MI study for investigating the true MEC.

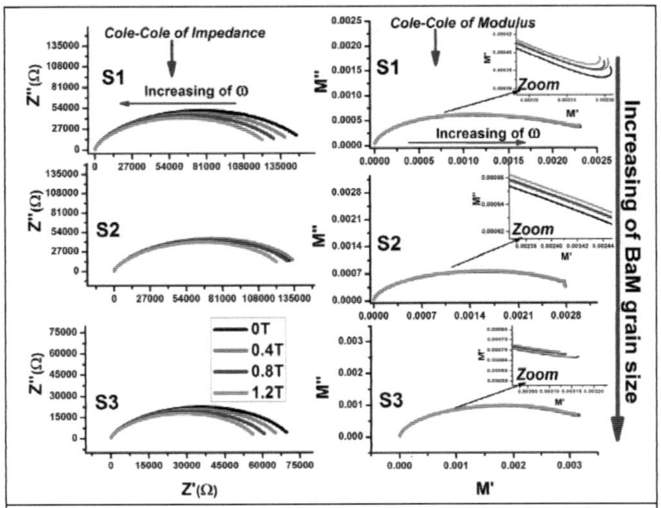

Figure 4.cole-cole of impedance and modulus for different magnetic field at room temperature.

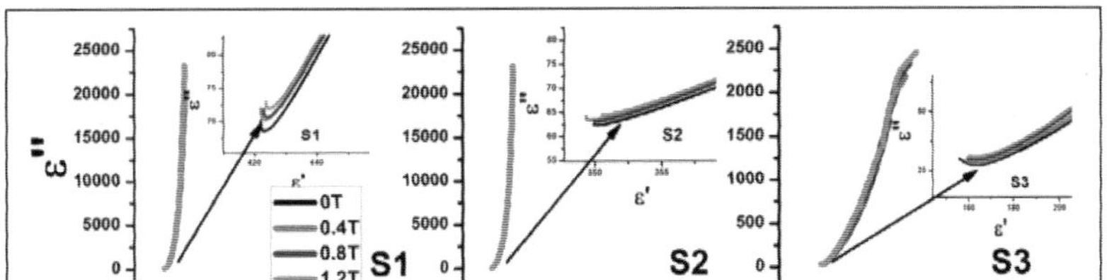

Figure 5. cole-cole of dielectric permittivity for different magnetic field at room temperature.

It was well known that, cole-cole of impedance and modulus can extract the intrinsic and extrinsic relaxation behaviour of polycrystalline system in the form of semicircles[33]. Figure 4 shows the cole–cole of impedance and modulus for different magnetic field at room temperature (RT). Both the cole-cole plots were significantly deviated at magnetic fields having a single semicircle, so we assume that the deviation of resistance and capacitance in magnetic field is due to extrinsic nature. But as we have MF behaviour at RT we could not deny the absence of MEC for these systems. So we have studied further the cole-cole of dielectric permittivity (ε'vs. ε'') which can extract the information about dipolar relaxation (which is intrinsic) present in the system (Figure 5)[31].

From the figure 5 it was estimated that all the cole-cole of dielectric permittivity could not able toshow the relaxation behaviour (absent even at higher frequency side- refer insight of figure 5). Therefore, at last, we assigned the single semicircle actions of impedance is effect of due to from interfaces (grain boundaries) only which follow the magneto resistance effect (MR) principle.

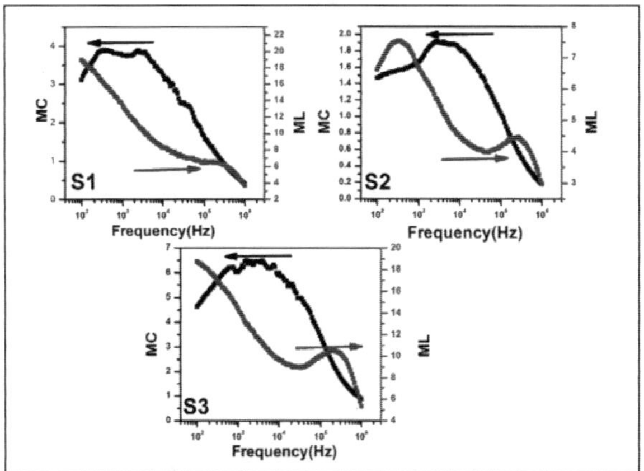

Figure6. Variation of magneto-capacitance (MC) and magneto loss (ML) with frequency for three composite systems.

But *P. Mandal et al. [34]* proposed a novel method by which one can predict the true (intrinsic) magneto electric coupling (MEC) for a system. Considering the Mandal et al.'s method we calculated and drawn Magneto-capacitance $MC(\%) = \frac{\varepsilon(H)-\varepsilon(0)}{\varepsilon(0)} \times 100$ and magneto-loss $ML(\%) = \frac{D(H)-D(0)}{D(0)} \times 100$ with variation of frequency (Figure6). From figure6 we were concluded that, all the systems could not able to fulfil the condition for MEC in the studied frequency range. From multiferroics data and magnetoelctric data we were confirmed that all the systems have both intrinsic and extrinsic magnetoelctric effect. The extrinsic magnetoelctric effect was observed from cole-cole of impedance where as the intrinsic magnetoelctric effect (MEC) may be able to be predictable from the variation of PE loop of composite systems (Figure 3).

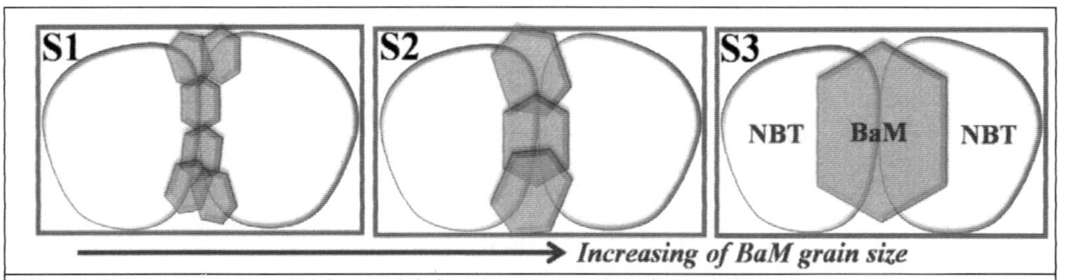

Figure 7. A proposed model for explaining the magteoelectric coupling.

As we know, the MEC for theses composite systems was based upon strain mediated coupling between two phases through interfaces and also depends on connectivity of both the phases, for predicting the intrinsic MEC we were consider a simple model which was supports to our results (Figure7). From the figure7 (model) we can perceive that, as BaM grain size increases the connection of amount of surface area of NBT system with BaM phases (which is better for S3 system than others) was continuously increases. Therefore the probability of reveres strain from BaM phases due to magnetostrictive effect will be increase for NBT phase with increasing of BaMgarin size. As the result we were obtained the increasing of P_S with increasing of BaM grain size (the largest P_S obtained for S3 system).

4. Conclusion

Three novel [90 wt% $Na_{0.5}Bi_{0.5}TiO_3$ (NBT)-10 wt% $BaFe_{12}O_{19}$ (BaM)] magnetoelctricmultiferroic composite systems were successfully fabricated by solid state reaction technique with increasing BaM grain size. All the composite systems were shown significant PE and MH loop (multiferroicbehavior) at room temperature. It was found that as BaM grain size increases the saturation polarization ($\boldsymbol{P_S}$) and saturation magnetization ($\boldsymbol{M_S}$) increases. The magneto impedance results (in a restricted frequency range) were revealed about the presence of dominating extrinsic magnetoelctric effect for these composite systems. The evidence towards the presence of intrinsic MEC for S3 system (largest BaM grain size) was concluded from the variation of MC and ML with frequency. To explain/predict the variation of intrinsic MEC we have proposed a simple mechanical model from which the important of connectivity scheme for a composite system was genuinely detected. The proposed model was well supports to the experimental results for these composite systems. In future we can also expect that, the proposed model may be able to explain the intrinsic MEC for other composite systems having different type of FE and FM phases.

Acknowledgment
The authors want to acknowledge Prof. P. N. Vishwakarma, Department of Physics and Astronomy, NIT Rourkela for magnetic-capacitance data. To BiswanathSamantaray, Saha Institute of Nuclear Physics, 1/AF Bidhannagar, Calcutta - 700 064, India, for collection of magnetic data and Ministry of Human Resource Development (MHRD), Government of India for providing scholarship.

References

[1] Vaz C A F, Hoffman J, Ahn C H and Ramesh R 2010 Magnetoelectric coupling effects in multiferroic complex oxide composite structures *Adv. Mater.***22** 2900–18

[2] Wang K F, Liu J-M and Ren Z F 2009 Multiferroicity: The coupling between magnetic and polarization orders *Adv. Phys.***58** 321

[3] Bedekar V, Bichurin M I, Ivanov S N, Pukinski Y J, Priya S, Bedekar V, Bichurin M I, Ivanov S N, Pukinski Y J and Priya S 2010 Metal-ceramic laminate composite magnetoelectric gradiometer Metal-ceramic laminate composite magnetoelectric gradiometer *Rev. Sci. Intruments***81** 33906

[4] Lam K H, Lo C Y, Chan H L W, Lam K H, Lo C Y and Chan H L W 2010 Frequency response of magnetoelectric 1 – 3-type composites Frequency response of magnetoelectric 1 – 3-type composites *J. Alloys Compd.***107** 93901

[5] Dong S, Zhai J, Xing Z, Li J and Viehland D 2005 Extremely low frequency response of magnetoelectric multilayer composites *J. Appl. Phys.***86** 102901

[6] Wang Y, Li J and Viehland D 2014 Magnetoelectrics for magnetic sensor applications : status , challenges and perspectives *Mater. Today***17** 269–75

[7] Palneedi H, Annapureddy, Venkateswarlu Priya S and Ryu and J 2016 Status and

Perspectives of Multiferroic Magnetoelectric Composite Materials *Actuators***5** 2–31

[8] Eerenstein W, Mathur N D, Scott J F and F J Scott 2066 Multiferroic and magnetoelectric materials *Nature***442759** 759–65

[9] Newnham R E, Skinner D P and Cross and L E 1978 CONNECTIVITY AND PIEZOELECTRIC-PYROELECTRIC COMPOSITES *Mater. Res. Bull.***13** 525–36

[10] Wang Y and Weng G J 2015 Magnetoelectric coupling and overall properties of multiferroic composites with 0-0 and 1-1 connectivity *J. Appl. Phys.***118** 174102

[11] Nan C W 1994 Magnetoelectric efFect in composites of piezoelectric and piezomagnetic phases *Phys. Rev. B***50** 6082

[12] Nan C W, Bichurin M I, Dong S, Viehland D and Srinivasan G 2008 Multiferroic magnetoelectric composites: Historical perspective, status, and future directions *J. Appl. Phys.***103**

[13] Rani R, Juneja J K, Singh S, Raina K K and Prakash C 2013 Dielectric , ferroelectric , magnetic and magnetoelectric properties magnetoelectric composites *Ceram. Int.***39** 7845–51

[14] Zhang R F, Deng C Y, Ren L, Li Z and Zhou J P 2013 Dielectric, ferromagnetic and maganetoelectric properties of BaTiO 3-Ni0.7Zn0.3Fe2O4 composite ceramics *Mater. Res. Bull.***48** 4100–4

[15] Sharma A, Kotnala R K and Negi N S 2014 Observation of multiferroic properties and magnetoelectric effect *J. Alloys Compd.***582** 628–34

[16] Yang P, Zhao K, Yin Y, Wan J G and Zhu J S 2006 Magnetoelectric effect in magnetostrictive / piezoelectric laminate composite Terfenol- D Li Nb O 3 [(z x t w) 129 ° 30 °] Magnetoelectric effect in magnetostrictive / piezoelectric laminate composite *Appl. Phys. Lett.***88** 172903

[17] Li T, Ã S W O, Lai H and Chan W 2006 Magnetoelectric effect in a parallel sandwich of magnetostrictive pseudo-1 – 3 composite and piezoelectric 2 – 2 composite *J. Magn. Magn. Mater.***304** 442–4

[18] Xu Y, Li J, Ma J and Nan C 2011 Magnetoelectricity of lateral 1-3 type composites with CoFe 2 O 4 ferromagnetic microstrips embedded in (K , Na) NbO 3 -based piezoceramic substrate *J. Appl. Phys.***110** 44104

[19] Wang J, Zhao F, Wu W and Zhao G 2011 Unusual temperature dependence of the magnetic anisotropy constant in barium ferrite $BaFe_{12}O_{19}$ Unusual temperature dependence of the magnetic anisotropy constant in *J. Appl. Phys.* 96107

[20] Rao B N, Datta R, Chandrashekaran S S, Mishra D K and Sathe V 2013 Local structural disorder and its influence on the average global structure and polar properties in $Na_{0.5}$ $Bi_{0.5}$ TiO_3*Phys. Rev. B***88** 224103

[21] Guerra J D S , Pal M, Portugal R J , Cótica L F , Santos I a., Guo R and Bhalla a. S 2013 Multiferroism and magnetoelectric coupling in $(PbZr_{0.65}Ti_{0.35}O_3)0.97–(BaFe_{12}O_{19})0.03$ ceramic composites *J. Appl. Phys.***114** 224113

[22] Srinivas A, Gopalan R and Chandrasekharan V 2009 Room temperature multiferroism and magnetoelectric coupling in $BaTiO_3 – BaFe_{12}O_{19}$ system *Solid State Commun.***149** 367–70

[23] Babu S N, Min S G, Malkinski L, Babu S N, Min G and Malkinski L 2011 Electric field tunable magnetic properties of lead-free $Na_{0.5}Bi_{0.5}TiO_3/CoFe_2O_4$ multiferroic composites *J. Alloys Compd.***109**07D911

[24] Islam R A and Priya S 2008 Effect of piezoelectric grain size on magnetoelectric coefficient of $Pb(Zr_{0.52}Ti_{0.48})O_3$-$Ni_{0.8}Zn_{0.2}Fe_2O_4$ particulate composites *J. Mater. Sci.***43** 3560–8

[25] Sheikh A D and Mathe V L 2009 Composition dependent phase connectivity , dielectric and magnetoelectric as piezoelectric phase *Mater. Res. Bull.***44** 2194–200

[26] Durkan C and Welland M E 2000 Size effects in the electrical resistivity of polycrystalline nanowires *Phys. Rev. B***61** 14215–8

[27] Etier M, Schmitz-Antoniak C, Salamon S, Trivedi H, Gao Y, Nazrabi A, Landers J, Gautam D, Winterer M, Schmitz D, Wende H, Shvartsman V V. and Lupascu D C 2015 Magnetoelectric coupling on multiferroic cobalt ferrite-barium titanate ceramic composites with different connectivity schemes *Acta Mater.***90** 1–9

[28] Ounnunkad S 2006 Improving magnetic properties of barium hexaferrites by La or Pr substitution *Solid State Commun.* **138** 472–5

[29] Pattanayak R, Muduli R, Kumar R, Dash T, Sahu P, Raut S and Panigrahi S 2016 Investigating the effect of multiple grain – grain interfaces on electric and magnetic properties of [50 wt% $BaFe_{12}O_{19}$–50 wt% $Na_{0.5}Bi_{0.5}TiO_3$] composite system *Phys. B Phys. Condens. Matter* **485** 67–77

[30] Catalan G 2006 Magnetocapacitance without magnetoelectric coupling *Appl. Phys. Lett.* **88** 2013–6

[31] Schmidt R, Ventura J, Langenberg E, Nemes N M, Munuera C, Varela M, Garcia-Hernandez M, Leon C and Santamaria J 2012 Magnetoimpedance spectroscopy of epitaxial multiferroic thin films *Phys. Rev. B - Condens. Matter Mater. Phys.* **86** 1–10

[32] Kotnala R K, Gupta R and Chaudhary S 2015 Giant magnetoelectric coupling interaction in $BaTiO_3/BiFeO_3/BaTiO_3$ trilayer multiferroic heterostructures *Appl. Phys. Lett.* **107** 82908

[33] Pattanayak R, Panigrahi S, Dash T, Muduli R and Behera D 2015 Electric transport properties study of bulk $BaFe_{12}O_{19}$ by complex im- pedance spectroscopy *Phys. B Phys. Condens. Matter* **474** 57–63

[34] Mandal P, Bhadram V S, Sundarayya Y, Narayana C, Sundaresan A and Rao C N R 2011 Spin-reorientation, ferroelectricity, and magnetodielectric effect in $YFe_{1-x}Mn_xO_3$ ($0.1 \leq x \leq 0.40$) *Phys. Rev. Lett.* **107** 1–5

Effect of rapid thermal annealing on the structural and electrical properties of RF sputtered CCTO thin film

N Tripathy[1], K C Das[1], S P Ghosh[1], G Bose[2] and J P Kar[1*]
[1]Department of Physics and Astronomy, National Institute of Technology, Rourkela, India 769008
[2]FST, IFHE University, Hyderabad, India 501203

E-mail: karjp@nitrkl.ac.in, Phone : +91-9438532157

Abstract: $CaCu_3Ti_4O_{12}$ (CCTO) thin films have been deposited by RF magnetron sputtering on silicon substrates at room temperature. As-deposited thin films were subjected to rapid thermal annealing (RTA) at different temperatures ranging from 850°C to 1000°C. XRD and capacitance - voltage studies indicate that the structural and electrical properties of CCTO thin film strongly depend upon the annealing temperature. XRD pattern of CCTO thin film annealed at 950°C revealed the polycrystalline nature with evolutions of microstructures. Electrical properties of the dielectric films were investigated by fabricating Al/CCTO/Si metal oxide semiconductor structure. Electrical properties were found to be deteriorated with increasing in annealing temperature.

Keywords: RTA, X-ray diffraction, FESEM, C-V measurement

1. Introduction

Materials with high dielectric constants are drawing a great deal of interest among researchers. Recently, $CaCu_3Ti_4O_{12}$ (CCTO) a cubic perovskite material is under active research due to its giant dielectric permittivity [1]. It shows a negligible temperature dependence on dielectric constant [2]. From neutron diffraction, no change in phase was found room temperature to 300°C [1, 2], which suits its application in microelectronics [3]. Miniaturization of electronic devices demands CCTO thin film to be integrated to semiconductor technology. Many research groups have deposited CCTO thin film by different deposition techniques like pulse laser deposition [4], sputtering [5], wet chemical methods [6]. However, sputtering has several advantages like low cost, uniformity, ease of synthesis, low temperature processing, non toxic nature. Post deposition annealing plays a significant role in the morphological and electrical properties of thin films. Rapid thermal annealing (RTA) has several advantages over conventional annealing like low thermal budget, less chance of dopant distribution, rapid grain growth [7]. In this work, an attempt has been made to investigate structural, morphological and electrical properties of CCTO thin film annealed with different RTA temperature.

2. Materials and Methods

CCTO sputtering target have been fabricated by adopting solid state route. Stoichiometric amount of $CaCO_3$, CuO, TiO_2 were mixed by using ball milling in acetone medium. The powder was calcined at 1050°C for 10 hours. Formation of single phase CCTO was confirmed by XRD studies. Thereafter, the target was fabricated by mixing the calcined powder with 5wt.% PVA solution and was pressed into a pellet of 2 inch dia and 3mm thickness using uniaxial compression with pressure 40MPa. Finally, the target was sintered at 1100°C for 8 hours. The details of target fabrication process have been reported elsewhere [8]. CCTO thin films of thickness 100 nm have been deposited on p-Si (100) substrates by RF magnetron sputtering technique with RF power of 105 W and deposition pressure of 5×10^{-3} mbar [9]. Post-deposition RTA was carried out at various temperatures ranging from 850°C-1000°C for 10min in air atmosphere using MTI (OTF- 1200X) system.

The structural and morphological properties were studied by X-Ray diffractometer (Rigaku ultima IV) and field-emission scanning electron microscope (FESEM: NOVA-FEI). For electrical measurements, MOS capacitors were fabricated using thermally evaporated aluminum film as top and bottom electrodes. The electrical properties of the Al/CCTO/Si metal oxide semiconductor (MOS) structures were studied by Agilent E4980A precision LCR meter and Keithley 6487 Picoammeter/voltage source.

3. Results and discussion

Fig. 1. shows the XRD pattern of CCTO thin films annealed at different temperature. The films, annealed at 950°C, show distinct (220) and (400) XRD peak, which depicts the polycrystalline nature of CCTO. The ratio of maximum intensity peak (200) of thin film annealed at 1000°C and 950°C $I_{1000^\circ C}/I_{950^\circ C} = 0.60$ which clearly indicate that there is a reduction in crystallinity for thin film annealed at 1000°C. This may be due to the creation of defects in the CCTO films including incorporation of impurity, variation of morphology and density, formation of unsaturated bonds at higher temperature. This increases the density of disorder due to which there is reduction of crystallinty at high temperature. Fig. 2. shows the FESEM micrographs of the CCTO thin films annealed at various annealing temperature. Thin films, annealed at 850°C and 900°C, depict the formation of microstructures with wide range of size variation (500nm - 4μm). On the other hand, uniform distributions of microstructures were appeared for the films annealed at 950°C and 1000°C. This may be due to the availability of adequate amount of thermal energy to the microstructure to form uniformly on the substrate.

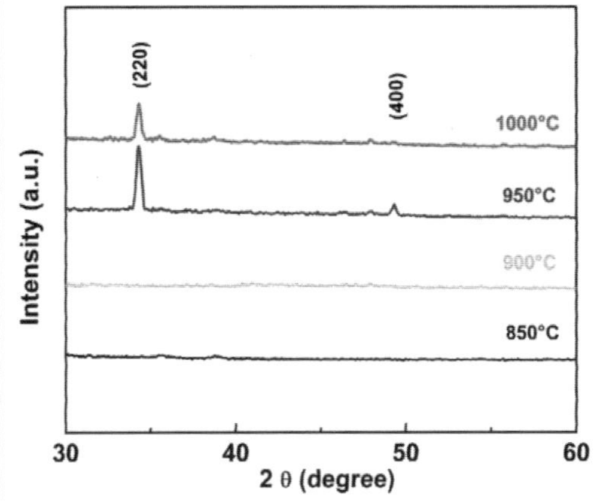

Figure 1. XRD pattern of CCTO thin films annealed at different temperatures

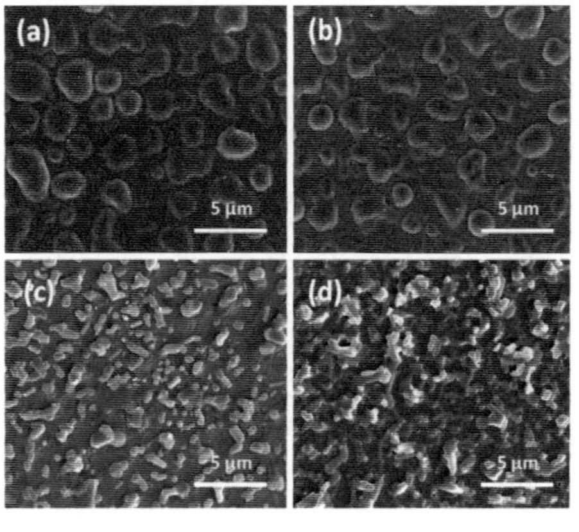

Figure 2. FESEM images of CCTO thin annealed at (a) 850°C, (b) 900°C, (c) 950°C and (d) 1000°C. Scale bar represents 5 μm

Capacitance-voltage (C-V) measurements were carried out for the films annealed at 950°C and 1000°C after the confirmation of CCTO phase. Fig. 3. shows the normalized high frequency (1MHz) (C-V) curves. Three regions of the C-V curves namely accumulation, depletion and inversion are distinctly visible in all the curves. The oxide charge density (Q_{ox}) was calculated from the shift in flatband voltage. Q_{ox} was found to be 0.14×10^{11} cm^{-2} and 1.81×10^{11} cm^{-2} for thin films annealed at 950°C and 1000°C, respectively. High Q_{ox} value for 1000°C annealed sample can be correlated to the formation of defects in CCTO thin films and therefore, possess large amount of trapping centres. The CCTO/Si interface charge density (D_{it}) is calculated using Hill- Coleman method [10]. Thin film annealed at 1000°C have comparatively high D_{it} (3.6×10^{11} $eV^{-1}cm^{-2}$) as compare to the 950°C annealed film (2.0×10^{11} $eV^{-1}cm^{-2}$) which may be due to the presence of dangling bonds at the interface and also due to interfacial stress, defects, nonstoichiometry of few initial deposited layers for high temperature annealing [11]. Leakage Current characteristics for CCTO thin films annealed at different temperature is plotted in Fig. 4. Thin film annealed at 1000°C shows high leakage current, which may be due to structural defects and/or incorporation of impurities due to high temperature annealing.

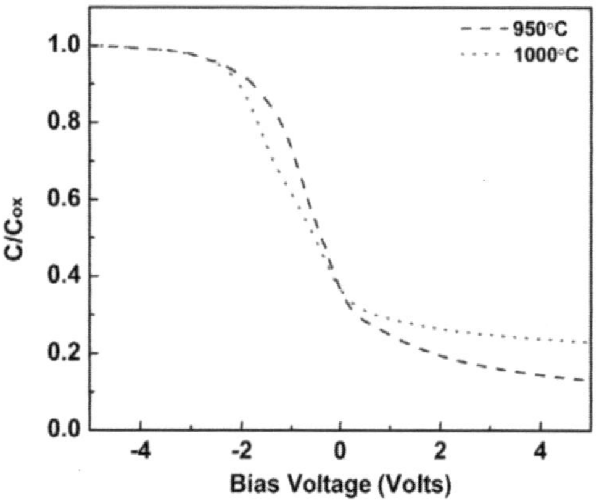

Figure 3. Capacitance - voltage characteristics of annealed CCTO films

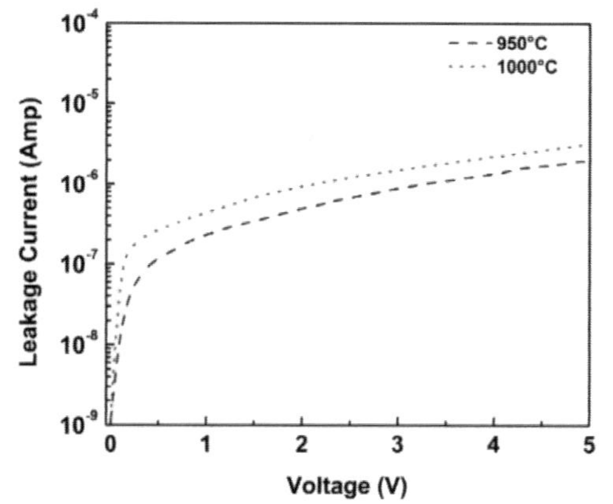

Figure 4. Current - voltage characteristics of annealed CCTO films.

4. Conclusions
CCTO thin film has been deposited by RF sputtering on silicon substrates. Post deposition annealing was carried out with different temperatures. Evolution of CCTO polycrystalline phase was confirmed from XRD patterns. Thin film annealed at 950°C have shown better crystallinity, uniform microstructures, lower interface and oxide charge density and also lower leakage current.

5. Acknowledgement
This work was supported by Science and Engineering Research Board (SERB), DST project (SR/FTP/PS-099/2012) under Fast Track Scheme for Young Scientist.

6. References

[1] Subramanian M, Li D, Duan N, Reisner B, Sleight A (2000) *J Solid State Chem.* **151** 323–325.

[2] Ramirez A, Subramanian M, Gardel M, Blumberg G, Li D, Vogt T (2000) *Solid State Commun.* **115** 217–220.

[3] Sinclair D C, Adams T B, Morrison F D, West A R (2002) *Appl. Phys. Lett.* **80** 2153.
[4] Deng G, Yamada T, Muralt P (2007) *Appl. Phys. Lett.* **91** 202903.
[5] Foschini C, Tararam R, Simões A, Cilense M, Longo E, Varela J (2013) *J. Alloys Compd.* **574** 604–608.
[6] Liu J, Smith RW, Mei WN (2007) *Chem. Mater.* **19** 6020–6024.
[7] Li J, Huang J H, Zhang Y, Yang Y, Song W, Li X M (2011) *J. Electroceram.* **26** 84–89.
[8] Tripathy N, Das K C, Ghosh S P, Bose G, Kar J P (2016) *IOP Conf Ser: Mater Sci Eng* **115** 012022.
[9] Tripathy N, Das K C, Ghosh S P, Das B, Kumar P, Kar J P (2016) *J Mater Sci: Mater Electron* doi:10.1007/s10854-016-5846-x.
[10] Hill W, Coleman C (1980) *Solid-State Electron* **23** 987–993.
[11] Kar J P, Bose G, Tuli S (2006) *Vacuum* **81** 494–498.

Effect of Moisture Absorption on the Mechanical Properties of Ceramic Filled Jute/Epoxy Hybrid Composites

Priyadarshi Tapas Ranjan Swain* and Sandhyarani Biswas
Department of Mechanical Engineering, National Institute of Technology, Rourkela-769008, Odisha, India

*E-mail: swainptr@gmail.com

Abstract. The present work emphasizes on the mechanical properties such as micro-hardness, flexural and impact strength of jute fiber and Al_2O_3 filler based polymer composites at dry and wet conditions. Composite samples reinforced with different wt.% of fibers and filler were prepared by hand lay-up technique. To improve the mechanical properties, jute fiber was hybridized with Al_2O_3 filler. The maximum flexural strength of 72.94 MPa and impact strength of 1.902 J is obtained for composites with 30 wt.% fiber content and 10 wt.% of filler content. The hardness of composite increases with increase in fiber and filler loading i.e 40 wt.% fiber content and 10 wt.% of filler content. The maximum hardness value is obtained 29.9 Hv. The effect of water absorption on mechanical properties of jute reinforced hybrid polymer composites is also investigated. To determine the influence of water absorption on the mechanical properties, specimens were immersed in distilled water for 10 days before testing. For reference purpose, dry specimens were tested. It is observed that the rate of water absorption depends on the fiber content as well as filler content. All the mechanical properties of composites are decreased after water absorption. Scanning electron microscopy (SEM) is used to characterise the microstructure and failure mechanisms of dry and wet jute fiber reinforced polymer composites.

Keywords: Jute fiber, Water absorption, Flexural strength, Impact strength, Scanning electron microscope.

1. Introduction

Recently, natural fiber composites pay attention for making environmental suitable engineering materials. The main advantages of natural plant fibers over synthetic fibers are acceptable as good specific strengths and modulus, cost-effective, low density and bio-degradability [1, 2]. The use of natural fibers reduces the weight by 10% and reduces the required energy to generate 80% while the cost of this component is 5% less than the comparable component fiber reinforced glass. Natural plant fiber-reinforced polymeric composites; also have some drawbacks such as the unsuitability between the hydrophilic natural fibers and hydrophobic thermoplastic and thermoset matrices. However, requiring the suitable use of physical and chemical treatments to enhance the adhesion between fiber and the matrix [3]. It is necessary that the composites retain their mechanical properties with less degradation in the water or moisture environments. Jute fibers are produced in large quantities, but jute easily absorbs water. The hydrophilic nature of the jute fibers makes the composites to water absorption sensitive, which in turn reduces the interfacial bonding between the fibers and the resin and also plasticizes the resin matrix. Poor processing to induce voids leading to the water intake. Furthermore, the mechanical characteristics of composites can be improved by the inclusion of inorganic fillers to the matrix which aimed the cost reduction, improvement in performance of polymer composites and some cases improve process ability. Several researchers have studied that addition of ceramic fillers in polymer composite enhances the mechanical properties of the resulted composite [4, 5]. Aluminium oxide (Al_2O_3) commonly referred to as alumina, is the most cost effective and broadly used material in the family of engineering ceramics and has the potential to be used as filler in various polymer matrices. With an excellent arrangement of properties and a reasonable price, it is no wonder that fine grain technical grade alumina has a very wide range of applications. It is essential that the composites retain their mechanical properties with less degradation in the water or moisture environments. Also, It is believed from the water environment that water molecules will enter quickly the interphase of the composites between the fiber and the resin because of the capillarity. Moisture absorption leads to the degradation of fiber-matrix interface region creating poor stress transfer efficiencies resulting in the reduction of mechanical and dimensional properties.

Therefore, the present study investigates the effect of Al_2O_3 filler on the water absorption behaviour of jute fiber-reinforced epoxy composite with a different percentage. However, according to the best knowledge of authors, no research was reported about the effect of water absorption on the mechanical properties of jute/Al_2O_3 reinforced hybrid polymer composites. Comparative analyses of composites were studied before and after water absorption.

2. Materials and methods

In the current study woven jute fiber and Al_2O_3 is taken as reinforcement and epoxy is taken as matrix material. The jute fiber is collected from local sources and the Al_2O_3 is obtained from NICE Ltd India in a range of 80-100 micron. The epoxy resin and the corresponding hardener HY-951 is brought from Ciba Geigy India Ltd.

2.1. *Composite fabrication*

The fabrications of the composite slab were carried out by conventional hand layup technique. The low-temperature curing epoxy resin and corresponding hardener were mixed in a ratio of 10:1 by weight as recommended by the manufacturer. Composites with four different fiber loading (10, 20, 30 and 40 wt.%) with three different filler loading (0, 5 and 10 wt.%) were fabricated and subjected to post-curing at room temperature for 24 h. The detail designation and composition of composites are given in Table 1.

Table 1. Detailed composition of the composites

Composites	Compositions
J1	Fiber (10%)+Filler(0%)+Epoxy(90%)
J2	Fiber (20%)+ Filler(0%)+Epoxy(80%)
J3	Fiber (30%)+ Filler(0%)+Epoxy(70%)
J4	Fiber (40%)+ Filler(0%)+Epoxy(60%)
J5	Fiber (10%)+ Filler(5%)+Epoxy(85%)
J6	Fiber (20%)+ Filler(5%)+Epoxy(75%)
J7	Fiber (30%)+ Filler(5%)+Epoxy(65%)
J8	Fiber (40%)+ Filler(5%)+Epoxy(55%)
J9	Fiber (10%)+ Filler(10%)+Epoxy(80%)
J10	Fiber (20%)+ Filler(10%)+Epoxy(70%)
J11	Fiber (30%)+ Filler(10%)+Epoxy(60%)
J12	Fiber (40%)+ Filler(10%)+Epoxy(50%)

2.2. *Water Absorption*
Moisture absorption studies were performed according to ASTM D 570-98 standard test method for moisture absorption of plastics. The samples were taken out periodically and after wiping out the water from the surface of the sample weighted immediately using a precise balance machine to find out the content of water absorbed. The specimens were weighed regularly at 24, 48, 72, 96, 120, 144, 168, 192, 216, 240, 264 hours. The moisture absorption was calculated by the weight difference. The weight gain in percentage of the samples was measured at regular time intervals of time by using the following equation:

$$\text{Water absorption } (\%) = \frac{W_n - W_d}{W_d} \quad (1)$$

Where W_n is the weight of composite samples after immersion and W_d is the weight of composite samples before immersion.

2.3. *Mechanical testing*
The flexural strength of a composite is determined by the maximum tensile stress that it can withstand during bending before getting the breaking point. The flexural strength measured in three-point bend test to understand the flexural behaviour of composites using the universal testing machine (Instron 1195). The test is conducted as per ASTM standard D2344-84.

$$F.S = \frac{3PL}{2bt^2} \quad (2)$$

Where P represent maximum load, b width of specimen, t thickness of specimen and L span length of the specimen.
Impact is a single point test that measures the resistance of material to impact from a fluctuation pendulum. This test can be used as a fast and easy quality control check to determine if a material meets specific impact properties or to compare materials for general toughness. For composite specimens low velocity instrumented impact tests are carried out. The tests are done as per ASTM-D 256 using an impact tester.
Leitz micro-hardness tester is used to found out micro-hardness measurement. A diamond indenter, in the form of a right pyramid with a quadrangular base and an angle 136° between opposite faces, is hard-pressed into the material under a load F. The two diagonals X and Y of the indentation left on the surface of the material after removal of the load are measured and their arithmetic mean L is calculated. In the current study, the load considered F = 24.54 N and Vickers hardness number is calculated using the following equation.

$$H_V = 0.1889\frac{F}{L^2} \text{ and } L = \frac{X+Y}{2} \quad (3)$$

Where F is the applied load *(N)*, L is the diagonal of square impression (mm), X is the horizontal length (mm) and Y is the vertical length (mm).

Results and discussion
The results obtained from the experimental study can be divided into two parts. The first part reflects water absorption behaviour of jute/Al_2O_3 polymer composites and the second evaluates the effects of water absorption at room temperature on the mechanical properties.

3.1. *Water absorption behaviour*

The effect of fiber loading on the water absorption behaviour of jute-epoxy composites filled with Al_2O_3 particulate filler with an increase in immersion time is shown in Figure 1. It is evident from the figure that the rate of water absorption of the composites increases with increase in immersion time. After a certain value of water absorption a saturation point will reached where no more water absorption takes place. It has been reported by the researchers that the water absorption rate increases with the increase in fiber loading [6]. The reason may be due to the fact that as the content of natural fiber increases in the composite, the number of free OH groups of cellulose increases and therefore, the moisture absorption increases. These free OH groups come in contact with moisture and form hydrogen bonding and results in weight gain in the composites. From Figure 1, it is also observed that J12 composite (i.e. with 40 wt.% fiber loading and 10 wt.% filler content) has maximum water absorption. It is due to the presence of a large amount of fibers and voids which absorbs more water in the composites. In comparison to all the composites fabricated, composite J1 (having 10 wt.% fiber loading and 0 wt.% filler loading) exhibits minimum water absorption rate due to the presence of less fiber and voids in the composite.

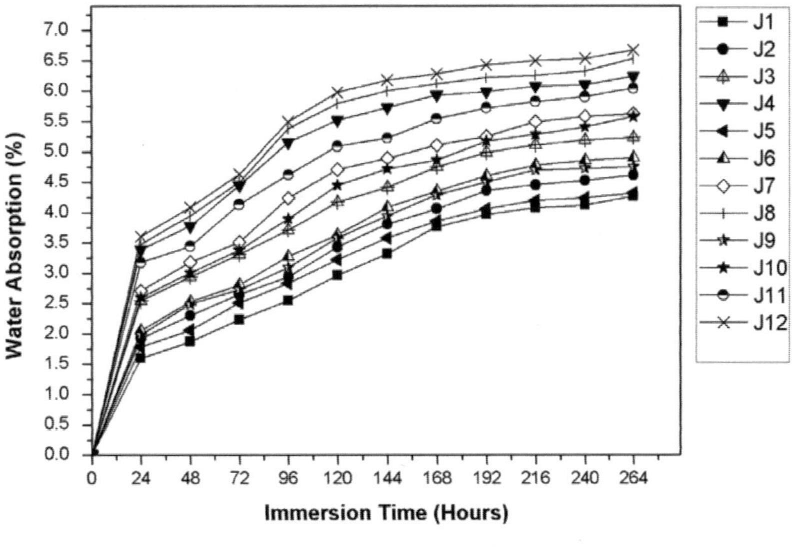

Figure 1. Water absorption behaviour of jute/Al_2O_3 composites

The main disadvantage in the use of natural fibers as reinforcement is that the fibers are more sensitive to water which will increase the dimension of the composites and also will reduce the mechanical properties [7, 8]. Basically, moisture absorption of the composites would lead to swelling and degradation at the fiber-matrix interface which may result in poor stress transfer resulting in poor mechanical properties and dimensional stability [9].

3.2. *Effect of water absorption on mechanical properties*

The effect of water absorption on the mechanical properties of jute/Al_2O_3 polymer composites was investigated after placing specimens in water at room temperature and comparing them with samples of the same composites kept in dry conditions.

3.2.1. *Flexural strength*

The effect of fiber and filler content on the flexural strength of dry jute/Al_2O_3 polymer composites is presented in Figure 2. In the dry condition, flexural strength increased as fiber content increased at 30 wt.% of fiber loading and 10 wt.% of filler loading. The flexural strength of jute/Al_2O_3 Polymer increased from 48.1 to 72.94 MPa due to the addition of Al_2O_3. This enhancement in flexural strength of jute/Al_2O_3 Polymer composites is may be due to the ability of natural fiber to resist bending forces and good stress transfer from the matrix resulting in improve strength properties [10]

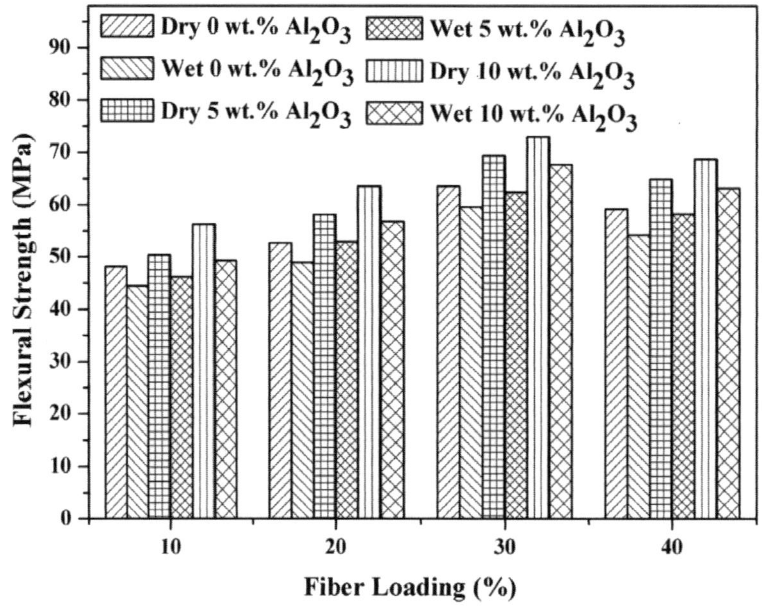

Figure 2. Flexural strength of jute/Al_2O_3 polymer composites in dry and wet conditions

The effect of water absorption on flexural strength of jute/Al_2O_3 polymer composites is also presented in Figure 2. It can be seen that the flexural strength of composites decreased after water absorption. Wet composites characteristics become decrease as compared to dry composites. This may due to the fact that the immersion of the composite samples in water affects the interfacial adhesion between fiber and matrix and makes de-bonding, leading to a decrease in mechanical properties of composite. When the fiber matrix interface was accessible to moisture in the environment the jute fibers expanded. Development of shear stress at the interface led to the ultimate de-bonding of the fibers, delamination and damage of structural integrity [11].

3.2.2. *Impact strength*

Impact strength is an important property that provides a sign of overall material toughness. Impact strength of fiber/filler-reinforced polymer is governed by the matrix–fiber/filler interfacial bonding, and the properties of both matrix and fiber/filler. After composites endure a sudden load, the impact energy is dissipated by the combination of fiber/filler pull-outs, fiber rupture and matrix deformation

[12]. Generally impact strength increases as fiber content increases in fiber-reinforced polymer composites because of the increase in fiber pull out and fiber breakage [13].

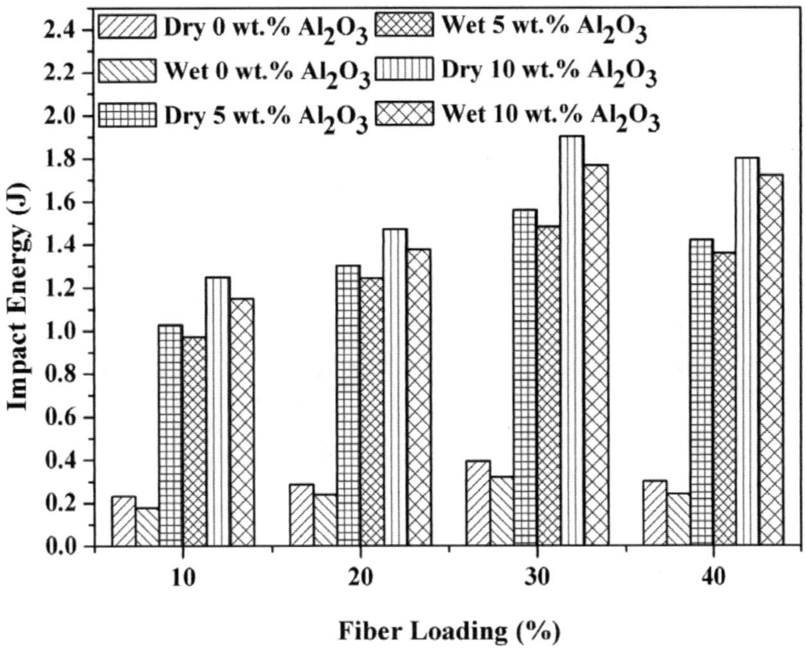

Figure 3. Impact strength of jute/Al_2O_3 polymer composites in dry and wet conditions

It can be seen from Figure 3 that impact strength rapidly increased upto fiber content 30 wt.% and filler content 10 wt.% in dry composites. The presence of Al_2O_3 the matrix increases the ability of these composites to absorb impact energy. In dry conditions, the addition of Al_2O_3 with fillings of 0, 5 and 10 wt.% and jute fiber with filling of 10, 20 and 30 wt.% increases the impact strength from 1.25 and 1.902 J respectively. After 30 wt.% of fiber loading and 10 wt.% of filler loading it decreases may be the matrix failure or deboning of fiber and filler loading.

Nevertheless, impact strength is unfavourably affected by water absorption. The decrease in impact properties after water immersion can be linked to the weak fiber–matrix interface, which resulted in a reduction of the mechanical properties and dimensional stability of composites [14].

3.2.3. *Hardness*

The effect of jute fiber and Al_2O_3 filler contents on the hardness of the polymer composites is presented in Figure 4. The hardness of polymer composites reinforced with 0, 5 and 10 wt.% filler loading and 10, 20 and 30 wt.% of fiber loading increased from 21.4 to 29.9 Hv, respectively. This enhancement in hardness is caused by the distribution of the test load on the fibers, which reduced the penetration of the test ball on the surface of the composite material and accordingly improved the hardness of this material [15]

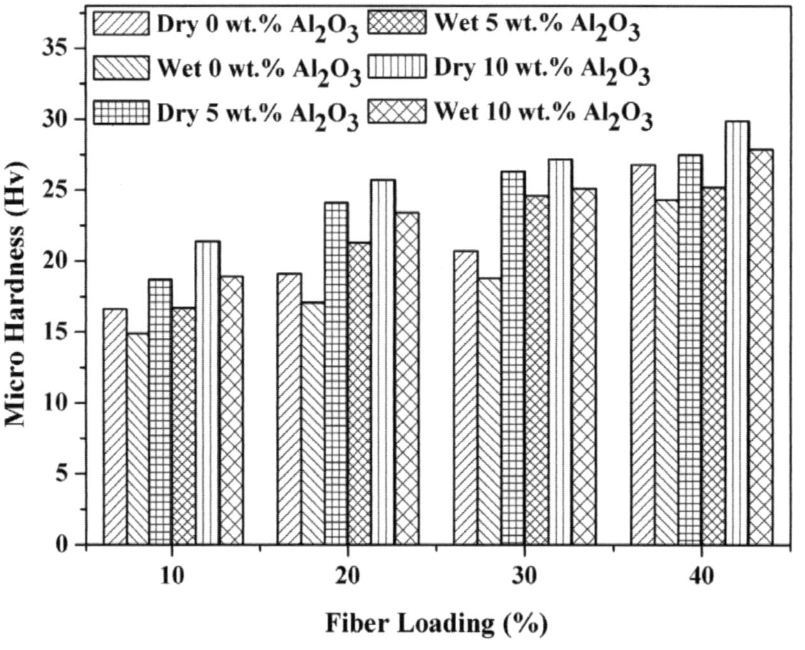

Figure 4. Micro-hardness of jute/Al_2O_3 polymer composites in dry and wet conditions

However, hardness is affected by water absorption, as shown in Figure 4. Hardness decreases in all cotton fiber-reinforced samples in wet condition, and is associated with the weakening of inter-face between the polymer matrix and the jute fiber caused by the water absorption. This type of decrease has been founded by other researchers working with natural fiber-based composites. As water absorption increased, the hardness of flax fiber-reinforced composites decreased, and found that the deformation depth increased for water-immersed specimens compared to dry ones, due to the hydrophilic nature of the fibers, and eventually led to the formation of a weak fiber–matrix interface [16].

3.2.4. *Morphology study*

Figure 5 a-c shows the micrographs of fractured specimen after flexural test in both dry and wet condition. SEM micrograph specifies the phenomenon of 'pull-out' occurred to a greater extent causing the failure of material. The image analysis also shows the formation of voids due to fiber pull-out, de-bonding, breakage. Overall, larger amount of broken fibers were observed if compared to the number of voids associated to the fiber pull-out. At some region weak fiber-matrix adhesion were observed, which can be attributed to low compatibility between hydrophilic nature of fiber and hydrophobic nature of matrix used.

Figure 5 (a) and (b) represents the wet fractured specimen, whereas Figure 5(c) shows the dry fractured specimen after the flexural test. It has been witnessed from the figure that more void created in wet type of fractured specimen. The fiber breakage, tearing of fibers, and pull-out were detected from fractographs. The voids and fiber entanglement were also found due to poor blending quality. These defects may act as crack-tips leading in matrix cracking portents or, in the second thing, resin reach areas that leading failure of the composite bond. In the case of dry fractured test, fewer voids occur. Overall, it is witnessed that due to water absorption, the fiber strength is reduced and the swelling is more when compared to other composite, breakage of fiber is also observed along with fiber pull-out due to flexural loading.

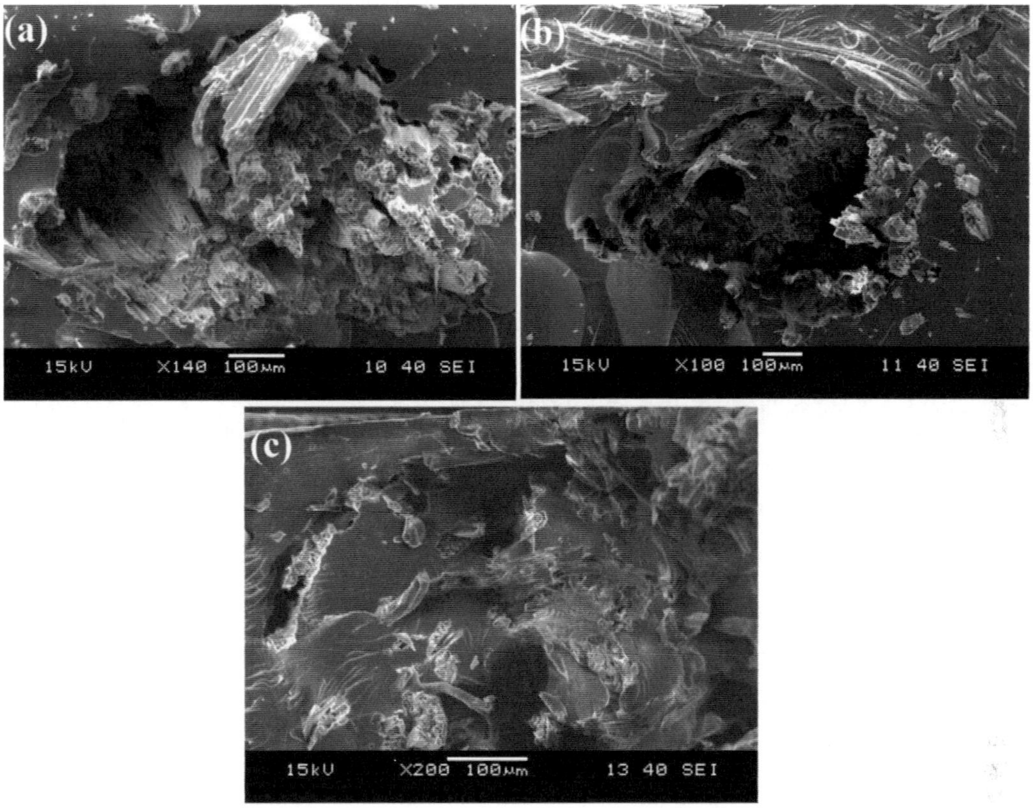

Figure 5 (a-c). SEM of jute/Al_2O_3 polymer composites after flexural test in case of dry and wet condition

4. Conclusion

In the present investigation, the effect of moisture absorption characteristics and its effect on mechanical properties of jute fiber and Al_2O_3 filler reinforced epoxy composite were studied. The incorporation of jute fiber and Al_2O_3 filler significantly increases the mechanical properties (e.g. flexural strength, impact strength, micro-hardness) of the composite.

However, jute fibers are hydrophilic in nature and hence have a poor resistance to water absorption. The water absorption of jute-reinforced polymer composites at room temperature was found to increase with increasing fiber content. Introduction to moisture for an extended period causes a reduction in flexural strength, impact strength, and hardness. A reasonable clarification for this would be that bonding at the fiber–matrix interfaces is degraded as a result of water absorption. After water absorption, SEM images of the composites confirm that jute has a great tendency of swelling and absorb more moisture.

Reference

[1] Bledzki A K and Gassan J 1999 *Prog Polym Sci*; **4:** 221-274.
[2] Mohanty A K, Mishra M and Drzal L T 2002 *J Polym Environ*; **10**: 19-26.
[3] Gassan J and Gutowski V S 2000 *Compos Sci Technol*; **60**: 2857-2863.
[4] Swain P T R and Biswas S 2014 *Int J Current Engg Tech*; 67-71.
[5] Swain P T R and Biswas S 2014 *Appl Mech Mater*; **592**: 729-733.
[6] Deo C and Acharya S K 2010 *J Reinf Plast Compos*; **29:** 2513-2521.
[7] Thiruchitrambalam M and Shanmugam D 2012 *J Reinf Plast Compos*; **31**: 1400-1414.
[8] Abdelmouleh M, Boufi S, Belgacem M N and Dufresne A 2007 *Compos Sci Technol*; **67**: 1627-1639.
[9] Dhakal H N, Zhang Z Y and Richardson M O W 2007 *Compos Sci Technol*; **67**: 1674-1683.
[10] Ghosh R, Reena G, Krishna A R and Raju B L 2011 *Int J Adv Sci Eng Inf Technol*; **4**: 89-91.
[11] Girisha C, Sanjeevamurthy N and Srinivas G R 2012 *Int J Eng Technol*; **2**: 166–170.
[12] Wambua P, Ivens J and Verpoest I, 2003 *Compos Sci Technol*; **63**: 1259-1264.
[13] Mishra S, Mohanty A K, Drzal L T, Misra M, Parija S, Nayak S K, and Tripathy S S, 2003 *Compos Sci Technol*; **63**: 1377-1385.
[14] D. Nabi Saheb, J.P. Jog, 1999 *Adv Polym Technol*; **18**: 351-363.
[15] Al-Mosawi A I 2009 *Al-Qadisiya J Eng Sci*; **2**: (2009) 14–24.
[16] Dhakal H N, Zhang Z Y, Bennett N, Lopez-Arraiza A and Vallejo F J 2014 *J Compos Mater*; **48**: 1399-1406.

Fabrication and characterization of theAl6063/5%ZrO_2/5%Al_2O_3 composite

Munmun Bhaumik and Kalipada Maity
Department of Mechanical Engineering, NIT Rourkela, Orissa- 769008, INDIA
Email: munmunbhaumik@gmail.com, kpmaity@gmail.com

Abstract. Metal matrix composites (MMCs) combine the desirable characteristics of metals (thermal conductivity and ductility) and ceramics (high stiffness, high hardness, and low thermal conductivity). Aluminium hybrid composites are new generations of metal matrix composites that have the potential to satisfy the recent demands of advanced engineering applications. In the present study, Al-MMC has been fabricated by mixing the 5wt% ZrO_2 and Al_2O_3reinforcement into the Al6063aluminium alloy matrix. Stir casting method has been adopted for the fabrication of MMC.The prepared casted MMC has been characterized by scanning electron microscope(SEM), X-ray diffraction(XRD) analysis.The physical property (density) and mechanical property (hardness, tensile test, bend test, compression test) have been measured for the fabricated MMC. The fracture surface has been studied.The fracture of Al-MMC is found brittle in nature.

Keywords: Al6063aluminium alloy; metal matrix composite; stir casting; mechanical property; physical property.

1. Introduction

Metal matrix composite consists of greater properties such as high strength, high elastic modules, high stiffness, high electrical and thermal conductivity, high corrosion resistance, oxidation and wearcompared to conventional material. Aluminium alloy based metal matrix composites (AMMCs) are extensively used in various engineering field due to their low weight to density ratio. Al-MMC reinforced with particles and whiskers are widely used for high-performance appliances such as automotive, aerospace, military, and electricity industry because of their enhanced physical and mechanical properties [1]. The manufacturing of Al-MMC can be categorized into three types. They are solid phase processes viz. diffusion bonding and powder metallurgy, semi-solid process viz. spray, compo casting and rheo casting and liquid phase process viz. stir casting [2]. Stir casting or vortex is most commonly used method out of all the available method for metal matrix composites production and allow a very large size component to be manufactured [3]. Carbon (C), boron nitride (BN), silicon carbide (SiC), titanium carbide (TiC), aluminium oxide (Al_2O_3), Zirconia (ZrO_2) are most popular reinforcements used for fabricating metal matrix composite.

Two types of MMC have been developed for the last few decades: fiber reinforced MMCs and particle reinforced MMCs [4, 5]. Baghchesara et al. [6] have fabricated Al-ZrO_2 composites by using vortex method using ZrO_2 powder and studied the fracture surface of the fabricated MMCs. Kumar et al.[7] have adopted the liquid metallurgy technique and successfully prepare the Al6061-SiC and Al7075-Al_2O_3 composites. They have found the significance improvement of the physical and mechanical properties than their base material. Gautam et al. [8] have prepared Al-Al_2O_3 composite by the mechanical alloying method and studied the physical and mechanical behavior of the composite. Unlu [9] has made Al/Al_2O_3/SiC composite by casting and P/M method and studied the mechanical and tribological properties for both the method.

In this present study, Al6063 aluminium alloy composite has been fabricated by mixing ZrO_2 and Al_2O_3 reinforcement particle into the matrix. MMC has been characterized by scanning electron microscope, X-ray diffraction analysis. The physical property such as density and mechanical property like hardness, tensile test, bend test, compression test have been measured. The fracture surface has been studied using SEM.

2. Experimental details

The major raw materials used in this study are aluminium alloy, ZrO_2 and Al_2O_3 powder. Al6063 is taken as a matrix of the composite. Figure 1(a) and (b) shows the SEM image of Al_2O_3 and ZrO_2 which have used as reinforcement.

(a) (b)

Figure 1.SEM image of the as-received powders: (a) Al_2O_3 (b) ZrO_2

2.1 Fabrication of Al6063/ZrO$_2$/Al$_2$O$_3$MMC

For fabrication of hybrid metal matrix composite (h-MMC), stir casting technique was adopted.For fabrication commercially available Aluminium alloy (Al6063) was used as a matrix material and ZrO_2(~20μm)and Al_2O_3 (~30μm)were used as reinforced material.

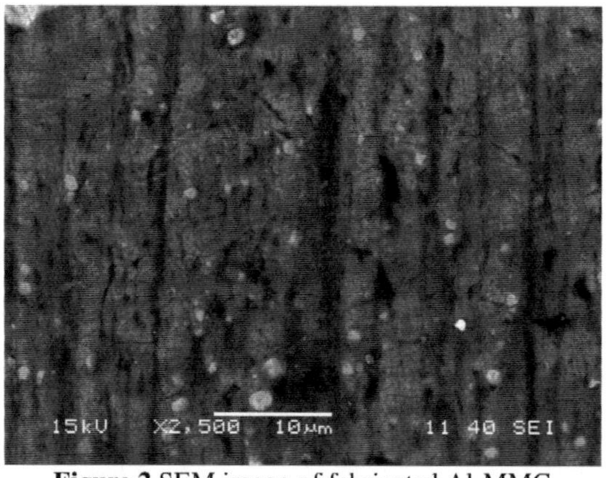

Figure 2.SEM image of fabricated Al-MMC

During stir casting route the matrix material was melted at 850^0C in a graphite crucible under the controlled argon environment in the resistance heated furnace. The specific amount of Al_2O_3 and ZrO_2 particles equal to 5wt% preheated upto 300^0C and 1200^0C and then mixed into the molten bath of Al6063. Reinforced particles were stirred by a stirrer at 700rpm. After complete stirring, the material mixture was again reheated into the furnace and then poured into the mold at 750^0C to 850^0C. The material was allowed to cool in open atmosphere. After cooling the casting material was machined to a required shape. Figure 2 shows the SEM image of Al-MMC fabricated by stir casting.

3. Result and discussion

3.1 X-ray diffraction (XRD) analysis

Figure 3 represents the XRD pattern obtained from the Al-MMC. Al, Al_2O_3, ZrO_2 are clearly identified as the main phase of the composite. Al_3Zr_4, Al_3Zr and Ti_9Al_{23} are also observed as a secondary phase.

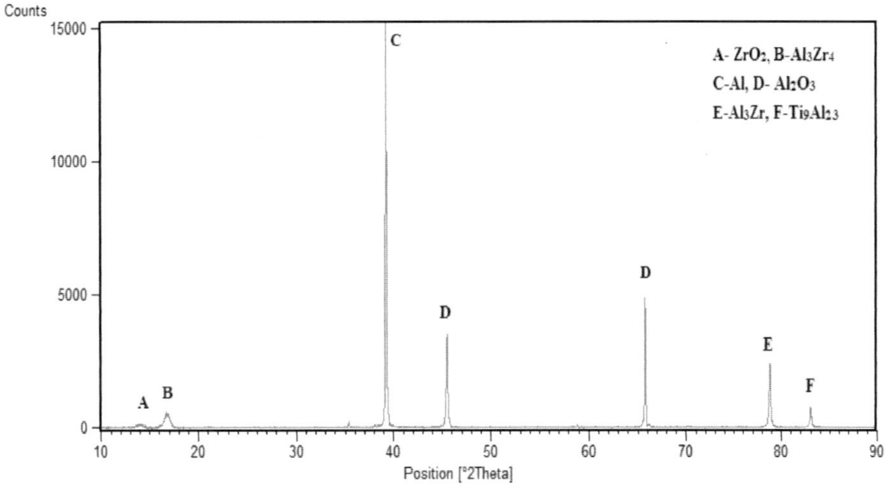

Figure 3.X-ray diffraction pattern of fabricated Al-MMC

3.2 Physical property

3.2.1Density

The fabricated composite specimen was considered for density measurement. The actual density was determined by Archimedes principle at room temperature and theoretical density was calculated by using "Rule of Mixture" formula. The theoretical and actual density obtained from the calculation are 2.89gm/cm^3 and 2.71gm/cm^{3}respectively. The relative density is 93.36% for this fabricated composite.

3.3 Mechanical property

3.3.1Hardness

The hardness of the Al-MMC was measured by Vickers hardness tester (LecoMicrohardness tester LM248AT). Before testing the sample was polished to a scratch free surface. During testing, theload was kept constant at 50gf for a dwell time of 5 seconds. A minimum five reading was takenfor determining the mean micro-hardness value. The hardness was found 91.1 HV for the Al-MMC.

3.3.2Tensile test

The tensile strength of the specimenwas investigated using Instron, UK/SATEC 600KN.The tensile specimen was grounded according to the ASTM B557. The maximum tensile strength has been foundout 78.51MPa. Figure4 shows the load vs.extension curve for the tensile test. The brittle phase of Al_2O_3 and ZrO_2 is the reason behind the low tensile strength.

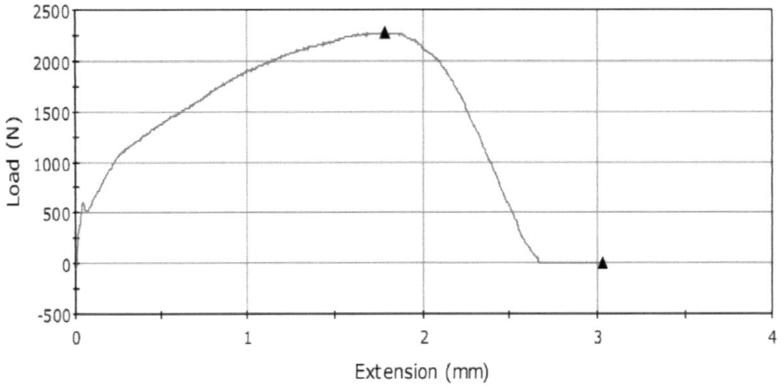

Figure 4. Tensile test of Al-MMC

3.3.3Compression test

Compression test was performed by universal testing machine (INSTRON/SATEC 600KN). Specimen dimension 10mm of diameter and 8mm height was taken according to ASTM Standard E9-89. The experimentwas performed under 28^{0}C, 50% humidity and at 1 mm/min strain rate. Compressive strength has been found 685 MPa for Al-MMC.Figure 5 shows the load vs. extension curve for the compression test.

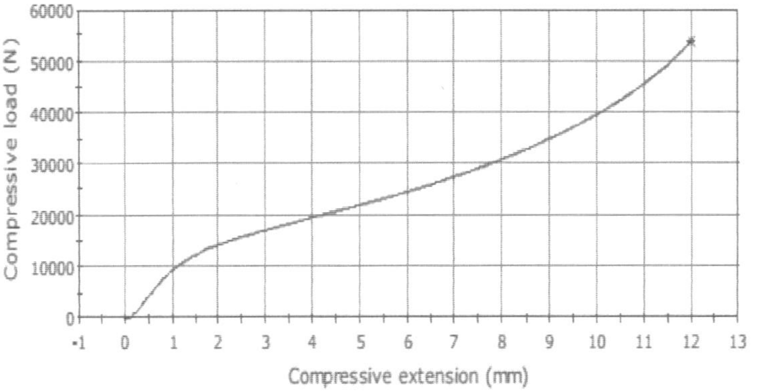

Figure 5. Compression testing of Al-MMC

The fabricated composite shows low strength under tensile mode compared to compressive mode. This may cause because of the presence of ceramic reinforcement in the metal matrix. The effects of elastic property of reinforcement inhibit the plastic deformation of aluminium matrix resulting increase in compressive strength [10].

3.3.43-point bend test

The flexural test is a combination of both tensile and compressive tests. Bend test was performed by universal testing machine (INSTRON 5967). In this test a rectangular cross section is placed on two supporting pins. The loading force is appliedto the middle portion using loading pin. Fig.6 depicts the details of thebending test.

The flexural stress is calculated by the Eq. 1.

$$\sigma_f = \frac{3PL}{2bd^2} \quad (1)$$

Where, σ_f is stress in outer surface at midpoint (MPa)

P is load at a given point (N), L is supported span (mm), b is thewidth of test beam (mm),and d is the depth of tested beam.

Specimen dimension is 30×10×5 mm^3. ASTM standard B925-08 was taken for this test. The flexural test was carried out in the universal testing machine (INSTRON-5967) maintaining crosshead velocity of 0.2 mm/min and a span length of 24 mm during the test. Figure 6 depicts the flexural test of Al-MMC.

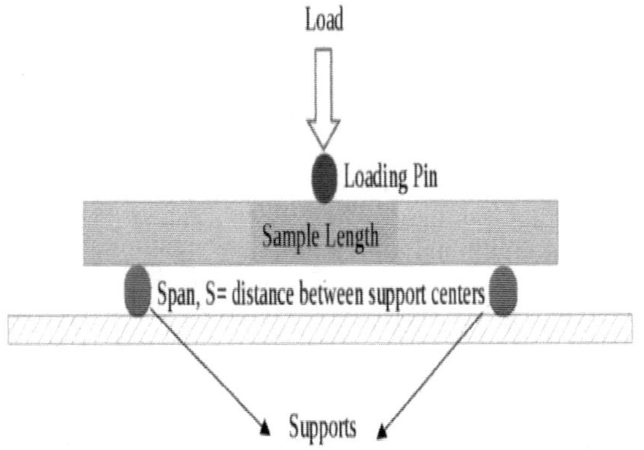

Figure 6. Details of bending test

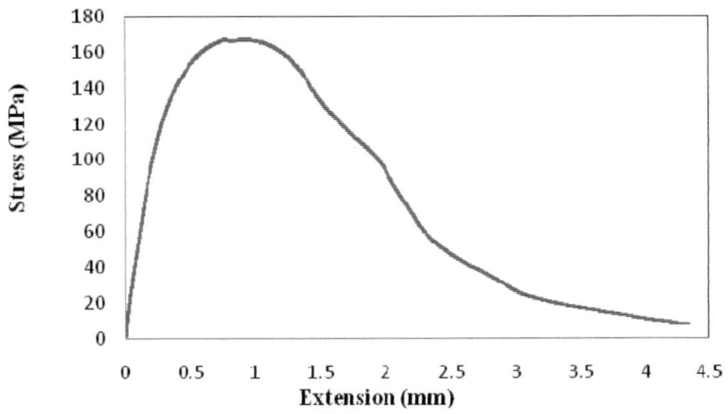

Figure 7. Flexural testing of Al-MMC

From Figure 7, it is observed that the maximum flexural stress is 167 MPa at 1.16 kN flexural break load.

3.3.5Fracture surface

To understand theunusual mechanical behavior of this composite the fractographywas performed on the specimen (Figure 8). Figure 8 (a) and (b) shows the SEM micrographs of fracture surfaces of the specimen containing 5% ZrO_2 and 5%Al_2O_3 produced by stir casting.

Fracture normally occurs as a result of one or the combination of the following mechanism

1. Fracture of reinforcing particle[11]
2. Partial debonding of reinforcement-matrix interface and nucleation of voids [12]
3. Development of the voids and ignition of cracks in the matrix [13]

Various types of factors are responsible for the composite fracture. They are the processing method, the applied stress, the post heat treatment, reinforcement distribution and morphology of the reinforcing particle, etc.

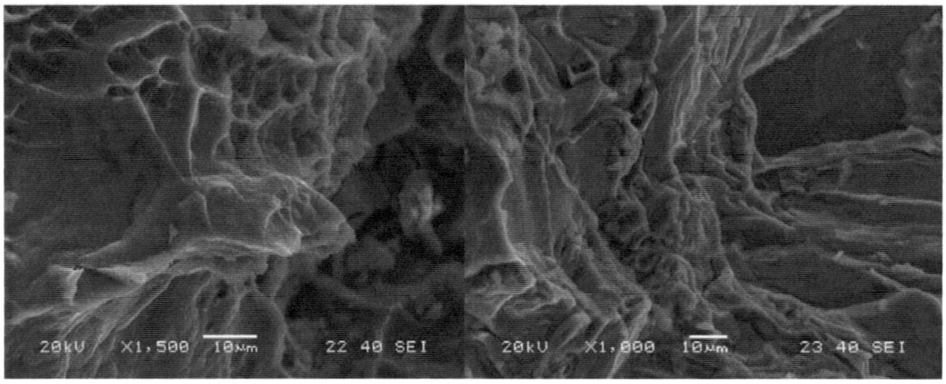

(a) (b)

Figure 8. Topography images of the fracture surface of the specimen

Figure 8 reveals the presence of dimplewith tear ridges, while eutectic phases are mostly present. The incoherency of matrix-particle particle has been observed extensively. Relatively rough fracture surface and presence of porosity have been observed. The reinforcement size isalso responsible for the failure.The presence of hard and brittle ZrO_2 and Al_2O_3 particle in the ductile aluminium matrix exerts constraints on the plastic flow of the matrix. When powder concentration and triaxiality of stress combines in the clustered particle regions, the composite undergoes brittle fracture [14].This indicates the brittle mode of failure of the Al-composite.Fracture of Al composite initiates from the matrix. In other words, composite fracture is directed by matrix fracture [15, 16].

4. Conclusions

Al6063 aluminium alloy reinforced with ZrO_2, Al_2O_3 powders is successfully fabricated by stir casting method.The mechanical (hardness, tensile test, bend test, compression test) and physical (density) properties have been studied for the Al-MMC.The relative density is found 93.36% for the fabricated MMC. The composite shows low strength under tensile load compared to compressive load.The topographic observation of the fracture surface revealed that the fracture of Al-MMC is brittle in nature, and the matrix-fracture is dominant.

References

[1] Seal S, Kuiry S C, Georgieva P, Agarwal A 2004 Manufacturing Nanocomposite Parts: Present Status and Future Challenges *MRS Bulletin* 16-21.

[2] Naher S, Brabazon D, Looney L 2004 Development and Assessment of a New Quick Quench Stir Caster Design for the Production of Metal Matrix Composites *Journal of Material Processing Technology***166** 430-439.

[3] Hashim J, Looney L, Hashmi M S J 1999 Metal Matrix Composites: Production by the Stir Casting Method *Journal of Material Processing and Technology***92** 1-7.

[4] Miserez A G T 2003 Fracture and toughening of high volume fraction ceramic particle reinforced metals (dissertation) Lausanne: EPFL.

[5] Tjong S C, Ma ZY 2000 Microstructural and mechanical characteristics of in situ metal matrix composites*Materials Science and Engineering***R29** 49–113.

[6] Baghchesara M A, Abdizadeh H, Baharvandi H R 2009 Fractography of Stir Casted Al-ZrO_2 Composites *Iranian Journal of Science & TechnologyB* **33(B5)** 453-462.

[7] Kumar G B V, Rao C S P, Selvaraj N, Bhagyashekar M S 2010 Studies on Al6061-SiC and Al7075-Al_2O_3 Metal Matrix Composites *Journal of Minerals & Materials Characterization & Engineering***9** 43-55.

[8] Gautam R K, Mohan A, Mohan S 2013 Fabrication and Characterization of Al-Al_2O_3 Composite by Mechanical Alloying *Materials Science Forum***736** 81-97.

[9] Zong B Y, Derby B J 1996Characterization of microstructural damage during plastic strain of a particulate-reinforced metal matrix composite at elevated temperature*Journal of Materials Science***31(2)** 297-303.

[10] Zhang Q, Zhang H, Gu M and Jin Y 2004 Studies on the fracture and flexural strength of Al/Si_p composite *Materials Letters* **58(27-28)** 35-45

[11] Syu D G C, Ghosh A K 1994 The Effect of Temperature on the Fracture Mechanism in 2014Al/15 vol%Al_2O_3 Composite *Materials Science and Engineering: A***184** 27–35.

[12] Christman T, Needleman A, Nutt S, Suresh S 1989 On microstructural evolution and micromechanical modelling of deformation of a whisker-reinforced metal-matrix composite *Materials Science and Engineering* A**107** 49-61.

[13] Lu Y X, Meng X M, Lee C S, RKY L, Huang C G 1998 Failure mechanisms of a SiC particles/2024 Al composite under dynamic loading *Phys. Status Solidi(a)***169** 49–55.

[14] Pakdel A, Farhangi H and Emamy M 2007 Effect of extrusion process on ductility and fracture behaviour of SiCp/aluminum-alloy composites*Proceedings of 8th International Fracture Conference 7 – 9 November 2007 Istanbul/TURKEY* 460-470.

[15] Kamat S V, Hirth J P, Mehrabian R 1989 Mechanical properties of particulate-reinforced aluminum matrix composites*Acta Metall.* **37** 95–402.

[16] Unlu B S 2008 Investigation of tribological and mechanical properties Al_2O_3–SiC reinforced Al composites manufactured by casting or P/M method*Materials and Design***29** 2002–2008.

Processing and characterization of epoxy composites reinforced with short human hair

Bishnu Prasad Nanda and Alok Satapathy*

Department of Mechanical Engineering, National Institute of Technology, Rourkela, 769008, India

bishnuprasad.nanda726@gmail.com, alok@nitrkl.ac.in

Abstract: Human hair is a biological fiber with well characterized microstructure. It has many unique properties like high tensile strength, thermal insulation, unique chemical composition, elastic recovery, scaly surface etc. But due to its slow decomposition, it creates many environmental problems. Although a number of utilization avenues are already in place, hair is still considered as a biological waste. In view of this, the present work makes an attempt to explore the possibility of fabricating a class of polymer composites reinforced with short human hair fibers. Epoxy composites with different proportions of hair fiber (0, 2, 4, 6 and 8 wt.%) are prepared by simple hand lay-up technique. Mechanical properties such as tensile, flexural and compressive strengths were evaluated by conducting tests as per ASTM standards. It was found out that with the increase in fiber content, the tensile and flexural strength of the composite were increasing significantly while the compressive strength improved marginally. Scanning electron microscopy was done on these samples to observe the microstructural features.

Key words: Epoxy composite, hair fiber, microstructure, biological waste

1. Introduction

Primary constituent of human hair fiber is keratin protein. Keratin protein is made up with long chain of amino acids like cytosine, serine, glutamine, threonine, glycine, leucine, valine and arginine. Hair fiber is composed of three main structures cuticle, cortex and medulla. The cuticle is the outermost part of hair fiber. Cortex is the cylindrical part after the cuticle, which occupies 75% of area. Medulla is the thin cylindrical layer located at the center of human hair. A no. of studies on the mechanical and physical characterization of human hair have been reported [1-2]. Aniruddha *et al.* [3] studied the nanomechanical and micromechanical properties of human hair. Bhusan *et al.* [4] studied the friction and wear behavior of human hair and skin at a range of loads, speeds and skin area. Carmen *et al.* [5] studied the tribological behavior of human hair and skin using atomic force microscopy. Guohua *et al.* [6] studied the nanomechanical properties of human hair including hardness, elastic modulus and creep using nanoindentation technique. Use of hair fiber in different possible field of applications also researched by many researchers [7]. Jain *et al.* [8] studied the effect of human hair on plain cement concrete on the basis of its compressive, crushing and flexural strength. Bagadiya *et al.* [9] studied the use of keratin waste, derived from human hair as short fiber for elastomer.

Numerous works have been reported on the characterization of human hair and its use on different fields. But its use as a reinforcing element in polymers has not been adequately researched. In view of this, the present work includes the investigation of the physical, mechanical and micro structural characteristics of epoxy composites reinforced with short hair fiber (SHF) in different proportions (0, 2, 4, 6, 8 wt.%).

2. Experimental details

2.1 Composite Fabrication

Raw hair fibers were collected from a nearby saloon. These fibers were thoroughly washed in detergent and dried in sunlight for few hours. Fibers were cut into approximate length of 3-4 mm with scissor. The low temperature curing epoxy resin (LY556) and corresponding hardener (HY 951) were mixed in the ratio of 10:1 by weight. The short hair fibers were then mixed with various weight percentages to prepare the composites by hand lay-up technique. The mixture was thoroughly mixed until uniform dough was formed. The dough was then slowly decanted into a rectangular mold of size 200 mm length, 100 mm width and 4 mm thickness, and also on a cylindrical mold of 10 mm diameter and 60 mm depth. The casting was put under load for 24 hours and after that composite specimens were taken out from the mold. Specimens of required size were cut for physical and mechanical characterization.

2.2 Density and Void Fraction

The actual densities of the composites were measured by using the Archimedes principle. According to this principle, when an object is immersed in a liquid, the apparent loss in its weight is equal to the weight of the liquid it displaces. To conduct the test distilled water was taken as the medium. This method is covered in ASTM standard D792.The density of the composite was obtained by using the following equation.

$$\rho_a = \frac{\rho_w W_a}{W_a - W_w}$$

Where,

ρ_a is actual density of composite,
ρ_w is the density of distilled water,
w_a is weight of the sample in air and,
w_w is weight of the sample in water.

The theoretical density (ρ_t) of the composite material can be determined by using the following equation given by Agarwal and Broutman [10].

$$\rho_t = \frac{1}{(w_f/\rho_f)+(w_m/\rho_m)}$$

Where

w_f is the weight fraction for fiber,
w_m is the weight fraction for matrix,
ρ_f is the density of fiber and,
ρ_m is the density of matrix.

Volume fraction of voids can be calculated by using the following equation [10]:

$$v_v = \frac{(\rho_t - \rho_a)}{\rho_t}$$

Where ρ_t and ρ_a represents the theoretical density and actual density of the composite, respectively.

2.3 Tensile Strength

Composite samples of size 150 mm×20 mm×3 mm were prepared to conduct tensile test as per the ASTM E 1309 standard. Uniaxial load was applied through both ends. The test was performed in universal testing machine Instron 1195 at a cross-head speed of 10 mm/min and the results were used to calculate the tensile strength of the composite samples. Three tests were conducted for each composition and the mean value of the three replications was reported as the tensile strength of the composite.

2.4 Flexural Strength

Three point bending test was used to obtain the flexural strength. Composite specimens of size 80 mm×20 mm×4 mm were prepared to conduct the test. The tests were conducted using an Instron 1195 universal testing machine while maintaining 40 mm span length and 2 mm/min cross-head speed. Flexural strength of each specimen was calculated using the following formula.

$$FS = \frac{3PL}{2bt^2}$$

Where,

P is maximum applied load,
b is the width of the specimen,
t is the thickness of the specimen and
L is the span length of the sample.

2.5 Compressive Strength

Composite samples of required dimensions (height 10 mm, diameter 10 mm) were used for the uniaxial compression test as per the ASTM D 695 standard. Static uniaxial compression tests were carried out on these specimens using Universal Testing Machine Instron 1195 while maintaining 1 mm/min crosshead speed.

2.6 Scanning Electron Microscopy

The micro structural features of the raw human hair and the hair fiber reinforced composite specimens were examined by Scanning Electron Microscope JEOL JSM-6480 LV. The specimens were mounted on stubs with silver paste. To improve the penetration of light and for better surface micrographs, a thin film of platinum was vacuum-evaporated onto the samples before the images were taken.

3. Results and Discussion

3.1 Density and Void Fraction

Table 1 shows the theoretical density, measured density and void fraction of epoxy-SHF composites for different weight percentages. It can be observed that the theoretical density was greater than the measured density. This can be attributed to the presence of voids and pores present in the composite during the fabrication process. It was also found that with the increase in SHF content, the measured density of the composite decreased from 1.09 gm/cc at 0 wt.% of SHF content to 1.059 gm/cc at 8 wt.% of SHF content, but the void fraction increased from 0.9% at 0 wt.% of SHF to 1.9% at 8 wt.% of SHF content.

Table 1. Theoretical and measured densities and void fraction of Epoxy-SHF Composites

SHF content (wt%)	Theoretical density (gm/cc)	Measured density (gm/cc)	Volume fraction of voids (%)
0	1.1	1.09	0.9
2	1.095	1.082	1.2
4	1.093	1.077	1.4
6	1.085	1.066	1.7
8	1.08	1.059	1.9

3.2 Tensile Strength

The tensile strength of epoxy composites with different wt.% of short human hair fiber are presented in Table 2.It was seen that the tensile strength of the composite increased with increase in the fiber content. With a reinforcement of 2 wt.% short hair fiber tensile strength of composite increases from 65 MPa to 81 MPa. Further, for the 8 wt. % of short hair fiber, the tensile strength of epoxy composite reached 127 MPa.

Table 2. Mechanical Properties of Epoxy-SHF Composites

SHF content (wt%)	Tensile Strength (MPa))	Flexural Strength (MPa)	Compressive Strength (MPa)
0	65	58	84
2	81	71	87
4	94	86	91
6	112	99	94
8	127	106	96

3.3 Flexural Strength

Effects of SHF loading on the bending strength of the composites have also been studied in this work. Table 2 presents the values of flexural strength for composite with different SHF content. A consistent rise in flexural strength of composite with increase in SHF content was recorded. It was found that the flexural strength of composite increased to 71 MPa with addition of just 2 wt% hair fibers. Similarly, with the addition of 8 wt % fiber, the flexural strength of composite attained a value of 106 MPa.

3.4 Compressive Strength

The compressive strength of epoxy composites with different wt.% of short human hair fibers are presented in Table 2. A marginal rise in compressive strength of composite with increase in SHF content was recorded. It was found that the compressive strength of composite increased to 87MPa with addition of 2 wt % hair fibers. Similarly, with the addition of 8 wt % fiber the compressive strength of composite attained a value of 96 MPa.

Due to the higher tensile strength of hair fibers major portion of the load transmitted from matrix to fiber during the tensile and bending test of the epoxy-SHF composite. However during the compression test of the composite specimen void fraction presents in the composite causes early failure of the composite. Mechanical properties such as tensile, flexural and compressive strengths of epoxy-SHF composite increases because of improved interfacial adhesion between the matrix and the fiber though the void fraction of the composite increases with the addition of SHF.

3.5 Scanning Electron Microscopy
Figure 1 presents some SEM images of raw hair fibers and the epoxy composite. Randomly oriented fibers are shown in Fig. 1-a.

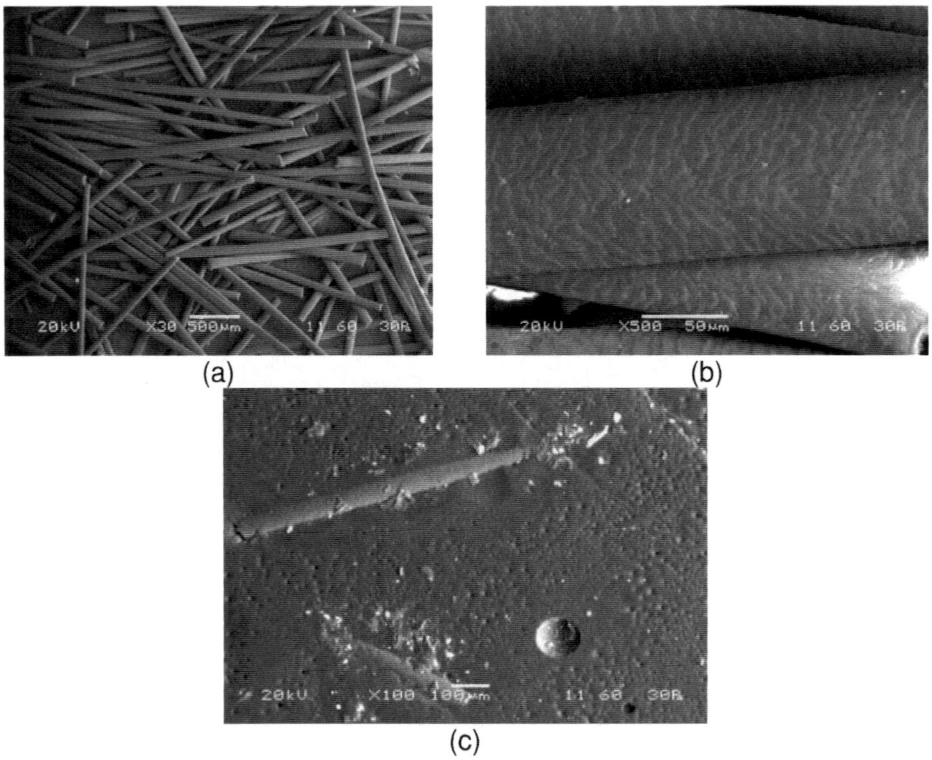

Figure 1. SEM images of raw hair fiber and epoxy-SHF composite specimen

The surface features of a single fiber body are seen in the magnified SEM image, given in Fig. 1-b. The scaly nature of fiber surface with ridges is evident from this image. Figure 1-c is the micrograph of a typical epoxy-hair composite with 2 wt% fiber content.

4. Conclusions

Successful fabrication of epoxy composites reinforced with short human hair fiber with simple hand lay-up technique is possible. Tensile strength and flexural strength of the composite can be largely modified with the incorporation of short human hair fiber. This work thus opens a new avenue for the utilization of bio waste like human hair fiber.

References:

[1] Velasco M V R, Dias T C S, Freitas A Z, Junior N D, Pinto C A S O, Kaneko T M, Baby A R 2009 BJPS **45** 153-161.

[2] Seshadri I P, Bhushan B 2008 Acta Mater. **56** 774–781.

[3] Samanta A, Bhattacharya M, Dalui S, Acharya M, Das P S, Chanda D, Acharya S D, Sivaraman S K, Nath S, Mandal A, Ghosh J, Mukhopadhyay A K 2016 J. Mech. Behav. Biomed. Mater. **56** 229-248.

[4] Bhushan B, Wei G and Haddad P 2005 Wear **259** 1012-1021.

[5] Torre C and Bhushan B 2005 Ultramicroscopy **105** 155-175.

[6] Wei G, Bhushan B and Torgerson P 2005 Ultramicroscopy **105** 248-166.

[7] Zheljazkov V D, Silva J L, Patel M, Stojanovic J, Lu Y, Kim T, Horgan T 2008 Horttechnology **18** 592-596.

[8] Jain D. and Kothari A. 2012 Res.J.Recent.Sci. **1** 128-133.

[9] Bagadiya M, Desai R N and Padhiyar S J 2016 IJARIIE **2** 2395-4396.

[10] Agrawal B D, Broutman L J, Chandrashekhara K 2006 John Wiley and Sons, Inc 3rd ed.

Effect of Hydrothermal immersion and Hygrothermal Conditioning on Mechanical Properties of GRE Composite

A.P. Chakraverty,[*, 1, 2] U. K. Mohanty, [2, 3] S. C. Mishra[2] and B. B. Biswal[4]

[1]Department of Applied Physics,
C. V. Raman College of Engineering, Bhubaneswar-752054, Odisha, India
[2]Department of Metallurgical and Materials Engineering,
National Institute of Technology, Rourkela-769008, Odisha, India
[3]Department of Mechanical Engineering
Institute of Technical and Educational Research (ITER),
SOA University, Bhubaneswar-751030, Odisha, India
[4]Department of Industrial Design,
National Institute of Technology, Rourkela-769008, Odisha, India
*Corresponding Author Email: anantac8@gmail.com

Abstract: Glass fibre reinforced epoxy (GRE) composite meet several degrading agents like moisture and temperature while its use in real time applications in civil infrastructures. Keeping this in mind, the short beam shear (SBS) specimens of GRE composite were exposed to such laboratory created stringent environment as a combination of moisture and elevated temperature for several periods. The environments are as: immersion in distilled water coupled with 65°C as hydrothermal conditioning and an ambience containing 95% relative humidity at 60°C as hygrothermal conditioning. Moisture treated SBS specimens were subjected to 3-point bend test to reveal inter laminar shear strength (ILSS), stress/strain at rupture and modulus values with periods of exposures. The concerned sample suffered 23% of degradation in ILSS values after 120 days of hydrothermal immersion and 25% after 90 days of hygrothermal conditioning. Samples at some optimum exposing conditions of both the exposures are thermally characterized by adopting differential scanning calorimetry (DSC) test. Glass transition temperature (T_g) of such representing samples were determined from the DSC thermograms. About 8 % reduction in T_g values was observed for the GRE composite sample, expectedly, due to moisture induced matrix plasticization and swelling. The fractographs as obtained through scanning electron microscope (SEM) revealed some causes of failures indicating the prime modes of failure of the treated GRE samples with optimum duration of both the exposures.

Key Words: GFRP Composite, Hygrothermal Conditioning, Inter Laminar shear strength (ILSS), Glass Transition Temperature (T_g), SEM Fractographs

1. Introduction

Glass Fibre reinforced epoxy (GRE) composites, in recent days, are considered as the most suitable alternatives for application in civil infrastructures [1], aerospace structures [2] and many more [3]. In such applications, metallic counterparts, expectedly, degrade due to corrosion and crack formation property with respect to any kind of mechanical hurdles [4]. Owing to some specific properties [5] like high strength to weight ratio, corrosion free, ease of fabrication and installation, higher fatigue performance, cost effectiveness, GRE composite system creates a special attention among materials expert community and of course be a material of choice for the applications under question. These materials need rigorous investigations pertaining to their exposure to various stringent environments.

GRE composites while its use in different civil structural applications can meet several environmental severities, out of which moisture is the prime agent combined with concurrent ambience like temperature [6]. Thus, an environment combined with moisture and temperature (elevated/ambient/sub-ambient) is the key point to be considered as one important conditioning exposure of such material.

Moisture is the chief detrimental factor for GRE composite being affected by some matrix dominating property likes Plasticization [7], swelling [8] and hydrolysis of polymer [9]. Though, a little amount of moisture is very helpful for this material in a way of creating better contact area between fibre and polymer layer via siloxane coating agent [10], still its attack to the concerned material with a huge amount can deteriorate considerably by delamination, fibre (glass) leaching, strength degradation and strain misfit between fibre and matrix as a result of unequal moisture induced expansion [11, 12].

In addition, temperature is one of the driving factors for moisture absorption in to the GRE composite body. Higher temperature can accentuate the rate of moisture absorption process, generally in a moisture laden ambience [13]. Moisture absorption at elevated temperatures may also cause cracking, blistering, chemical degradation and debonding, hydrolysis, oxidation and leaching of small molecules in the matrix/fibre, the processes being all irreversible [10, 14]. Loos and Springer [15] found that the maximum moisture content is a strong function of the relative humidity of the exposure environment while the diffusivity is a strong function of the temperature. An ambience laden with moisture is expectedly most degrading for the GRE composite, as unlike the hydrothermal exposure where the sample is actually immersed in water, here the interface between the composite and the ambience continuously changes giving a scope for higher rates of interactions causing greater uptake of moisture and accelerating the associated damage-intending processes [16]. On this aspect, epoxy resin with higher mechanical properties is very prone to moisture attack under the influence of higher temperature. Hence, epoxy resin, on this issue, need to be investigated for its better implementation as polymer matrix in FRP composite system.

Unlike some polymers like vinyl ester, poly-propylene, polyester, epoxy with double oxirane group (from Bisphenol type) is prone to polymerization process with a very limited curing interval and gains excellent strength and mechanical property [17]. Its brittle property demands for its use as matrix incorporation with some reinforcing agent like glass fibre, carbon fibre, Kevlar fibre and also some ceramic particles to enhance the toughness as required for different engineering applications. Glass fibre, on the way towards of it, can be a suitable choice on the basis of its low cost with some mechanical properties to be compromised with respect to carbon fibre [18-19]. Also, Glass fibre is not usually affected by moisture unless until the leaching induced damage will not affect the glass fibre subject to prolonged huge moisture attack [20]. Hence a safety zone below a certain amount of moisture pick up is to be claimed for the GRE composite system with compromised level of mechanical degradation to avoid the leaching induced damage of glass fibre.

The present investigation is an attempt of investigating the physical and mechanical response of E-glass fibre/epoxy composite exposed to prolonged continuous hydrothermal immersion and hygrothermal conditioning (moisture laden with some higher temperature) for various lengths of time. The mechanism of the moisture ingression and the damages caused specially the mechanism of failure under such exposures are examined at length on the basis of the results of the experiments conducted. The present paper can reveal the material for its suitable applications for civil structural applications subject to prolonged exposure to moisture of different stringent moist environment.

2. Materials and methods

The glass fibre reinforced epoxy (GRE) composite was fabricated by adopting the conventional hand lay-up method [6] considering epoxy resin (LAPOX L-12) as polymer matrix and woven E-glass fibre as reinforcing agent. Specific characteristic properties of the components are mentioned in our research paper [9]. Epoxy resin was mixed with Diamine (LAPOX K-6) as hardener/curing agent in the ratio of 10:1 by weight prior to its application on glass fibre. After 48 hours of curing, at ambient condition, short beam shear (SBS) specimens were prepared by the help of diamond cutter keeping the sample size as per the ASTM D 2344-84(1984) standards for 3-point bending test (Figure 1). One batch of such samples was immersed in distilled water at 65° C in an electric oven up to 120 days (referred as hydrothermal conditioning). Samples after completion of each 20 days of hydrothermal immersion were kept out of this concerned exposing condition to record the percentage of moisture gain data and for further characterizations. Another batch of GRE composite samples was exposed to a moisture laden environment i.e. hygrothermal environment (95% Relative humidity at 60°C) in a humidity cabinet for 90 days. After each 15 days of such conditioning period, group of samples were removed from the Hygrothermal chamber and moisture gain data for corresponding periods are claimed. The periods (in days) of exposure to hygrothermal and hydrothermal conditioning are different owing to the fact that moisture ingression rate in case of hygrothermal exposure is higher than that in the case of hydrothermal exposure [6, 21].

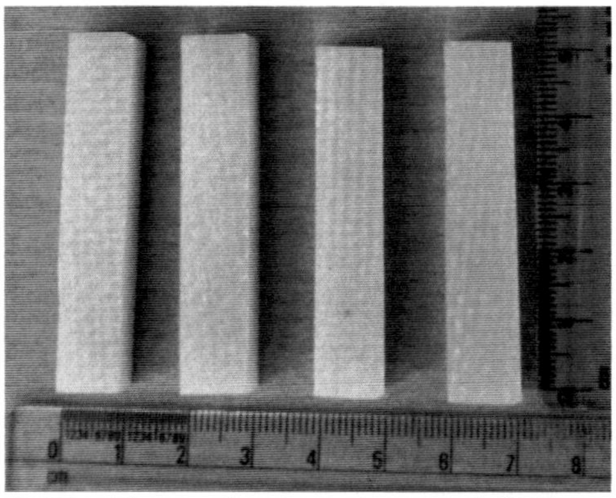

Figure 1. Short beam shear specimens of GRE composite

The percentage of moisture gain data of GRE composite samples subject to both type of conditioning environments were determined by the help of the weight gain data as a function of time of exposure, as [9]

$$M(t) = \frac{w_t - w_o}{w_o} \times 100 \qquad (1)$$

where, 'w_t 'is weight of the sample after concerned exposure (hydrothermal immersion and hygrothermal conditioning) and 'w_o ' is initial weight (fix weight) of the samples before conditioning. All such conditioned group of samples for concerned periods (20, 40, 60, 80, 100 and 120 days for hydrothermal immersion/ 15, 30, 45, 60, 75 and 90 days for hygrothermal conditioning) were subjected to 3-point bend test in INSTRON-1195 as one universal testing machine. The test revealed the data like load at rupture, stress at rupture, strain at rupture and modulus, all being recorded to monitor the mechanical properties.

Inter laminar shear strength (ILSS) data of the moisture treated samples were calculated by the formula [4-6]:

$$\text{ILSS} = 0.75\ {p_b}/{bt} \qquad (2)$$

where, p_b = breaking load (load at rupture in KN), b = width of the specimen in mm and t = thickness of the specimen in mm.

Samples after optimum duration of both the conditioning environments (hydrothermal immersion and hygrothermal conditioning) were allowed for glass transition temperature (T_g) measurement by differential scanning calorimetry (DSC) test by the help of Mettler Toledo-821 (ADSC) module compiled with STAR software. The DSC scanning was performed for the range 30-150° C at 10° C/min. The first change in slope in DSC thermogram reveals the glass transition temperature (T_g). For maximum period of both the conditionings, the fractured specimens after 3-point bend test were viewed by scanning electron microscope (SEM; JEOL; JSM-6480 LV and ZEISS, EVO 60) to reveal the mode of failures.

3. Result and Discussion

3.1 Moisture gain

Similar trend of moisture gain variation with respect to time of conditioning of hydrothermal immersion and hygrothermal exposure was observed. The percentage of moisture gain, as revealed in Figure 2, is increased with increase in conditioning time of both the exposures (hydrothermal and hygrothermal). Initially the slope assumes a higher value, flattens somewhat and then again increases. The maximum percentage of moisture gain by glass/epoxy composite after 120 days of hydrothermal immersion was found to be 1.642 and that after 90 days of hygrothermal conditioning to be 1.682.

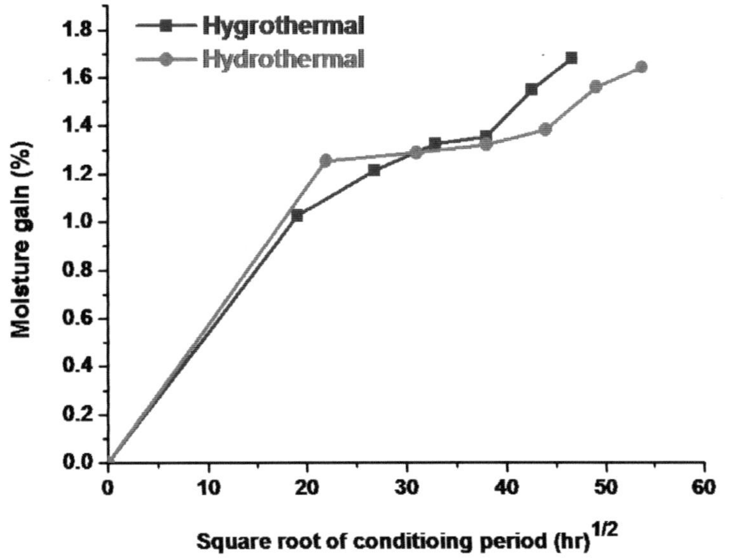

Figure 2. Percentage of Moisture gain of moisture conditioned GRE composite samples

Many investigators [22] have tried to explain the transport of moisture through the process of diffusion in resins and resin-glass composites. The initial continuing uptake trend in moisture gain is due to the difference in concentration of moisture between the composite body and the ambience. The reported rate of moisture absorption is also due to the enhanced rate of moisture pick up by the epoxy polymer; since epoxy resin is prone to moisture absorption due to the presence of –OH groups which can attract polar water molecules by hydrogen bonding [6]. Initially, the moisture absorption is concentration dependent and obeys Fick's 2nd Law i.e. the absorption of moisture is a direct function of the time of exposure. The process of moisture ingression can be explained due to the action of some driving force (higher exposing temperature; 65°C for hydrothermal immersion and 60°C for hygrothermal conditioning) associated with absorption, diffusion and permeation; all these phenomena being gradient driven [23].

The present reported result pertaining to moisture gain of GRE composite is true for an exposure to plane/distilled water or to an atmosphere laden with moisture i.e. under hygrothermal as well as hydrothermal exposures. The environment associated with hygrothermal conditioning is laden with moisture, and here the moisture lies in the form of molecular dimension where the group of water molecules being linked by hydrogen bonds to the polymer. Hence, the permeation of moisture in to the composite body is easier in hygrothermal process and the ingress is said to be more, as revealed in the plot of moisture gain (Figure 2). The higher extents of moisture ingression in the case of hygrothermally treated samples may be due to the dynamic interface between the ambience and the composite body.

The rate of moisture intake, after the moisture ingression reaches a threshold value obeying the Fick's Law [24], decreases and then again increases, as a consequence of exposure to longer periods. This is non-Fickian where the linearity of the plot between moisture gain and square root of time of exposure is disturbed. Finally, for the higher conditioning periods after attaining saturation in moisture content, the composite material behaves differently with added moisture ingression. The glass transition temperature (T_g) changes promoting plasticization and swelling causing reduction of modulus. The deteriorating process exhibited huge moisture ingression at later stages of hygrothermal conditioning/hydrothermal immersion.

Therefore, the huge increase in moisture ingression at the later stages of both conditioning environments indicating Non-Fickian nature of diffusion process could ameliorate the cause of deterioration in the composite material and this depends upon the relative rates at which polymer structure and property changes. In this case, the concentration gradient rate decreases with the ingress of more and more moisture. Here, the moisture absorption is anomalous, can't be explained by Fick's Law and is thus Non-Fickian. This is attributed to the creation of some defects like micro-cracks, delamination, debonding between fibre and polymeric matrix as a result of swelling of the matrix and also to a variation of absorption behaviour by alternation in T_g values (as revealed in DSC results).

3.2 Inter laminar shear strength (ILSS)

ILSS of the GRE composite samples continuously decrease with increase in duration of both hygrothermal and hydrothermal exposures (Figure 3) and also with increased moisture ingression (inserted figure of Figure 3).

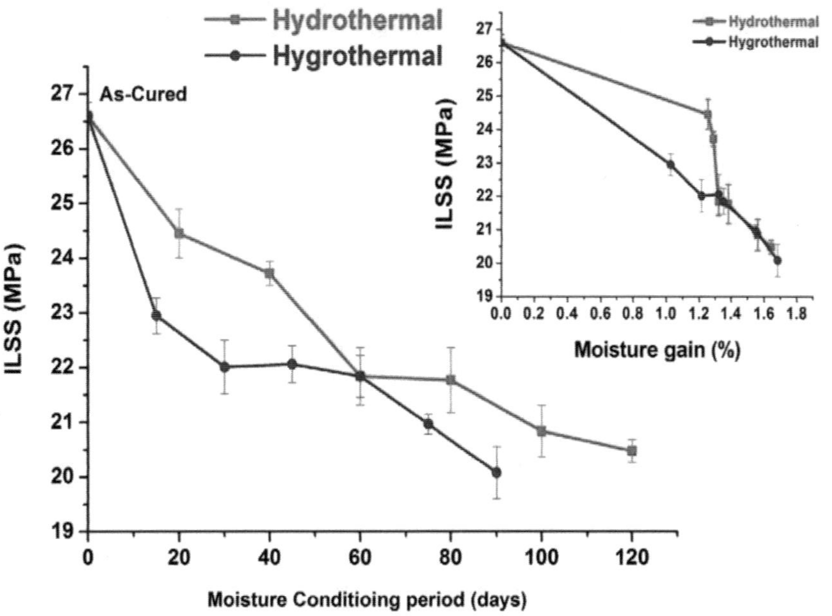

Figure 3. Variation of ILSS of GRE composite with moisture conditioning period (Inserted figure: Variation of ILSS with moisture gain percentage)

Hygrothermal conditioning causes 17% reduction in ILSS after 45 days of exposure and 25% after 90 days of same treatment. The continuous decrease in ILSS with increasing hygrothermal ageing duration establishes the fact that, prolonged moisture conditioning coupled with higher temperature causes greater mechanical degradation pertaining to creation of cracks, voids in the epoxy polymer of the composite assisting further moisture uptake and delaying the saturation level of moisture uptake [11, 25]. This result is also reflected in Figure 3 indicating decrease of ILSS values up to about 23% after 120 days of immersion in distilled water. The plot pertaining to the inserted figure of Figure 3 shows the decrease of ILSS with increased moisture gain during hygrothermal and hydrothermal treatments.

It may be inferred that, the continuous decrease in ILSS is due to the plasticization of the polymer matrix by continuous moisture absorption promoting break down of chemical bonds between polymer and fibre and/or secondary forces of attraction at the interfaces. In the present work, time and temperature have significant synergetic effect on inter laminar shear property of the composite body. With increasing time of exposure the laminar strength decreases up to a significant extent. Similar results were reported by Shivkumar *et al.* [26]; the ILSS of Glass/epoxy composite decreased with increasing time and temperature of hygrothermal atmosphere belonging 85% R.H. Also, the results reported by Kim *et al.* [13] are in agreement with the present trend; the tensile strength of E-glass fibre/vinyl ester composite decreased with increasing hydrothermal immersion period and with increasing temperature compared to as-cured sample.

3.3 Stress/strain at Rupture

As evident from Figure 4 and Figure 5, the composite samples require lesser amount of stress values for rupturing with increased hydrothermal as well as hygrothermal conditioning periods. Subsequently the strain at rapture values also assumes lower proportions for both the exposing environments. Stress at rapture for 90 days hygrothermal conditioning decreases to 21% of that for the as-cures sample and that for 120 days of hydrothermal immersion to be same amount (21%).

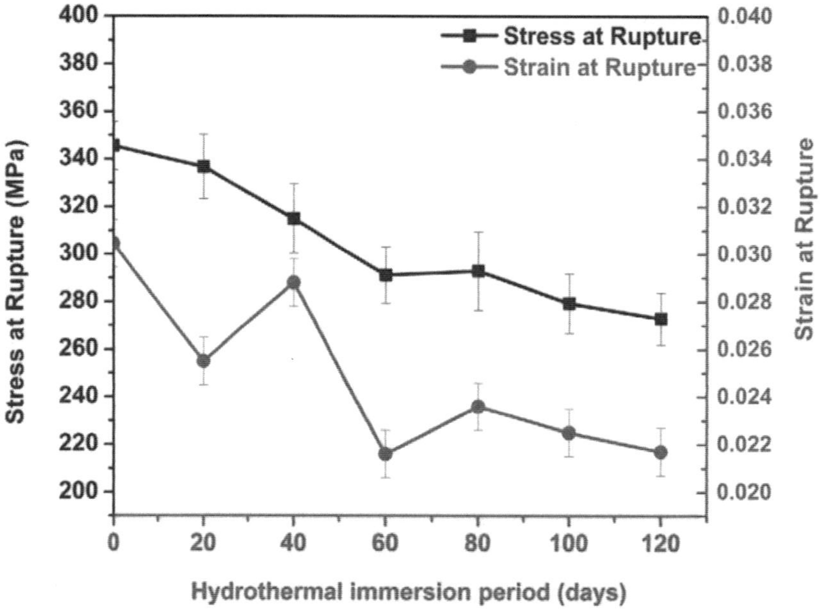

Figure 4. Variation of stress at rupture and strain at rupture values with hydrothermal immersion period

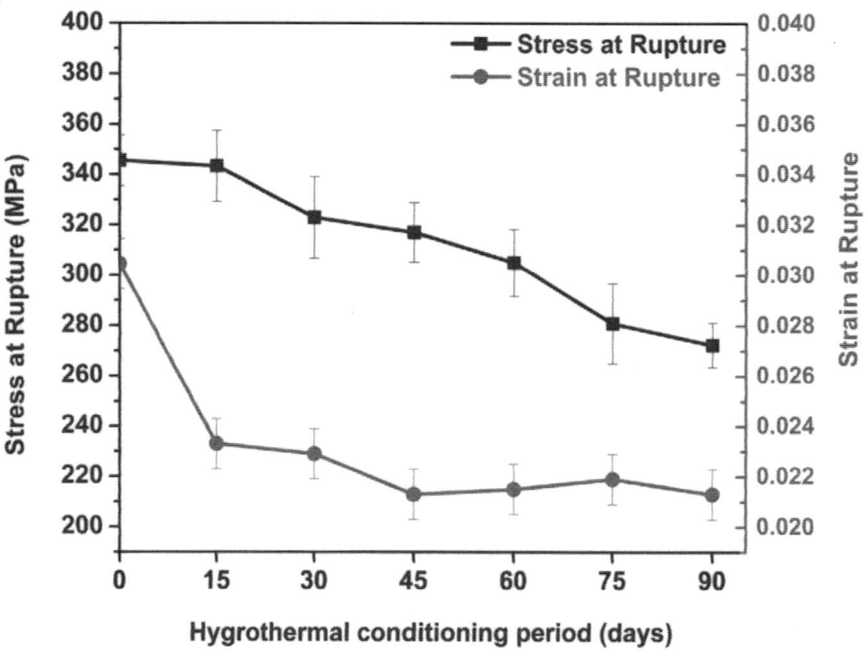

Figure 5. Variation of stress at rupture and strain at rupture values with hygrothermal conditioning period

As described in this paper, the continuous decrease in ILSS reveal the deterioration of the composite body, which thus can stand lesser amounts of stress values after prolonged hydrothermal immersion/hygrothermal conditioning. As a consequence strain at rupture values decrease severely, affecting the tolerance limit of the composite. The difference in the values of stress requirement for rupture and subsequent strain at rupture are quite less even a higher duration difference between both the conditioning environments. Thus, the deterioration of GRE composite is found more prominent due to hygrothermal conditioning of duration lower than the maximum duration of hydrothermal immersion.

The thermal expansion mismatch between fibre and polymer is responsible for the generation of the residual stress in the overall composite structure [6]. After prolonged hygrothermal/hydrothermal exposure, this residual stress in the composite body would not allow the composite to accommodate the applied external stress in the way of prohibiting such level of straining as being observed for as-cured specimen.

3.4 Modulus

Data pertaining to Figure 6 reveal the variations in Modulus of the GRE composite samples with respect to durations of hydrothermal immersion and hygrothermal conditioning. Around 6 % decrease in modulus is observed after 90 days of hygrothermal conditioning and about 9% after immersion in distilled water.

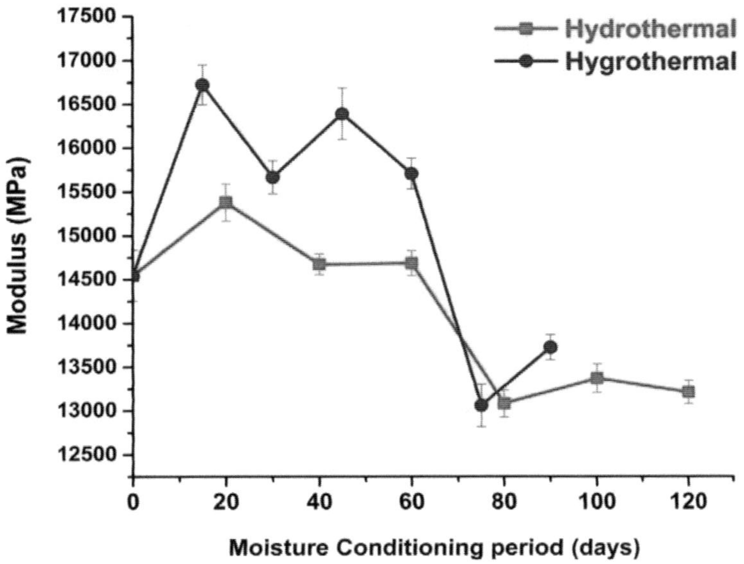

Figure 6. Variation of Modulus with periods of hydrothermal immersion and hygrothermal conditioning periods

Modulus values decreased irrespective of an initial increment with increasing conditioning period of both the exposures. The initial increase in modulus may be due to creation of better contact area between fibre and matrix due to weakening of hilly structure of Silane coupling agent of E-glass fibre.

This may happen due to initial small quantities of moisture absorption, which in turn may effectively decrease the attachment of Siloxane layer from glass fibre [10].

The result pertaining to decreasing trend in modulus is in agreement with the corresponding ILSS results of hygrothermally and hydrothermally treated GRE composite samples. The polymeric matrix during the concerned treatments is affected by plasticization and reduces the modulus and glass transition temperature. Modulus of layered FRP composites is analogous to the perfect fibre/matrix interfacial adhesion and of course rigidity of polymer structure, which in turn demand the better elastic limit during mechanical loading. The stress and strain variations indicating decreasing stress and straining amount with respect to increasing exposure period authenticates the modulus result by establishing a decreasing trend in elastic limit after prolonged hygrothermal and hydrothermal treatments.

3.5 Glass transition temperature (T_g)

Figure 7 and Figure 8 show the DSC curves for hygrothermally and hydrothermally treated composite samples, respectively. The samples with extreme conditions of both hygrothermal and hydrothermal treatments are chosen on the basis of time of exposure (15, 60 and 90 days for hygrothermal conditioning while 20, 80 and 120 days for hydrothermal immersion). The onset values of the glass transition temperature (T_g) as recorded from the DSC curves of the above representative samples are graphically illustrated in inserted figures of Figure 7 and Figure 8.

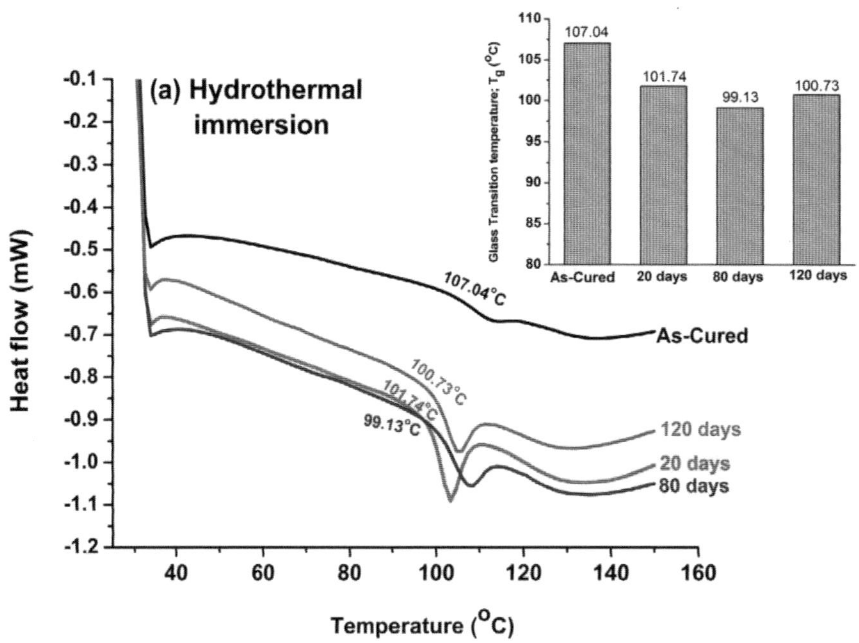

Figure 7. DSC thermograms of hydrothermally immersed GRE composite
(Inserted figure: T_g variation of GRE composite with respect to time of hydrothermal immersion)

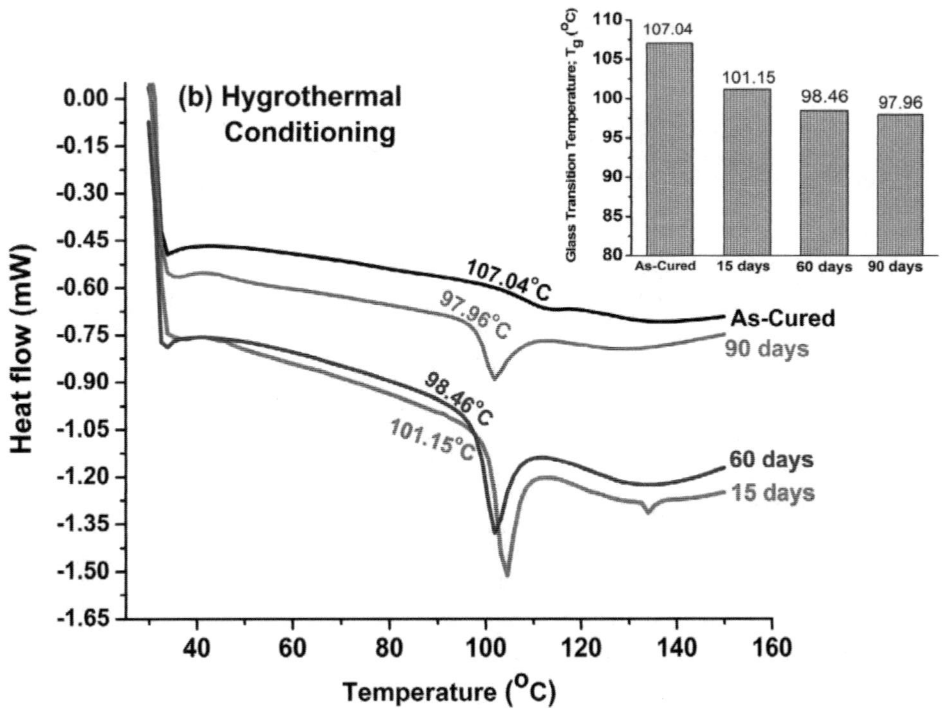

Figure 8. DSC thermograms of hygrothermally conditioned GRE composite
(Inserted figure: T_g variation of GRE composite with respect to time of hygrothermal conditioning)

The T_g values decreased with increasing period of hygrothermal and hydrothermal exposures. About 8% reduction in T_g is observed after 90 days of hygrothermal conditioning. About 6% reduction in T_g is observed after 120 days of hydrothermal immersion of the composite specimens. T_g of polymeric materials depends on the extent of cross-linking density [27]. Due to imperfect curing/cross-linking of the polymer, some voids could have been created in the polymer. These voids, referred as free volume, approximately occupy 1/40th of total polymer volume at/below glass transition temperature [6]. The T_g deterioration is affected when the extent of free volume/voids increase. Hence, it may be inferred that, on prolonged hygrothermal/hydrothermal treatment, polymer matrix could be affected by plasticization leading to chain scission/breaking by the action of hygro-elastic swelling stress. This may lead to increase in internal voids/free volumes in the polymer structure giving rise to early glass transition process. The greater percentage of T_g reduction for GRE composite subject to maximum duration of hygrothermal treatment could be due to more damaging effects concomitant with severe plasticization, swelling of epoxy polymer as a result of huge amount of hygro-elastic swelling stress. Hence, thermal stability of the material is more hampered after hygrothermal treatment of less duration compared to maximum duration of hydrothermal immersion.

3.6 SEM fractographs

As evident from Figure 9 and Figure 10, the SEM Micrographs for hydrothermally immersed and hygrothermally conditioned fractured specimens show some prime modes of failures like fibre breakage, fibre pull out, fibre-matrix debonding.

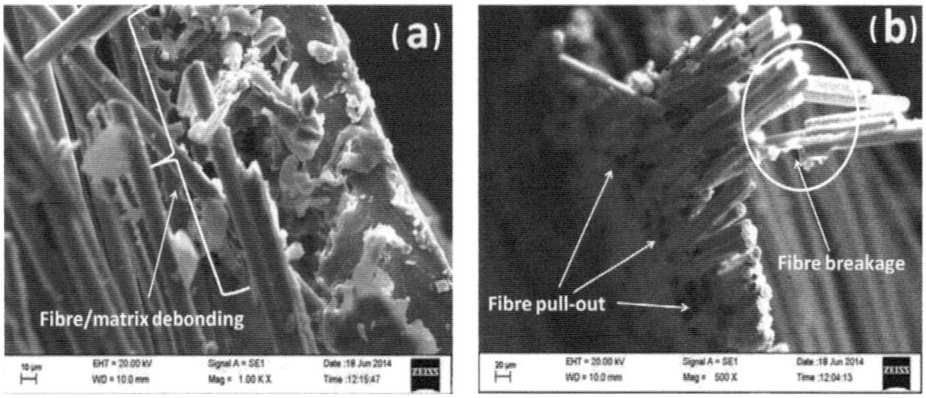

Figure 9. SEM fractographs of hydrothermallly immersed GRE composite samples
(a) 20 days (b) 120 days

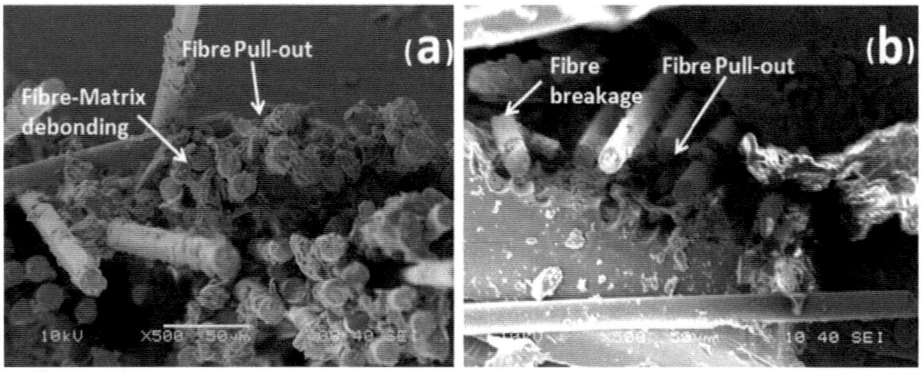

Figure 10. SEM fractographs of hygrothermallly conditioned GRE composite samples
(a) 15 days (b) 90 days

The causes of failure are found anyone of the modes or any combinations. The degrading phenomenon might have caused the irreversible damages in the composite body and the prolonged moisture treatment coupled with elevated temperature could have affected the polymer matrix to a great extent. Hence, fibre pull-out coupled with fibre breakage, matrix holes are created due to debonding of fibre from matrix material. Especially, fibre breakage is visible for the GRE composite samples after maximum duration of both the conditioning environment (hydrothermal and hygrothermal). Thus, the damage of GRE composite is quite comparable for both the conditioning exposures irrespective of variation of periods of exposures for both the cases.

4. Conclusion

The findings as established through the present experimentation indicate the following concluding remarks:

(i) Moisture ingression, whether due to hygrothermal or hydrothermal exposures is detrimental for the well being of the GRE composite.

(ii) Hygrothermal exposure is more effective towards deterioration of the properties of the composite.

(iii) Whether the exposure is hygrothermal or hydrothermal, time of exposure also plays an important role towards the mechanical stability of GRE composite.

(iv) Property degradation of GRE composite is comparable with respect to the both the exposures, irrespective of higher time of exposure for hygrothermal conditioning.

(v) Design stress of the GRE composite can thus, be made less than 25% of ILSS value of the as-cured sample so that the component can be safely used under hygrothermal and hydrothermal ambience.

References

[1] Hollaway L C 2010 *Constr Build Material* **24** 2419.
[2] Mishra G, Mohapatra S R, Behera P R, Dash B, Mohanty U K and Ray B C 2010 *Aircraft Eng Aeroso Tec* **82** 258 .
[3] Pinto M, Chalivendra V B, Kim Y K and Lewis A F 2016 *Compos Struct* **154** 333-337.
[4] Chakraverty, A P, Mohanty U K, Mishra S C and Biswal B B 2016 *IOP Conference Series: Materials Science and Engineering* **115** 012004.
[5] Ray B C, 2006 *J Colloid Interface Sc* **298** 111-117.
[6] Chakraverty A P, Mohanty U K and Biswal B B 2012 *Emer Mat Res* **1** 263.
[7] Fitriah S N, Abdul Majid M S, Ridzuan M J M, Daud R, Gibson A G and Assaleh T A 2017 *Compos Struct* **159** 350.
[8] Hawa A, Abdul Majid M S, Afendi M, Marzuki H F A, Amin N A M, Mat F and Gibson A G 2016 *Mater Des* **89** 455.
[9] Chakraverty A P, Mohanty U K, Mishra S C and Satapathy A 2015 *IOP Conference Series: Materials Science and Engineering* **75** 012029.
[10] Abdel-Magid B, Ziaee S, Gass K and Schneider M 2005 *Compos Struct* **71** 320.
[11] Nishizaki I and Meiarashi S 2002 *J Compos Constr* **6** 21.
[12] Berketis K S, Tzetzis D and Hogg P J 2008 *Mater Des* **29** 1300.
[13] Kim H Y, Park Y H, You Y J and Moon C K 2006 *KSCE J Civil Engg* **10** 291.
[14] Ellyin F and Rohrbacher C 2000 *J Reinf Plast Comp* **19** 1405.
[15] Loos A C and Springer G S 1979 *J Compos Mater* **13** 131.
[16] Bhuyan M K, Bhuyan M S, Rodriguez-Devora J I and Yanez M 2012 *J Compos Mater* **47** 3421.
[17] Zhou J, Jia S, Fu W, Liu Z and Tan Z 2016 *Materials Letters* **176** 228.
[18] Xu L R, Krishnan A, Ning H and Vaidya U 2012 *Composites: Part B* **43** 2480.
[19] Jang B Z 1994 *ASM International*, Materials Park, OH, USA.
[20] Charles R J 1958 *J Appl Phy* **29** 1549.
[21] Chakraverty A P, Mohanty U K and Biswal B B 2012 *Int J Curr Res* **4** 67.
[22] Feldman D 1990 *Prog Polym Sci* **15** 603.
[23] Odegard G M and Bandyopadhyay A 2011 *J Polym Sci* **49** 1695.
[24] Weitsmsn Y J 2000 in Kelly A and Zweben C of *Comprehensive Composite Materials, Elsevier* **2**
[25] Gautier L, Mortaigne B, Bellenger V and Verdu J 2000 *Polymer* **41** 2481.
[26] Shivkumar S and Shivarudraiah 2010 *Int J Adv Engg Tech* **1** 225.
[27] Hale A, Macosko C W and Bair H E 1991 *Macromolecules* **24** 2610.

Dielectric and conduction behaviour of H_2SO_4 doped conducting Polyaniline

J. Mohanty[1*], P. Behera[1], S. R. Mishra[1*], T. Badapanda[2], S.Anwar[3]

[1]Department of Chemistry, C.V. Raman College of Engineering, Bhubaneswar, Odisha, India-752054
[2]Department of Physics, C.V. Raman College of Engineering, Bhubaneswar, Odisha, India-752054
[3]Colloids & Materials Chemistry, Institute of Minerals and Materials Technology, Bhubaneswar, Odisha, India-751013
Corresponding author's email: sruti76@gmail.com (S. R. Mishra), jayashreemohanty7@gmail.com (J.Mohanty)

Abstract: We report the effect of H_2SO_4 doping on the dielectric and conduction behaviour of Polyaniline (PANI) samples. The PANI salt prepared by oxidising aniline hydrochloride in distilled water with the oxidant ammonium persulphate with continuous stirring at room temperature and PANI base is produced by subjecting PANI salt to a reaction with 0.5M NaOH. H_2SO_4 doped PANI is prepared by subjecting PANI base to reaction with 1M H_2SO_4 at room temperature under constant stirring for 1h. The synthesied PANI along with the doped samples were further washed with acetone to study the effect of acetone washing on the electrical behaviour. It is observed that the dielectric constant as well as the dielectric loss decreases with frequency in the entire studied sample. The frequency dependent AC conductivity at room temperature obeys the power law and the DC conductivity was obtained from the fitting parameter. It is found that the non acetone washed PANI doped in 1M H_2SO_4 shows highest dielectric constant and conductivity.

Key Words: PANI, H_2SO_4, dielectric study, Impedance study, Conductivity

1. Introduction

Conducting polymers have potential applications in many areas such as electrochemistry, electromagnetic systems, electronics, electromechanical systems, electroluminescence and sensors. Out of most semiconducting/conducting polymers, the charge transfer polymers like polypyrrole, polythiphene, polyparaphenylene and polyaniline, show excellent electrical and optical properties. Among these semiconducting/conducting polymers the most widely studied polymer is polyaniline due to its easy preparation method, environmental stability, low cost with applications in molecular electronics [1]. The conductivity in PANI arises due to the delocalisation of electrons in a continuously overlapped pi orbital in the polymer backbone. Because of this reason it has interesting optoelectronic properties and finds applications like photo-voltaic cells, organic light emitting devices, solid state laser, electromagnetic shielding, rechargeable battery electrodes, light emitting diode, transparent electrodes, inorganic semiconductors and sensors like gas sensors and humidity sensors [2-13]. There are several methods to improve the dielectric and condutivity properties of PANI by doping. The doping can be achieved either by chemical means involving partial oxidation or reduction of the polymer, electrochemical doping by applying a DC source, photo doping involving photogenerated charge carriers, charge-injection doping as well as non-redox doping carried out by proton doping of

PANI base with protonic acids [12-18]. Among them aid doping is found to be most suitable. The polymer in doped state shows semiconducting to metallic-like conductivity because doping incorporates charge carriers like polarons and bipolarons to the polymer backbone. With increasing the doping level, the localised polarons and bipolarons near to the dopant can overlap to produce new energy bands which lie in between the valence and conduction bands through which electron can flow.

The conducting polymers synthesis method and chemical modifications provides unlimited opportunities for their potential applications [19]. The preparation condition of PANI, dopant concentration, environmental parameters like humidity, use of various solvents and oxidants during preparation of polymer, presence of residual water creating hydrogen bonding between water and PANI during preparation, extent of disorder and cross linking in polymer, the reaction time can affect the electrical properties of PANI [20- 26]. Apart from this the incorporation of various functional groups on the polymer backbone, can increase the processibility in comparision to the pristine polymer. Besides the above mentioned parameters, the presence of short chain oligomers which lack polyconjugation and other reactive intermediates produced during polymerisation are undesirable for the electrical properties of the polymer. In order to get oligomers free highly conducting form of PANI, it is recommended to wash the polymer by acetone after polymerisation, because by acetone washing of the polymer, short chain oligomers and other reactive intermediates are removed [IUPAC]. However, to study regarding the electrical properties of conjugated polymers it is required to focus on the properties of more well defined oligomers to avoid problems such as defects in ideal polymer chain or disorder. In short oligomers molecules, the polaron wave functions are forced to overlap due to spatial confinement whereas in longer polymer molecule the polarons are free to move apart sufficiently and the wave functions do not overlap. In the non-degenerate ground state conjugated oligomers of varying length, in the absence of extrinsic effects bipolarons may be unstable than that of the formation of two independent polarons. Hence, the use of oligomers will increase both as test structures for studies of the formation of species such as polarons and bipolarons and their direct use in electronic applications making them to play the important roles in organic-based electronic applications. Overall, the bulk conductivity of conducting polymers consists of contributions from intra-chain, inter-chain and inter-domain electron transportation.
The charge transport mechanism in conducting polymers has been widely studied using different experimental techniques such as AC conductivity, dielectric relaxation and photoconductivity [27, 28]. Complex Impedance Spectroscopy (CIS) is a powerful technique for understanding the charge transport mechanism in disordered conducting polymer nanostructures. Its main advantage over other techniques is the use of wide frequency range [27]. It has been reported that the relaxation of electric field in a charge carrier system is due to the hopping of mobile charge carriers, which leads to both short-range AC conductivity and long-range DC conductivity [28, 29]. The frequency dependent conductivity and dielectric permittivity provide important information on the electronic transport mechanism in disordered materials. It gives an insight in to the molecular structure of the materials, since the localized electronic states within the material are created due to the presence of disorder in the molecular configurations [30]. In this manuscript we report the electrical behaviour (dielectric relaxation, impedance and conductivity) of PANI base, PANI doped in 1M H_2SO_4 with and without washed with acetone. Since the acetone washing removes various reaction intermediates as well as oligomers, hence the presence of these species on the PANI electrical properties has been analysed. The frequency dependences of the dielectric constant and the AC conductivity are analyzed in terms of a dipolar- or a conductivity-relaxation mechanism and pair approximation.

2. Experimental

The PANI salt was synthesised through the conventional method of oxidative polymerisation of aniline hydrochloride with the oxidant ammonium persulphate. Aniline hydrochloride (2.592gm) was dissolved in distilled water (50ml) to which ammonium persulfate solution (prepared by dissolving 5.704gm in 50 ml. of distilled water) was added slowly with continuous stirring at room temperature to prepare PANI salt. Then the reaction mixture was allowed to stand for three hour at room temperature followed by filtration. After filtration the washing of the samples was carried out in two different procedures. In the first procedure the residues were washed with 0.2M HCl followed by distilled water and in another procedure after washing the residue with 0.2M HCl and distilled water it was washed with acetone. The PANI samples prepared in the above experiments were air dried whole night followed by drying at 60^0C in an oven for one day. The resulting PANI salt samples were subjected to a reaction with 0.5M NaOH solution followed by air drying as well as drying at 60^0C in an oven for one day. The produced dried PANI base is powdered and doped in 1M H_2SO_4 by stirring PANI base with 1M H_2SO_4 for one hour at room temperature. The synthesized PANI base and 1M H_2SO_4 doped PANI samples with and without washed with acetone were structurally characterized by X-ray diffraction (XRD) using a Philips diffractometer model PW-1830 with Cu-K_α (k=1.5418 A °) radiation. In order to analyse the electrical properties, the samples were made in the form of circular disc having diameter 10mm and thickness 1mm using a hydraulic press. For electrical contact measurements the ends of the PANI pellets were coated with silver paints. The frequency dependent dielectric and impedance study was carried out by using a N4L-NumetriQ (model PSM1735) connected to a computer.

3. Result and Discussion

3.1 XRD Analysis

The X-ray diffraction pattern for PANI base and PANI doped in H_2SO_4, with and without acetone washing are given in Figure 1 (a,b). The XRD plot for PANI base shows a semicrystalline nature whereas the H_2SO_4 doped PANI samples show a more crystalline nature. The semicrystalline characteristics of PANI base is due to repetition of benzenoid and quinoid rings in the polymer chain whereas the crystallinity in these PANI samples may be due to the systematic alignment of polymer chains by chain folding or due to the formation of single or multiple helices [32]. Peaks appeared at positions $2\theta = 20^0$ corresponds to (020) plane of PANI which is due to the periodicity parallel to polymer chain and $2\theta = 25.6^0$ corresponds to (200) plane of the polymer and it is due to the occurrence of periodicity perpendicular to the polymer chain [34]. For acid doped PANI, the peak observed at $2\theta = 20^0$ is diminished and prominent peak is observed at $2\theta = 25.6^0$ suggesting the formation of PANI salt and it represents the extent of conjugation for delocalisation of π electrons in the polymeric chain structure.

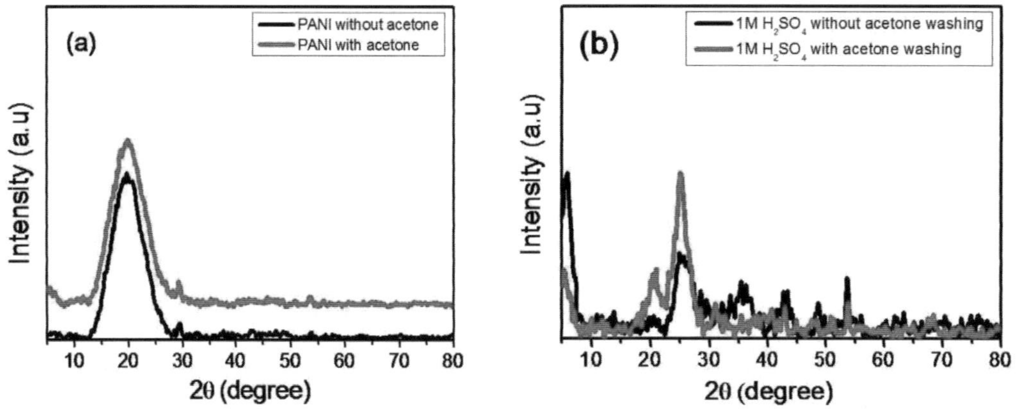

Figure 1. X-Ray diffraction pattern for the (a) PANI with and with acetone washing (b) 1M H_2SO_4 doped PANI with and with acetone washing

3.2 Frequency dependent dielectric study

The variation of dielectric permittivity at room temperature with frequency for PANI base and 1M H_2SO_4 doped PANI samples without and with acetone washing is shown in Figures 2(a,b) respectively. In all cases, a relatively high dielectric permittivity is achieved at low frequency region which is a characteristic of polymer as well other dielectric material and it is due to the space charge polarisation produced due to the free charges at the interface [14] and the nature of the dielectric constant ε' can be expressed as follows

$$\varepsilon^*(\omega) = \varepsilon_\infty + \frac{(\varepsilon_s - \varepsilon_\infty)}{[1 + (i\omega\tau)^{1-\alpha}]} \quad (1)$$

For very low frequency the charge carriers get sufficient time to move over macroscopic distances and build up at the interfaces between the sample and the electrodes within half a cycle of the applied AC field, resulting in high value of the dielectric permittivity [33]. After getting a high dielectric permittivity value at low frequency, the plot becomes asymptotic at higher frequency which is the characteristics of disordered conducting polymer like PANI and it occurs due to hopping of electrons between isolated polarons and bipolarons [34]. The decrease in the dielectric constant with increasing frequency is due to by dielectric relaxation, phenomena that tells that the charge carrier localization is not stable and frequency disturbances effect the charge carrier localization [35].

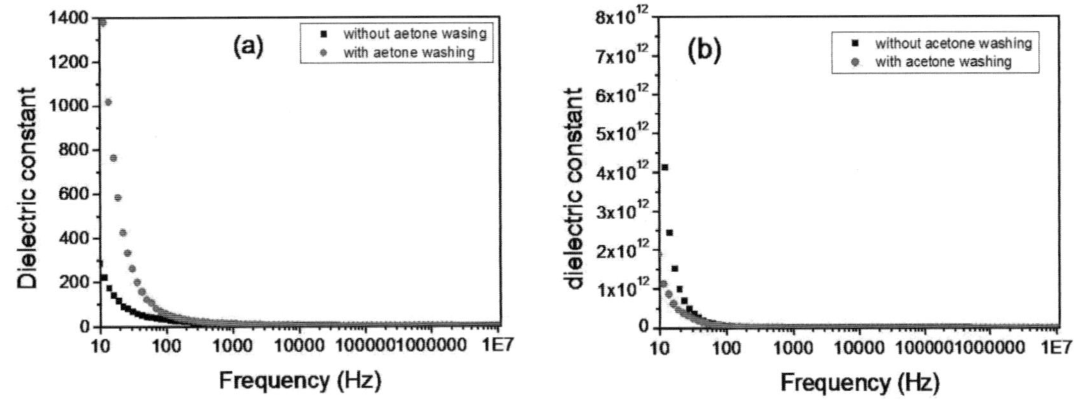

Figure 2. Frequency dependent dielectric constant of (a) PANI with and without acetone (b) 1M H_2SO_4 doped PANI with and with acetone washing

Figure 3 (a,b) shows the variation of dielectric loss with frequency for the PANI base and acid doped samples without and with acetone washing respectively. A high dielectric loss is observed in all the samples at low frequency signifies the increased conductivity occurs due to free mobility of the charge carriers in the polymer arising by the creation of percolation path by long conducting clusters present in the polymer [36]. Among the samples without acetone washed PANI base shows comparatively lower dielectric permittivity as well as loss. However, when the sample is doped with 1M H_2SO_4 the sample without washed with acetone has higher dielectric permittivity whereas for the same sample that is washed with acetone the loss factor is more in comparison to that of non acetone washed sample showing the decrease in capacitive nature and increased in conductivity in acetone washed acid doped sample. The higher dielectric permittivity of non acetone washed acid doped polymer may be due to the extra charge carriers contribution from the small chain oligomers present in non acetone washed sample.

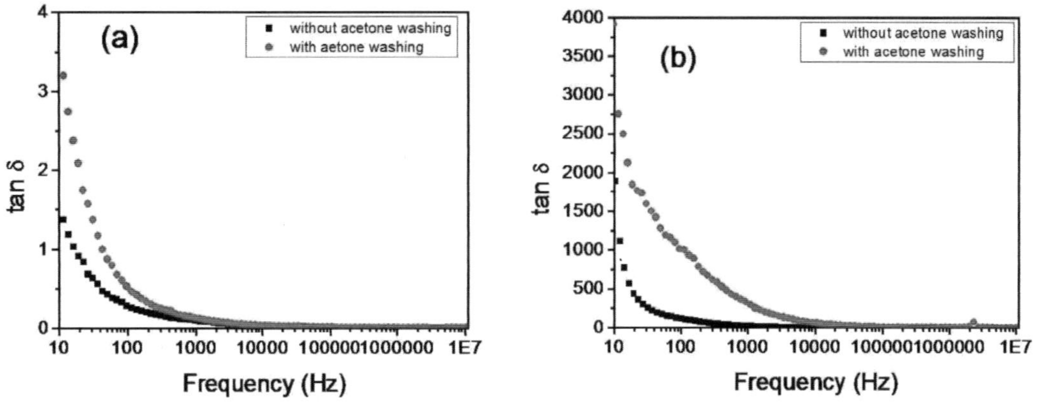

Figure 3. Frequency dependent dielectric loss of (a) PANI with and without acetone (b) 1M H_2SO_4 doped PANI with and with acetone washing

3.3 Complex Impedance Analysis

Figure 4 (a,b) shows the variation of real part of impedance with frequency for without acetone washed and with acetone washed PANI base and acid doped PANI samples respectively. It is observed that for PANI base (both without and with acetone washing) the magnitude of Z' decreases with increase in frequency and reaches an asymptotic value in the high frequency range. For acid doped samples the enhancement of impedance with frequency shows that the AC conductivity increases with frequency. A comparatively lower value of real part of impedance is observed in 1M H_2SO_4 doped PANI that is not washed with acetone than that of the doped polymer washed with acetone. Again, it is observed that the real part of impedance of acid doped in non acetone and acetone washed PANI sample is almost frequency independent which shows that the materials have no AC conductivity for a wide range of frequency.

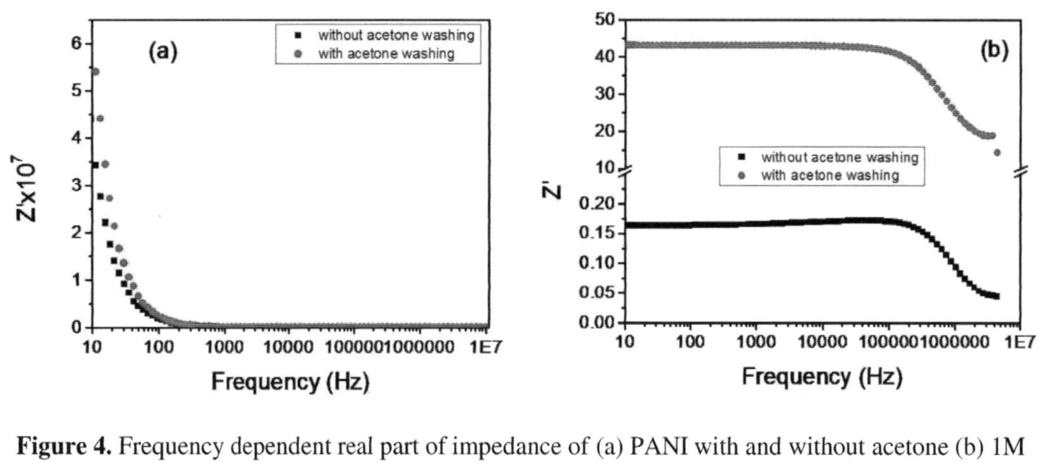

Figure 4. Frequency dependent real part of impedance of (a) PANI with and without acetone (b) 1M H_2SO_4 doped PANI with and with acetone washing

Figure 5 (a,b) shows the variation of imaginary part of impedance with frequency for without acetone washed and with acetone washed PANI base and acid doped PANI samples respectively. The curves show that the Z" values reach a maxima peak (Z''_{max}) in all the samples. The broadening of peaks suggests that there is a spread of relaxation times, i.e., the existence of electrical relaxation phenomenon in the material. The relaxation process may be due to the presence of immobile species. For PANI base in acetone washed sample the frequency corresponding to the Z''_{max} values move to the higher side than that of the non acetone washed sample which implies that the relaxation time decreases for PANI base when it is washed with acetone. For acid doped sample, the opposite observation has been found i.e. the relaxation time decreases for the sample when it is not washed with acetone. Again, maximum Z''_{max} is obtained for PANI base and least Z"max value is obtained for acid doped PANI that is not washed with acetone.

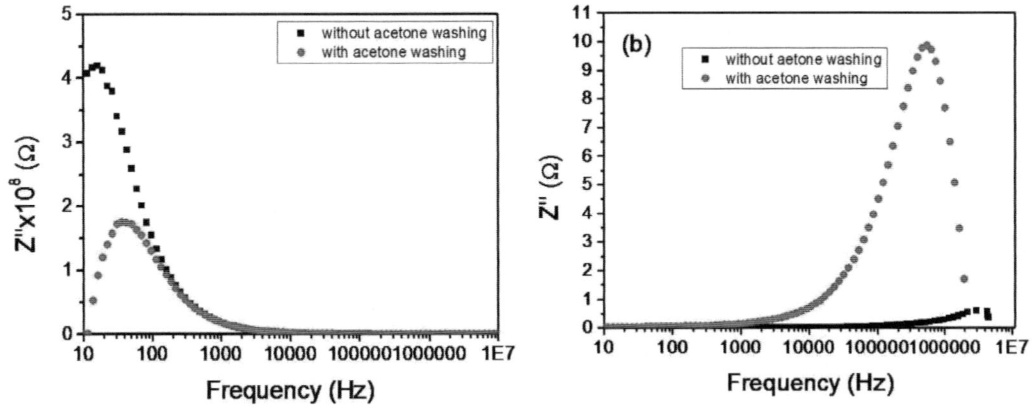

Figure 5. Frequency dependent imaginary part of impedance of (a) PANI with and without acetone (b) 1M H_2SO_4 doped PANI with and with acetone washing

3.4 AC conductivity studies

Figure 6 (a,b) shows the variation of AC conductivity with frequency at room temperature of all the samples without acetone and with acetone washing respectively. The AC conductivity is calculated using a relation $\sigma_{ac} = \omega\varepsilon\varepsilon_0 \tan\delta$, where ω is angular frequency, ε_0 is the dielectric permittivity of vacuum and tan δ is the loss factor. The AC conductivity σ(ω) obeys the Jonscher's power law [37] as given by $\sigma(\omega) = \sigma_{dc} + A\omega^n$ where n is the frequency exponent in the range of 0 = n = 1. Here A and n are temperature dependent parameters and indicates that the electrical conduction is a thermally activated process. Funke [38] explained that the value of n might have a physical meaning (i.e., $n \leq 1$ would mean that the hopping motion involved is a translational motion with a sudden hopping). On the other hand, value of n greater than 1 would mean that the motion involved is a localized hopping of the species with a small hopping without leaving the neighbourhood. The frequency at which change in slope takes place is known as hopping frequency (ω_p) of the polarons, which is temperature dependent. It is a function of both temperature as well as frequency, and corresponds to short-range hopping of charge carriers through trap sites separated by energy barriers of varied heights. The dispersion in conductivity at low frequency can be explained due to non-adiabatic hopping of charge carriers between impurity sites. In disordered polymers like PANI, various conduction mechanisms contribute to the total conductivity where there is movement of electrical charges when it is subjected into an alternating electric field [21].

The DC conductivity of the samples was obtained by fitting the frequency dependent AC conductivity with the Jonscher's power law. Basically, there are two types of charge carriers are present in case of PANI, one bound charges, responsible for dipole moment and another non-bound charges (polarons and bipolarons) responsible for DC conductivity [39]. In case of PANI base (both non acetone and acetone washed) there is no DC conductivity was observed with the measured frequency range and the conductivity increases with increase in frequency which shows that the materials are devoid of capacitive behaviour in the mentioned frequency range. This type of characteristics may be due to the presence of long conducting clusters present in the polymer [36]. In PANI base both without and with acetone washing, there is very less electrical conductivity and dielectric constant was observed which signifies the absence of proton in PANI Chain [22]. For acid doped samples, a comparatively higher conductivity was observed in case of sample that is not washed with acetone than that of the acetone washed sample. When the polymer is not washed with acetone it may contain some low molecular weight oligomers which can be fully doped by acids than that of the high molecular weight polymers [18], consequently increasing the charge carriers that may contribute to the PANI electrical properties with increase in electrical conductivity.

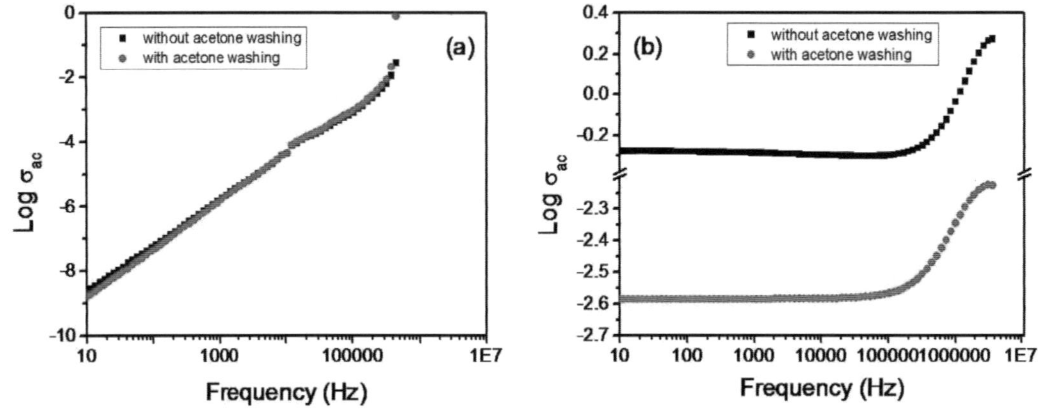

Figure 6. Frequency dependent AC conductivity of (a) PANI with and without acetone (b) 1M H2SO4 doped PANI with and with acetone washing

4. Conclusion

The PANI base was prepared by oxidising aniline hydrochloride with the oxidant ammonium persulphate with continuous stirring at room temperature. The samples were again washed with acetone to study the effect of acetone washing that removes the short chain oligomers on the electrical behaviour of the polymer. The incorporation of H_2SO_4 in the polymer increases the crystallinity as observed in the XRD study. The electrical properties were observed via dielectric and impedance measurement for all the samples and a comparative study with respect to acetone washing was presented. Among all acid doped samples, the sample without acetone washed shows higher dielectric as well as conductivity than that of acetone washed sample due to the stable charge carriers and extra electrical conductivity contribution from the small chain oligomers present in the polymer. Whereas for PANI base there is lowering in dielectric and conductivity value in without acetone washed sample than that of the acetone washed sample. From the above study, we can conclude that acetone washing has a significant role in the electrical behaviour of PANI especially in acid doped polymer. Hence, the presence of short chain PANI oligomers increases the electrical conductivity as well as dielectric permittivity of the acid doped polymer. The higher value of dielectric permittivity and electrical conductivity of non acetone washed acid doped PANI may find promising applications in various industrial applications.

References

[1] Chutia P and Kumar A 2014 *Physica B* **436** 200
[2] Salaneck W R, Friend R H and Bredas J L 1999 *Physics Reports* **319** 231
[3] Ameen S , Ali V , Zulfequar M, Haq M. M and Husain M 2007 *Current Appl. Phys* **7** 215
[4] Basavaraja C, Kim J K, Thinh P X and Huh D S 2012 *Polymer Composites* **33** 1541
[5] Cabezas A L, Feng Y, Zheng L R and Zhang Z B *Carbon* 2013 **59** 270
[6] Hu W, Chen S, Yang Z, Liu L and Wang H 2011 *J. Phys. Chem. B* **115** 8453
[7] John H, Thomas R M, Jacob J, Mathew K T and Joseph R 2007 *Poly. Comp.* **28** 588
[8] McCall R P, Scherr E M, MacDiarmid A G amd Epstein A J 1994 *Physical Review* **50** 5094

[9] Upadhyay P K and Ahmad A 2010 *Chin. J. Poly. Sci,* **28** 191
[10] Xue Y and Quesnel D J 2016 *RSC Adv.* **6** 7504
[11] Yoo JE, Cross JL, Bucholz TL, Lee KS, Espe MP and Loo YL 2007 *J. of Mater.Chem.* **17** 1268
[12] Howard R 1989 *Synthetic Metals* **30** 257–263
[13] Gospodinova N, Mokreva P and Terlemezyan L 1993 *Polymer* **34**, 2438–39
[14] Kim H M, Lee C Y and Joo J 2000 *J. Korean Phys. Soc.* **36** 371
[15] Jia QM, Shan SY, Jiang LH and Wang Y.M 2011 *Adv. Mater. Res.* **160-162** 1405
[16] Saravanan S, Joseph C M, Anantharaman MR, Venkatachalam S and Prabhakaran PV 2006 *J. Phys. Chem. Solids* **67** 1496
[17] Chang Q, Li J, Gu D, Yin P, Jiang H and Shen L 2015 *J. Macromolecular Science*, Part B: Physics **54** 381-92
[18] Yoo J.E, Cross JL, Bucholz TL, Lee KS, Espe MP and Loo YL 2007 *J. Materials Chemistry* **17** 1268
[19] Acevedo DF, Salavagione HJ, Miras MC and Barbero CA 2005 *J.Braz. Chem. Soc.* **16(2)** 259
[20] Limelette P, Schmaltz B, Brault D, Gouineau H, Autret-Lambert C, Roger S and Grimal V, Tran VF 2014 *J. Applied Physics* **115** 033712
[21] Matveeva ES 1996 *Synthetic Metals* **79** 127
[22] Novak M, Kokanovic I, Babic D and Bacani M 2010 *J. Non-Crystalline Solids* **356** 1725
[23] Oka O and Kiyohara O 1993 *Synthetic Metals* **55-57** 999
[24] Sengupta PP and Adhikari B 2007 *Mater. Sci. Eng. A* **459** 278
[25] Sapurnia YI and Stejskal J 2008 *Poly. Inter.* **57** 1295
[26] Stejskal J and Gilbert RG 2002 *Pure Appl. Chem* **74(5)** 857-67.
[27] McCall RP, Scherr EM, MacDiarmid AG and Epstein AJ 1994 *Physical Review B*, **50,** 5094
[28] Epstein AJ and MacDiarmid AG 1991 *Macromol. Chem. Macromol. Symp.* **51** 11
[29] Heeger AJ 1989 *Faraday Discuss. Chem Soc* **88** 203-211
[30] Jain H 1993 *Experimental Techniques of Glass Science, The American Ceramic Society*, Ohio
[31] Epstein AJ, Ginder JM, Zuo F, Woo HS, Tanner DB, Richter AF, Angelopoulos M, Huang WS and MacDiarmid AG 1987 *Synth. Met.* **21** 63
[32] Sharama BK, Khare N, Dhawan SK and Gupta HC 2009 *J. Alloys and Compounds* **477** 370
[33] Ray DK, Himansu AK and Sinha TP 2007 *Indian Journal of Pure and Applied Physics* **45** 692
[34] Mrlik M, Moucka R, Iicinova M and Stejskel J 2014 *Synthetic metals* **192** 37-42
[35] Saafan SA, El-Nimr MK and El-Ghazzawy EH 2006 *J.Appl. Polym. Sci.* **99** 3370
[36] Liang C, Gest J, Leory G and Carru JC 2013 *J. Applied Physics* **114** 124106
[37] Jonscher AK 1983 *Dielectric Relaxation in Solids (Chelsea Dielectric Press)* London
[38] Funke K 1993 *Prog. Solid state chem.* **22** 111
[39] Nithyaprakash D, Chandrasekharan J, Punithaveni B and Sasikumar L 2014 *Optik* **125** 5343

Fabrication of nano ZrO_2 dispersed novel $W_{79}Ni_{10}Ti_5Nb_5$ alloy by mechanical alloying and pressureless sintering

R. R. Sahoo, A. Patra[1], S. K. Karak

Department of Metallurgical and Materials Engineering, National Institute of Technology, Rourkela-769008, Orissa, India

Corresponding author Email : patraa@nitrkl.ac.in/anspat.met@gmail.com

Abstract: A high energy planetary ball-mill was employed to synthesize tungsten (W) based alloy with nominal composition of $W_{79}Ni_{10}Ti_5Nb_5(ZrO_2)_1$ (in wt. %) for 20 h with chrome steel as grinding media, toluene as process control agent (PCA) along with compaction at 500 MPa pressure for 5 mins and sintering at 1500°C for 2 h using Ar atmosphere. X-ray diffraction (XRD), Scanning electron microscopy (SEM), Energy dispersive spectroscopy (EDS), elemental mapping and Transmission electron microscopy (TEM) was used to study the phase formation, microstructure of both milled powder and consolidated alloy. The crystallite size of W in $W_{79}Ni_{10}Ti_5Nb_5(ZrO_2)_1$ powder was 37 nm, 14.7 nm at 10 h and 20 h of milling respectively and lattice strain enhances to 0.54% at 20 h of milling. The crystallite size reduction is more at 10 h of milling and the rate drop beyond 10 to 20 h of milling. The intense improvement in dislocation density was evident upto 10 h of milling and the rate decreases between 10 to 20 h of milling. Increase in the lattice parameter of tungsten in $W_{79}Ni_{10}Ti_5Nb_5(ZrO_2)_1$ alloy upto 0.09% was observed at 10 h of milling owing to severe stress assisted deformation followed by contraction upto 0.07% at 20 h of milling due to formation of solid solution. The large spherical particles at 0 h of milling transformed to elongated shape at 10 h of milling and finer morphology at 20 h of milling. The average particle size reduced from 100 μm to 4.5 μm with the progress of milling from 0 to 20 h. Formation of fine polycrystallites of W was revealed by bright field TEM analysis and the observed crystallite size from TEM study was well supported by the evaluated crystallite size from XRD. XRD pattern and SEM micrograph of sintered alloy revealed the formation of NbNi, Ni_3Ti intermetallic phases. Densification of 91.5% was attained in the 20 h milled and sintered alloy. Mechanical behaviour of the sintered product was evaluated by hardness and wear study. $W_{79}Ni_{10}Ti_5Nb_5(ZrO_2)_1$ alloy showed increase in hardness with decreasing load. The wear rate increases with increasing load due to higher abrasion effect at higher load.

1. Introduction

Highest melting point (3410°C), density (19.3 g/ml), excellent mechanical strength at elevated temperature, tensile elastic modulus (411 GPa at 20°C) and hardness of 9.8 GPa of tungsten (W) is fruitful for electrical, electronic, nuclear and space vehicle application [1]. Reduced workability and high ductile-brittle transition temperature (DBTT) (200-500°C) of tungsten imposes serious concern in its applications. Fabrication of tungsten based alloys in recent times is attributed to widen the application window and therefore to enhance the properties of W such as lowering of DBTT.

[1] patraa@nitrkl.ac.in/anspat.met@gmail.com (Prof. A. Patra)

Nanocrystalline tungsten alloys are potential candidates in this regard owing to enhanced mechanical properties [2, 3]. Mechanical alloying is a superior low temperature synthesis method of nanostructured materials by crystallite size refinement to improve the mechanical properties as compared to melt spinning and similar rapid quenching methods [4]. High-energy ball milling is a convenient alloying method due to easiness, mass production, improved productivity, comparatively economical equipment, and ability to fabricate a wide range of materials [5]. Ni addition is beneficial in improving the plastic flow property by alteration of dislocation core symmetry and also accelerates liquid phase sintering in case sintering temperature higher as compared to the melting point of Nickel (1455°C) [1]. Nb enhances the high temperature response of W [1] and Ti eases the densification of W, restricts the grain growth, however it increases the DBTT of W [6]. Dispersion of ultrafine particles at the grain boundary of W inhibits the grain boundary migration and subsequently improves the strength by dispersion strengthening mechanism [7].
The objective of the present research work is to reduce the sintering temperature of $W_{79}Ni_{10}Ti_5Nb_5(ZrO_2)_1$ alloy by nanostructuring via mechanical alloy before sintering. Milled powder and sintered $W_{79}Ni_{10}Ti_5Nb_5(ZrO_2)_1$ alloy was characterized by various characterization method.

2. Materials and methods

Elemental W, Ni, Nb, Ti (Sigma Aldrich, 99.5% purity, size: 50-100 μm), ZrO_2 (Sigma Aldrich, 99.8% purity, size: <100 nm) powders were milled in a planetary ball mill (Fritsch Pulverisette P5) with 300 r.p.m mill speed, chrome steel as grinding media and ball to powder weight (BPR) ratio of 10:1. Milled powders of 1, 5, 10, 15, 20 h were collected from the vial for characterization. The alloy composition and milling parameters are provided in table 1.

Table 1. Particulars of alloy composition and milling parameters

Composition (wt.%)	Grinding Medium	Ball to powder weight ratio	Mill Speed (rpm)	Milling Duration (h)	Milling Medium
$W_{79}Ni_{10}Ti_5Nb_5(ZrO_2)_1$	Chrome Steel	10:1	300	20	Toluene

X-ray diffraction pattern (XRD) of the mechanically alloyed powders of various milling times and the sintered alloy was investigated by high resolution X- ray diffractometer (Make: Rigaku Japan, Model: Ultima IV) by using Cu-K_α radiation (λ=1.541874 Å) with scan speed of 10°/min and step size of 0.05°. XRD analysis data and JCPDS data bank was matched to identify the appearance of phases during mechanical alloying [8]. Peak position and peak broadening from the XRD pattern was used to evaluate the crystallite size and lattice strain [9].
The dislocation density induced in the milled powder was measured by:

$$\rho_{\mathrm{d}} = 2\sqrt{3}\frac{\left(\varepsilon^2\right)^{1/2}}{D\,X\,b} \qquad (1)$$

Where, b is the burgers vector of dislocations, b = (a√3)/2 for the bcc structure, a= cell parameter = lattice parameter, D = crystallite size, ε = lattice strain, [10].

Precise lattice parameter calculation method was used to measure the lattice parameter from XRD pattern after stripping of $K_{\alpha 2}$ of XRD pattern [9]. The mechanically alloyed powder morphology at various milling time and the microstructure of the sintered alloy was studied by scanning electron microscope (SEM) (Make: JEOL, Model: JSM-6084LV). Powders milled for 20 h was dissolved in acetone in a test tube and intensely stirred by ultrasonic vibrator. Few drops of acetone containing powders were placed on a carbon grid prior to TEM study. Crystallite size of W in 20 h milled $W_{79}Ni_{10}Ti_5Nb_5(ZrO_2)_1$ powder was revealed by high resolution transmission electron microscope (HRTEM) (Make : JEOL, Japan, Model- JEM 2100) study. A uniaxial press was employed to compact the 20 h milled powders into pellets of 10 mm diameter by 500 MPa pressure and 5 mins dwell time for uniform pressure distribution throughout the sample cross section. The compacted pellets were subjected to sintering at 1500°C for 2 h with constant Argon injection (rate: 100 ml/min) to avoid any chance of oxide formation. Microhardness of the sintered product was recorded by a Microhardness tester (Make: Leco, Model: LM248AT) under load of 50 gf, 100 gf, 500 gf, 1000 gf with10 sec of dwell time. The wear depth of sintered alloy was measured by a ball on plate wear tester (Make: Ducom, Model: TR-208-M1) under 20 N and 30 N load with sliding speed and time of 25 r.p.m and 10 mins respectively.

3. Result and Discussion

3.1. X-ray Diffraction (XRD) study of milled powder

XRD patterns of milled $W_{79}Ni_{10}Ti_5Nb_5(ZrO_2)_1$ powder at various milling times (1, 5, 10, 15, 20 h) is shown in Figure 1. Increase in broadening (full-width at half-maximum) and decrease in the intensity of W diffraction peaks with progress of milling as evident from XRD pattern refers to crystallite size reduction of W and lattice strain build up [11]. Nb (BCC) forms continuous solid solution with W (BCC) with increase in milling time whereas Ni (FCC) and Ti (Hexagonal) shows limited solubility owing to difference in crystal structure, electronegativity and atomic radius as revealed from the presence of Ni, Ti peaks at 20 h of milling. Broadening was evaluated by Scherrer formula by eliminating the contributions the strain and instrumental broadening error [8].

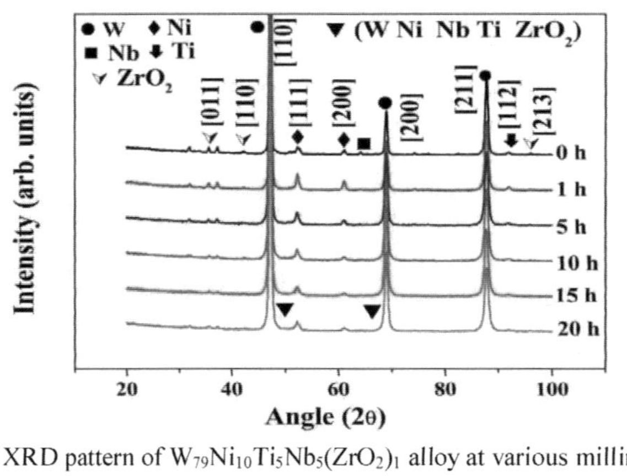

Figure 1. XRD pattern of $W_{79}Ni_{10}Ti_5Nb_5(ZrO_2)_1$ alloy at various milling time.

Williamson–Hall equation was used to measure the crystallite size and lattice strain [12] as:

$$\beta \cos \theta = \frac{0.94\lambda}{D} + 4\eta \sin \theta \qquad (2)$$

where, β is the full width half maxima (FWHM) or broadening, D is the crystallite size and η is the lattice strain. The intercept and the slope from the plot of β cos θ and sin θ is evaluated. The crystallite size (D) is measured after incorporating the λ value (1.541874 Å) in the intercept, whereas the slope corresponds to lattice strain. The change in crystallite size and lattice strain with milling time is presented in Figure 2. The minimum crystallite size and maximum lattice strain is recorded 14.7 nm and 0.54% respectively at 20 h of milling. The crystallite size reduction rate is substantially higher at 10 h of milling and the rate slows down between 10 h to 20 h of milling. The crystallite size refinement and increase in lattice strain is higher as compared to recently investigated $W_{80}Ni_{10}Nb_{10}$ alloy due to the presence of Ti which incorporates brittleness into the system and higher surface energy associated with nano ZrO_2 particles [13]. Presence of Ti at W/W grain boundary contributes in crystallite size reduction by alteration of grain boundary cohesive energy [14].

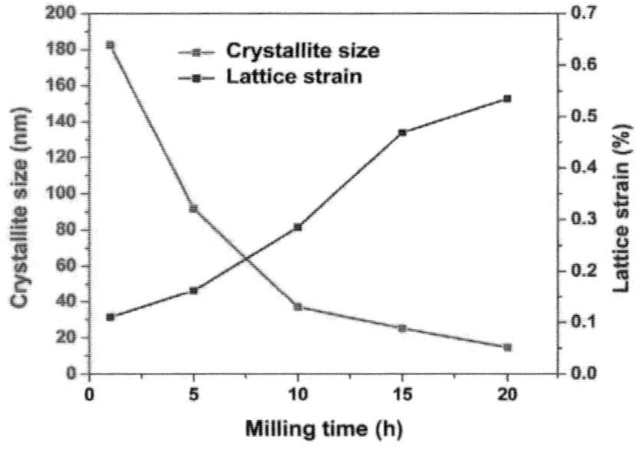

Figure 2. Crystallite size and lattice strain variation of W with milling time in $W_{79}Ni_{10}Ti_5Nb_5(ZrO_2)_1$.

The dislocation density present in $W_{79}Ni_{10}Ti_5Nb_5(ZrO_2)_1$ alloy shows rapid enhancement at 10 h of milling, though the rate of increase is restrained beyond 10 h of milling (figure 3(a)). Figure 3(b) shows that the lattice parameter of W in $W_{79}Ni_{10}Ti_5Nb_5(ZrO_2)_1$ alloy enlarges upto 0.09% at 10 h and reduces upto 0.07 % at 20 h of milling. Pabi et al. have reported similar trend of Nb lattice enlargement in mechanically alloyed pure Nb [15]. The present investigation also displays a similar trend. Nb and Ni form substitutional solid solution with W matrix. The atomic radius of W (0.193 nm) is larger than Ni (0.149 nm) and Ti (0.176 nm) and smaller than Nb (0.198 nm). Therefore the ensuing atomic radius displays a drop after 10 h of milling [16]. Similar contraction of W lattice in W-Ni-Mo alloys is reported in recent literature [17]. The stress required for dislocation generation (Peierls–Nabarro stress) rises with crystallite size reduction which results in decrease in the rate of crystallite size reduction and dislocation density enhancement. Grain boundary softening beyond 10 h of milling also corresponds to decrease in dislocation multiplication. Higher stress aided deformation is responsible for initial expansion of W lattice at 10 h of milling. Formation of substitutional solid solution majorly by Nb with W and Ni, Ti with W to a lower degree results in contraction of W lattice beyond 10 h of milling.

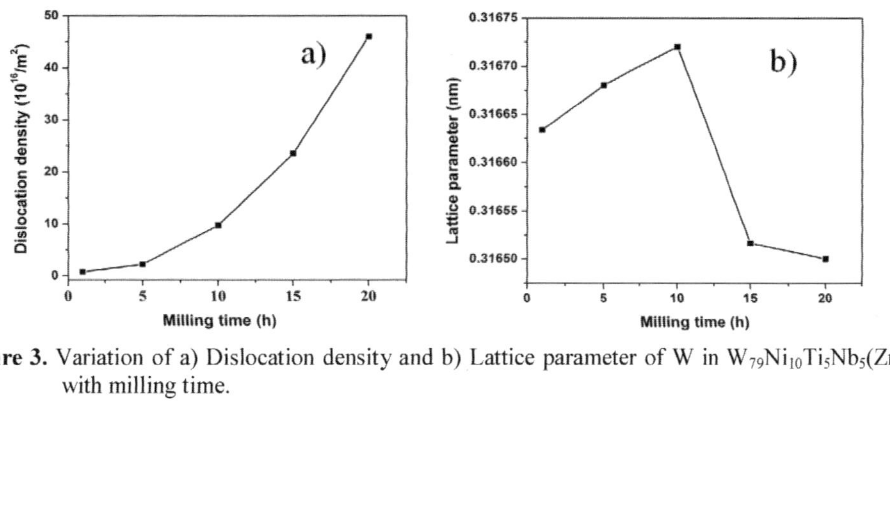

Figure 3. Variation of a) Dislocation density and b) Lattice parameter of W in $W_{79}Ni_{10}Ti_5Nb_5(ZrO_2)_1$ with milling time.

3.2. Microstructure study of milled powder

The particle morphology and size variation of the milled powder with increase in milling time is displayed in Figure 4 (a-d). The size of milled powder particles varies from ~100 μm to ~4.5 μm with the progress in milling. The powder particles are constantly work hardened, flattened and fractured as the milling time increases. The larger, spherical particle shape at 0 h transforms to elongated and finer morphology at 10 h and 20 h of milling respectively. The interparticle distance between W and solute atoms reduces with increase in milling time which results in enhanced atomic diffusion for the formation of alloy. Particle agglomeration to some extend is also observed at 10 h of milling (figure (4c)). The elongated particle morphology at 10 h of milling is related with reduced flowability owing to enormous interparticle friction which contributes in lower green density [17].

Figure 4. SEM images of powder morphology of $W_{79}Ni_{10}Ti_5Nb_5(ZrO_2)_1$ alloy at different milling times: (a) 0 h, (b) 5 h, (c) 10 h, and (d) 20 h.

Figure 5 (a) shows the bright field TEM image of $W_{79}Ni_{10}Ti_5Nb_5(ZrO_2)_1$ powder milled for 20 h. The size of the crystallites of W at 20 h of milling is in the range of 10 to 20 nm which is nearly identical with the measured crystallite size from XRD analysis (figure 2). Formation of fine polycrystallites is evident from figure 5 (b). Indexing of SAD pattern (figure 5 (b)) shows the presence of W (222), Ni (222), Nb (321), Ti (203), ZrO_2 (400) planes. The measured interplaner spacing (d) values of BCC-W is 0.091 nm for (222), Ni : 0.09 nm for (222), Nb : 0.089 nm for (321), Ti: 0.099 nm for (203), ZrO_2: 0.0909 nm for (400) respectively. The interplaner spacing (d) values matches well with the standard values of BCC-W from JCPDS data files.

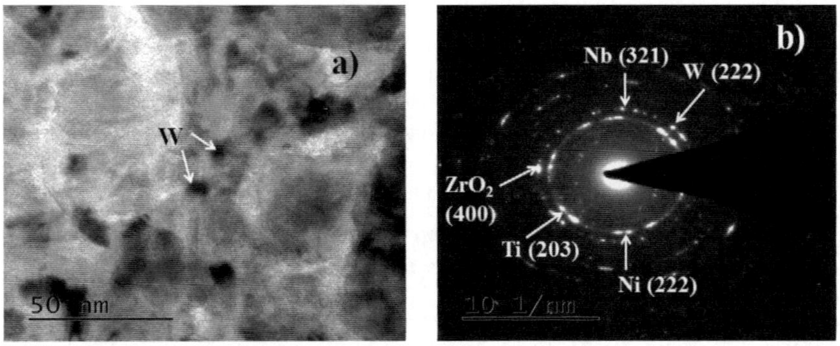

Figure 5. Bright field TEM image of $W_{79}Ni_{10}Ti_5Nb_5(ZrO_2)_1$ powder milled at 20 h: (a) Bright field TEM image and (b) corresponding SAD pattern.

3.3. Phase evolution analysis of sintered alloy

The phase formation in sintered $W_{79}Ni_{10}Ti_5Nb_5(ZrO_2)_1$ alloy is investigated from the XRD pattern (figure 6). Formation of NbNi and Ni_3Ti intermetallic is evident from the XRD study. The strengthening of the matrix is contributed by the hard NbNi and Ni_3Ti phases as well by the presence of ZrO_2 particles. Similar phenomenon of NbNi intermetallic formation is also reported in recent literature [18]. However the intermetallic induces brittleness in the alloy which is responsible for lowering the ductility of the W matrix. (211) diffraction peak of W in milled powder (diffraction angle : 87.6713°) is moved to higher diffraction angle (87.6738°) in sintered alloy due to W lattice distortion primarily by Nb and to lesser degree by Ni and Ti. The broadening of XRD pattern decreases in the sintered alloy as compared to milled powder owing to increase in particle size and reduction in lattice strain during sintering. Any other unwanted phase formation is not observed from the XRD pattern.

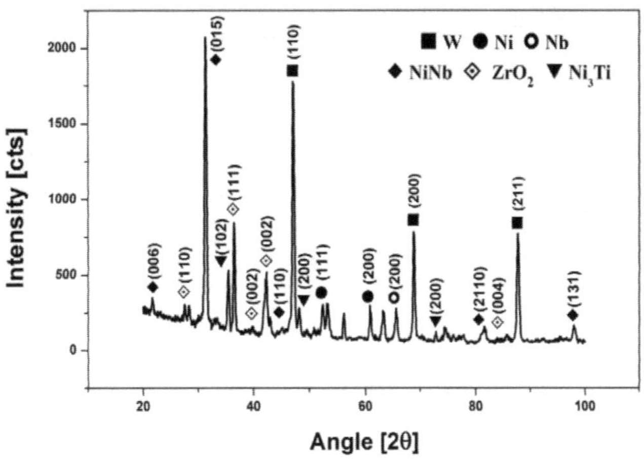

Figure 6. XRD pattern of $W_{79}Ni_{10}Ti_5Nb_5(ZrO_2)_1$ alloy milled for 20 h and sintered at 1500°C for 2 h.

3.4. Microstructure study of sintered alloy

Figure 7 (a) shows the SEM micrograph of $W_{79}Ni_{10}Ti_5Nb_5(ZrO_2)_1$ alloy sintered at 1500°C for 2 h. The micrograph displays the existence of Ni (gray phase) and W matrix (bright phase). The intermetallic phase formation (NbNi and Ni_3Ti : dark phase) is related melting point of Ni (1455°C)

IOP Conf. Series: Materials Science and Engineering **178** (2017) 012015 doi:10.1088/1757-899X/178/1/012015

which is lower than the sintering temperature (1500°C). However the melting point of Ti (1668°C) is higher as compared to the selected sintering temperature it is expected that size reduction of brittle Ti during mechanical alloying decreases the melting point below 1500°C and contributes in liquid phase formation along with Ni. Evolution of intermetallic decreases the weaker W/W area fraction (contiguity). The identified phases are confirmed from EDS analysis (figure 7(b)) and elemental mapping (figure 7(c-h)) and the phases are in good agreement with phases revealed from the XRD study.

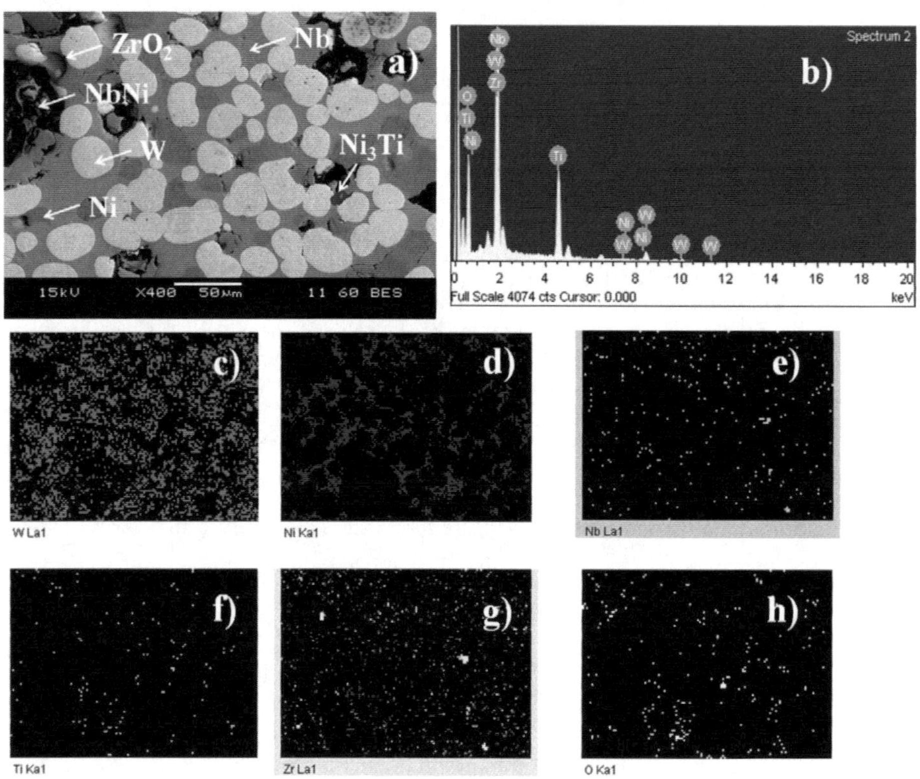

Figure 7. (a) SEM micrograph (b) EDS pattern (c, d, e, f, g, h) elemental maps of W, Ni, Nb, Ti, Zr and O of 20 h milled $W_{79}Ni_{10}Ti_5Nb_5(ZrO_2)_1$ alloy sintered at 1500°C for 2 h.

3.5. Densification study

Degree of densification is evaluated from the percentage ratio of sintered alloy density to the theoretical density of the sintered alloy. The density of the sintered alloy is measured by Archimedes' principle [19] as:

$$\rho_s = \frac{W_a}{(W_{sat} - W_{susp})} X \rho_w \frac{gm}{cm^3} \quad (3)$$

Where W_a is weight of the sintered sample in air. W_{sat} is the weight of the sample with all the open porosity saturated with water, W_{susp} is the weight suspended in water. ρ_w is the density of water. The %densification is measured from the percentage ratio of sintered density to theoretical density. The calculated densification of the sintered alloy records 91.5% (porosity: 8.5%) by compacting at 500

MPa pressure with 5 mins of dwell time. Table 2 shows the measured weight of the alloy at different condition for evaluation of %densification. The liquid phase formation during sintering improves the densification by closing of the open pores owing to enhanced grain boundary diffusion and capillary action of the liquid phase.

Table 2. Measured %densification of $W_{79}Ni_{10}Ti_5Nb_5(ZrO_2)_1$ alloy

Alloy Composition	Weight in air (gm)	Soaked weight in water (gm)	Suspended weight (gm)	Sintered density (gm/cm3)	Theoretical density (gm/cm3)	% Densification
$W_{79}Ni_{10}Ti_5Nb_5(ZrO_2)_1$	0.8246	0.8249	0.760	12.705	13.88	91.5

3.6. Hardness study

The microhardness has been measured by the following equation as [20]:

$$HV = 1.8544 \frac{P}{d^2} \qquad (4)$$

P is the applied load and d is the diagonal length of the indentation.

Change in hardness across the cross section of the $W_{79}Ni_{10}Ti_5Nb_5(ZrO_2)_1$ alloy with different load is shown in figure 8. Increase in hardness with reduction in applied load is due to considerable elastic recovery at reduced load and measurement error of the diagonal length owing to smaller size of indentation at reduced loads [21]. The hardness is measured on the W matrix phase, however the deviation in hardness value at a definite load is attributed to compositional heterogeneity across the cross section. The mean hardness at all loads is lesser than reported hardness of nanostructured W, as W is substituted by less harder Nb, Ni and Ti addition [1]. The measured hardness is comparatively less than Y_2O_3 dispersed W based ODS alloy owing to enhanced high temperature stability, hardness and strength imposed by Y_2O_3 dispersion [22].

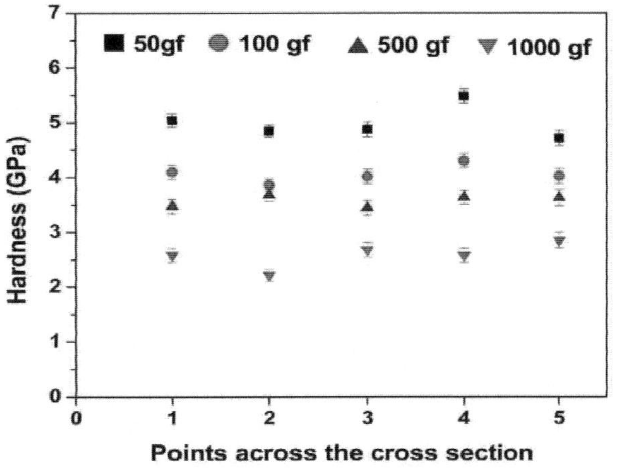

Figure 8. Variation of hardness of $W_{79}Ni_{10}Ti_5Nb_5(ZrO_2)_1$ alloy milled for 20 h and sintered at 1500°C for 2 h.

3.7. Wear study

Figure 9 shows the change in wear depth with sliding distance traversed by the indenter on the sample surface. The sliding distance can be calculated as [23]:

$$\text{Sliding Distance (S.D)} = \left(\frac{R}{60}\right) \times t \times 2\pi r \quad (5)$$

R is the number of rotation per minute traversed by the indenter on sample surface, t is time in sec (600 s), r is track radius (4 mm) measured from the center of the sample to the track. The wear depth increases with increasing load as revealed from Figure 9. The phenomenon may be owing to removing out the intermetallic and ZrO_2 particles from matrix and W/W interface to the wear track throughout the abrasion and results in increased wear. The variation in wear depth profile throughout the contour is attributed to compositional heterogeneity and variance in hardness (figure 8). The increase in width of the wear track with increase in load as evident from figure 10 can be attributed to the process of three body motion which contributes to increased wear owing to enhanced abrasion at higher load [24].

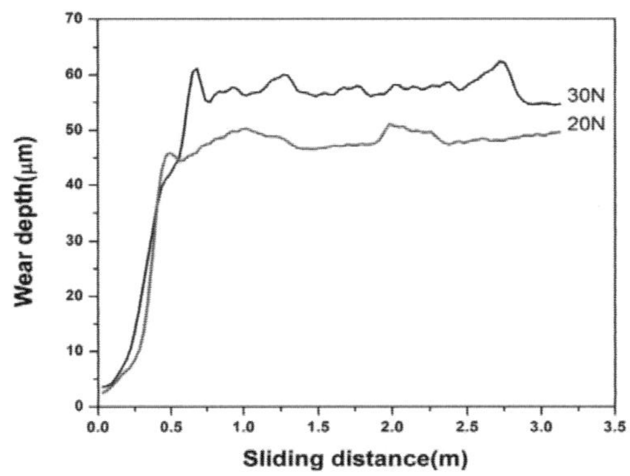

Figure 9. Variation of wear depth of $W_{79}Ni_{10}Ti_5Nb_5(ZrO_2)_1$ alloy milled for 20 h and sintered at 1500°C for 2h.

Wear rate is defined as volume of wear per unit load and sliding distance. Archard equation has been used to measure the wear rate at highest wear depth for both 20 N and 30 N load [25]. The measured wear rate at maximum wear depth has been recorded 4.10 ($\times 10^{-15}$) m^3/N m for 20 N load and 5.71($\times 10^{-15}$) m^3/Nm for 30 N load respectively.

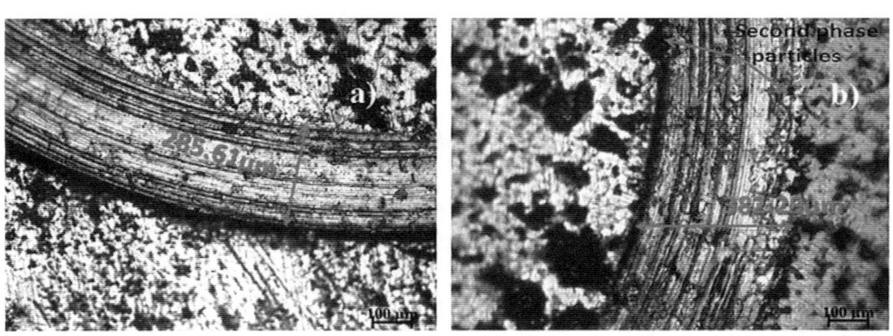

Figure 10. Micrograph of wear track of $W_{79}Ni_{10}Ti_5Nb_5(ZrO_2)_1$ alloy at a) 20 N b) 30 N load.

4. Conclusion

The conclusions derived from the synthesis, assessment of microstructure, physical, mechanical behavior of the investigated alloy are as follows:

- Mechanical alloying is a potential method for fabrication of $W_{79}Ni_{10}Ti_5Nb_5(ZrO_2)_1$ alloy powder.
- Crystallite size of W in $W_{79}Ni_{10}Ti_5Nb_5(ZrO_2)_1$ alloy reduces to 14.7 nm at 20 h of milling.
- Lattice parameter of W in $W_{79}Ni_{10}Ti_5Nb_5(ZrO_2)_1$ alloy increases upto 0.09% at 10 h and followed by contraction of 0.07% at 20 h of milling.
- Formation of nanocrystalline BCC W of 10-20 nm in size at 20 h of milling is confirmed from bright field TEM image and corresponding SAD pattern.
- Evolution of NbNi, Ni_3Ti intermetallic phase after sintering is attributed to the formation of Ni and Ti base liquid phase at present sintering temperature.
- Increase in microhardness with decrease in load is related with lower indentation size and measurement error of the smaller indentation diagonals at reduced loads.
- Increased abrasion at higher load results in enhanced wear rate.

References

[1] Patra A 2012 *Microscopy and Analysis* **26 (5)** 14-15.
[2] Glieter H 2001 *Acta. Mater* **48** 1.
[3] Koch C 1997 *Nanostruct. Mater* **9** 13.
[4] Suryanarayana C 1998 *Mechanical Alloying* ASM Handbook **7** 80.
[5] Suryanarayana C 2001 *Prog. Mater. Sci* **46** 1-184.
[6] Aguirre M V, Martin A, Pastor J Y, LLorca J, Monge M A and Pareja R 2009 *Metall. Mater. Trans. A* **40(10)** 2283-2290.
[7] Lassner E and Schubert W D 1999, *Tungsten-properties, chemistry, technology of the element, alloys, and chemical compounds* (Kluwer Academic/Plenum Publishers: New York) 255-268.
[8] Cullity B 1978 *Elements of X-ray diffraction* **2** (Addison Wesley: Philippines) 86-88.
[9] Williamson G and Hall W H 1953 *Acta. Mater* **1** 22.
[10] Zhao Y H, Sheng H W and Lu K 2001 *Acta. Mater* **49** 365–375.
[11] Nuthalapati M, Karak S K, Majumdar J D and Basu A 2014 *Metall. Mater. Trans. A* **45A** 3748-3754.
[12] Jenkins R and Snyder R L 1996 *Introduction to X-ray Powder Diffractometry* (J Wiley & Sons Inc: New York).
[13] Saxena R, Patra A, Karak S K, Pattanaik A and Mishra S C 2016 *Proc. Nat. Conf. on Processing and Characterization of Materials (Rourkela)* vol 115 (Bristol: UK/IOP Science) 012026.
[14] Chookajorn T and Schuh C A 2014 *Acta Mater.* **73** 128–138.
[15] Chattopadhayay P, Nambissan P M G, Pabi S K and I Manna 2001 *Appl. Surf. Sci* **182** 308-312.
[16] Singh V 2002 *Physical Metallurgy* **1** (Standard Publisher Distributor: New Delhi) 63-64.
[17] Patra A, Karak S K and Pal S 2015 *Proc. Nat. Conf. on Processing and Characterization of Materials (Rourkela)* **75** (Bristol: UK/IOP Science) 012032.
[18] Li X, Cheng Z, Hu K, Chen H and Yang C 2015 *Vacuum* **114** 93-100.
[19] Ozkan N and Briscoe B J 1994 *J. Eur. Ceram. Soc* **14** 143-151.
[20] Ayers L J 2012 *The Hardening of Type 316 L Stainless Steel Welds with Thermal Aging* Massachusetts Institute of Technology.
[21] Dieter G 1928 *Mechanical Metallurgy* SI Metric Edition (McGraw-Hill Co: Singapore) 189-93.
[22] Patra A, Saxena R and Karak S K 2016 *Int. J. Refract. Met. Hard Mater* **60** 131-146.
[23] Patra A, Karak S K and Pal S 2015 *Microscopy and Analysis* **29(4)** 18-22.
[24] Nuthalapati M, Karak S K and Basu A 2015 *Materials Today: Proceedings* **2** 1109-1117.
[25] Archard J F 1953 *J. Appl. Phys* **24 (8)** 981-988.

Elastic behavior of CNT-reinforced polymer composites with discontinuities in CNT configurations

Puneet Kumar, Dr. J.Srinivas
Department of Mechanical Engineering
National Institute of Technology Rourkela, Odisha, India
Email: puneetahalawat2009@gmail.com

Abstract: A numerical study has been made towards the effective elastic properties estimation of carbon nanotubes and carbon nanotube reinforced composite using finite element modelling (FEM). First, the elastic properties of Carbon nanotube (CNT) were predicted by considering that carbon atoms as nodes and carbon-carbon bonds as beam elements with linear and isotropic behaviour. It was observed that elastic properties of CNT predicted by FE analysis were in good agreement with previous data. Carbon atom vacancy defects were also included to investigate the adverse effect on elastic modulus of SWCNTs. To explore the macroscopic elastic behaviour of CNT in a finite densely packed polymer resin, a representative volume element (RVE) was selected instead of whole composite material in which the polymer resin was modelled as continuum material while CNT as an equivalent long fibre. FE results of RVE manifest that the CNT volume fraction and waviness have significant effect on elastic modulus of CNT reinforced polymer composite. An analytical formulation in terms of elastic properties and waviness ratio was also introduced in this study for waviness analysis. Moreover, the elastic properties of wavy CNT reinforced composite was compared with analytical outcomes. We extended present RVE model to incorporate the effects of CNTs agglomeration on the elastic behaviour of CNT-reinforced polymer composites. It was observed that anticipated elastic results not only depended on the volume fraction of CNTs, but also on the CNTs geometry, waviness and agglomeration.

Keywords: Agglomeration, elastic properties, finite element modelling, RVE, waviness

1. Introduction

Carbon nanotubes (CNT) are one of the famous reinforcing elements in several materials including polymers and hybrid matrices. Using only few weight percentage of this carbon allotrope [1], it is found the mechanical performance of the conventional composites enhances tremendously [2,3]. Over the past one decade, several investigations were carried-out for characterization of mechanical and elastic properties of confined isolated CNTs as well as CNT-reinforced polymer composites using micromechanical, molecular and experimental based approaches [4–6]. Atomistic simulation and experimental investigations stipulated Young's modulus of CNTs 0.6 to 1 TPa [7,8] and Poisson's ratio varying between 0.25 and 0.28 which are affected by their diameter, length, chirality, type of defects, sample synthesis, measurement techniques and computational theory. But utilization of these CNT properties as reinforcement data for polymer composites leads to several errors as there is a sharp decline in CNTs mechanical and geometrical properties during composite fabrication. In order to achieve the desired properties, recently, main focus is given on investigation of stiffening ability of CNT reinforced polymer composites considering the effects of geometric features of carbon nanotubes such as length, waviness, agglomeration and curvature. Formation of waved CNTs is mainly due to extremely high length to diameter ratio and low stiffness in transverse direction. Few studies reported in literature shows that the waviness of CNTs diminishes the overall axial stiffness of CNT-composites but has no significant change in transverse directions [9–13]. The earlier micro-mechanical models for demonstrating the waviness and agglomeration of CNTs were based on Mori Tanaka method [14,15]. Recent studies [16,17] illustrated continuum mechanics-based approach and micromechanical model to investigate the alignment and randomly oriented wavy CNT effects on effective elastic properties of composite with a parallel FE model generated for wavy CNT embedded representative volume element (RVE). Since experimental study of CNT reinforced polymers is an involved task at the nano scale, numerical modeling takes a vital part. In this line, various numerical and theoretical approaches

[18–24] outlined the combined effects of CNT distribution, dispersion, agglomeration, waviness, interfacial bonding deformation mechanisms on the elastic properties of CNT-reinforced composites.

In spite of several efforts in modeling CNT polymer composites, limited works were found related to the effect of the fiber geometric parameters on overall elastic behavior of composite using numerical modeling. The realistic molecular structure of CNT needs special focus and its integration within the polymer matrix require a careful study due to several complications like waviness, agglomeration, vacant sites etc. In present work, a finite element analysis based approach is proposed to predict the effects of CNT content, CNT waviness and agglomeration on effective elastic properties of composite. A two-dimensional representative element model is utilized for studying the waviness effect, whereas a three-dimensional representative volume element (RVE) model is developed by accounting the agglomeration effect using four CNT fibers within the matrix. The CNT fiber volume fraction and waviness factor are varied and the effects on elastic properties are reported. In this study, analytical expressions are employed to provide a better understanding of carbon nanotubes (CNTs) curvature on the overall behavior of nanocomposites. The waviness of CNT is modeled as a sine wave form and the transformed physical characteristics are applied to micromechanical framework. In addition, the predictions are compared with analytical data of the CNT-reinforced nanocomposites to assess the reliability of the proposed method.

2. Numerical Modeling

Analysis is divided into two parts. In the first part CNT modelling is presented using finite element discretization, followed by modelling of reinforced polymer composite by considering waviness and agglomeration.

2.1 Modeling of CNT geometry

Three types of CNTs are observed during experimental analysis: SWCNTs, DWCNTs and MWCNTs. A SWCNT can be viewed as a rolling up form of graphene sheet, which is composed of hexagonal carbon rings. The chiral vector (C_h) and chiral angle (θ) are two main influencing factors for atomic structure of carbon nanotubes. The chiral vector can be represented in terms of chiral indices and diameter of CNT and can be estimated as [25]

$$D_{cnt} = \frac{\sqrt{3}}{\pi} d_{cc} \sqrt{n^2 + m^2 + nm} \tag{1}$$

where (*n, m*) is the chiral indices and is the $d_{c\text{-}c}$ carbon-carbon bond length. Further CNTs are classified based on chiral indices: (*n, n*) armchair and (*n, 0*) zigzag respectively. In this section, modeling of SWCNTs, DWCNTS and MWCNTs is presented using finite element approach. First, a MATLAB code was developed to generate the coordinates of each carbon atom present in tube. Zigzag and armchair CNTs are considered in this analysis and corresponding atoms coordinates are generated shown in Figure 1. The carbon atom coordinates generated in MATLAB are imported in the commercial available software ANSYS and copied in axial direction to get desire length of CNTs. A finite element model of CNT is developed by considering the carbon atoms as nodes and carbon-carbon bond as beam element (Beam 188). In Figure 2 a finite element model of (50, 50) SWCNT is represented.

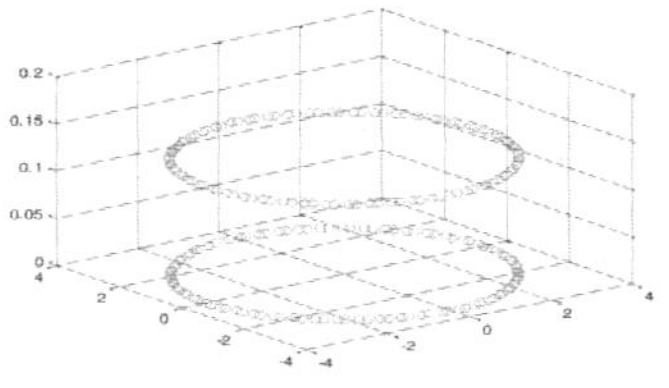

(a)

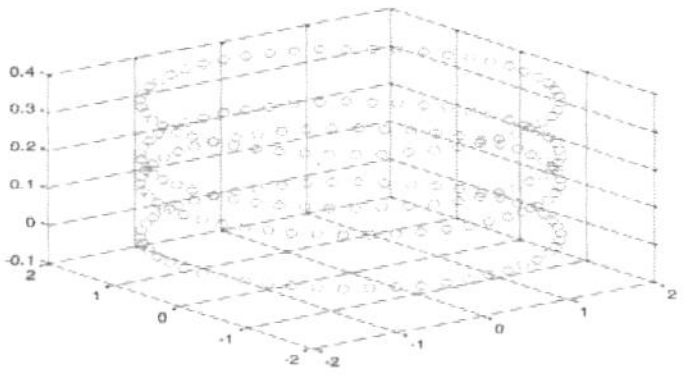

(b)

Fig, 1 Carbon atom coordinates generated in MATLAB for (50, 50) and (50, 0) CNT

In order to define the material model of beam elements, previous studies [26–28] results are employed in present modeling approach. Based on molecular mechanics and continuum theories, materials properties are obtained from equivalency theorem of molecular and structure energy. The material and geometric properties utilized in ANSYS APDL model are adopted from literature [26]. Figure 2 depicts the typical FE model of armchair (50, 50) and zigzag (50, 0) nanotube with and without vacancy defects for calculating the Young's modulus and shear modulus. The Young's modulus of a material is the ratio of normal stress to normal strain as obtain from a uniaxial tension test. In this regards, one end of the tube is remaining to be fixed in all directions while a small finite displacement is applied on other end. Then, following equation can be used to calculate the elastic modulus of CNT.

$$E_{cnt} = \sigma/\varepsilon = \frac{FL_{cnt}}{\pi D_{cnt} t \Delta L} \tag{2}$$

where t (=0.34 *nm*) is the wall thickness and D_{cnt} is the diameter of the SWCNT, L_{cnt} and ΔL are the initial tube length and elongation in axial direction. F is the reaction force estimated from ANSYS results. Similarly, shear modulus of CNT can be estimated by applying the initial small rotation at one end and keeping other end fixed. The following relation is used

$$G_{cnt} = \frac{TL_{cnt}}{\theta J_\theta} \tag{3}$$

where θ is the initial torsional angle applied at one end of the tube, J_θ is the polar moment inertia of the cross-section area calculated as $J_\theta=(\pi/32)[(D_{cnt}+t)^4-(D_{cnt}-t)^4]$. T represents the torque estimated from the ANSYS results. Double walled CNTs are modeled as two concentric SWCNTs connected with weak Vander Waals interactions. In multi walled CNTs case, three concentric tubes are considered. Similar analysis was carried for DWCNTs and MWCNTs as described in case of SWCNTs. Additionally, the interaction between two tubes is modeled using LINK 180 element with stiffness and cross-sectional area equal to 3.7×10^{-9} N/nm and 0.433 nm^2 respectively [29]. A three walled CNT [(5, 5) (10, 10) (20, 20)] and LINK elements between the tubes are represented in Figure 3.

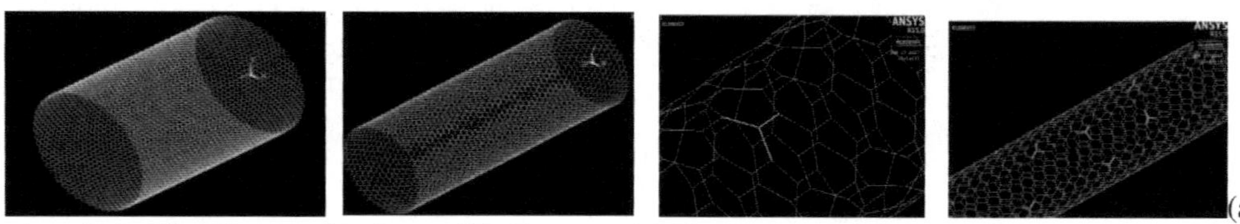

(a) Armchair (50, 50) (b) zigzag (50, 0) (c) Single vancy defect (d) Multi vacancy defect

Fig.2 Finite element models of SWCNTs

(a) Sectional view (b) longitudinal view

Fig.3 Inter atomic interaction between CNTs through link elements in 3 walled CNT

Agglomeration and waviness are responsible for the absence of homogeneous dispersion inside the polymers and bring about insufficient stress transfer leading to poor overall mechanical properties. In order to investigate the effect of these parameters, finite element analysis is employed to discretize a representative region of the composite in next section.

2.2 *Modeling of CNT reinforced polymer*

Naturally, interaction between reinforced CNT and its surrounding polymer resin takes place through Vander Waals bonds. Some investigators attempted to improve the load transfer efficiency between CNT and its surrounding polymer by utilizing functionalization technique which provides covalent bonds between and polymer and CNT. Functionalization has a main drawback that it distorted the structure of CNT which can diminish the overall mechanical properties of CNTs significantly. So Vander Waal's (vdW) interaction can be consider for including interphase effect in modeling and

IOP Conf. Series: Materials Science and Engineering **178** (2017) 012016 doi:10.1088/1757-899X/178/1/012016

represented using Lennard-Jones potential. Furthermore, vdW interaction can be ignored when the inter-atomic separation is equal or greater than 0.85 nm. For simplicity, CNT and the interphase can be modeled as an equivalent long fiber, a solid cylinder with diameter 1.424 nm. In previous literature [30,31] equivalent properties of long fiber have been investigation using inverse rule of mixture. An armchair SWCNT with chiral index of (10, 10) is picked here for both instances of RVE with waviness and agglomeration, thickness of CNT is chosen as 0.34 nm. Till to date lots of investigation have been made considering interphase effect but waviness and agglomeration effect is less in consideration, in this particular study waviness agglomeration phenomenon is described with a perfect bonding between CNT fiber and polymer matrix.

2.2.1 Simulation of CNT waviness

Two-dimensional representative model is employed for CNT/polymer composites as it has already been used by many researchers [32], for obtaining the effective elastic properties of wavy CNTs reinforced composites. A CNT of length L_{cnt} is embedded inside a polymer matrix and forms a RVE of length L_{Rve}, where $L_{rve}=L_{cnt}$. RVE dimensions used were fixed to W_{rve} = 120 nm and L_{rve} = 100 nm, based on amplitude of wavy CNT. It is important to mention that for a selected volume faction of CNT as 5%, the developed equivalent fiber has 7.5 % volume fraction. This difference is due to the hollow shape of CNT and interphase thickness while equivalent long fiber has a solid shape as shown in Figure 4. Simulating polymers chain at atomic scale obliges a huge measure of elements and calculations. Since atomic chains of polymer are significantly tighter than atomistic structure of CNT, it can be displayed as a continuum medium as a satisfactory simplification in modeling method. The Young's modulus of the simulated isotropic resin is considered 10 GPa and Poisson's ratio 0.3 [30].

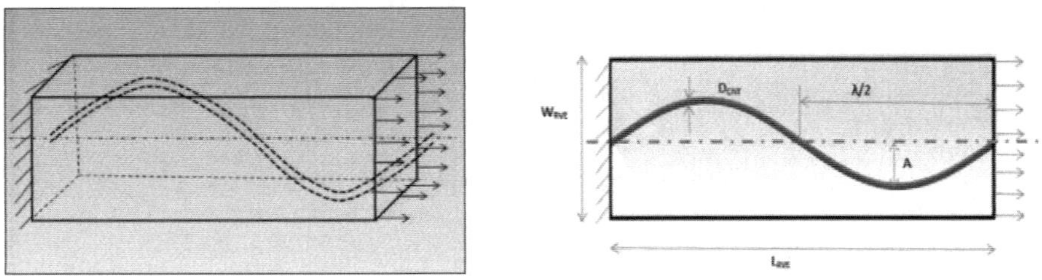

Fig.4. A 3-D and 2-D rectangle RVE with wavy CNT

A 2-D model is adopted to simulate the behavior of nanocomposites consist of waved carbon nanotube. Polymer matrix is treated as linear elastic and isotropic and perfect bonding is assumed between the nanotube and matrix along with consideration of equivalent fiber approach instead of nanotube and interphase. FEA is used to compute Elastic properties of CNT reinforced polymer composite. Commercial simulation software (ANSYS 15.0) was utilized for the analysis. Both CNT and polymer resin were modeled utilizing two-dimensional structural solid elements (''PLANE 83'') with 8-node elements (one at middle and one at each corner node) and two degrees of freedom at every node (translational in x and y-direction) configured in plane strain. The RVE was developed utilizing model of around 2000 elements, with 600 elements being connected with the CNT and the rest of the polymer matrix. In view of the equivalent fiber model representation which takes into account the CNT hollow structure with interphase as equivalent solid structure. To compute the longitudinal and transverse components of elastic modulus, the RVE was subjected to a strain in the x or y-direction as shown in Figure 5, depending on the component of interest. For computing longitudinal modulus, the nodes at the left side edge of the RVE (x = 0) were constrained to have zero displacement in the x-direction while small strain was applied to the right side edge ($x = L_{rve}$). An analogous procedure was performed for the calculation of transverse modulus, but constraining now nodes at y = 0 to have a zero displacement condition in the y-direction and imposing a uniform displacement in the y-direction on the nodes located at y = W_{rve}. Thus, a pure tensile testing is performed in both cases and bending was excluded by the applied displacement boundary conditions.

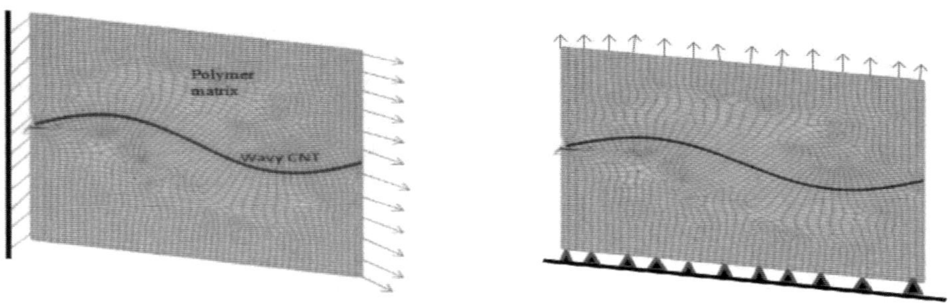

(a) Axial loading (b) Transverse loading

Fig.5 Finite element model with loading conditions for (a) axial and (b) transverse case.

In linear elastic range, effective Young's modulus of CNTs reinforced composite is defined as follows:

$$E_i = \frac{\sum F_i \times L_{RVE}}{A \times \Delta L} \qquad (i=x,y) \tag{4}$$

where $\sum F_i$ is the summation of reaction force on the constrained edge, A is the edge area of the RVE ($A=W_{rve} \times 1$) and ΔL is a very small displacement. Since wavy amplitude has direct effect on Young's modulus majorly depends on a parameter called waviness factor ($w=A/\lambda$). In this investigation wavelength of wavy CNT (λ) kept constant and is equal to the length of RVE and waviness factor varied from 0.01 to 0.1 with help of the amplitude of Wavy CNT. As result point of view first we varied the waviness and then secondly volume fraction of CNTs (as area fraction in 2-D RVE model) varied by varying the width of RVE (W_{rve}).

A two-dimensional micromechanical model [16] is employed here for predicting effective elastic properties of CNTRC with aligned wavy CNT. An appropriate RVE is considered and upper and lower bounds for effective stiffness properties are calculated using energy principles. Upper bound is calculated using the minimum potential energy principle and iso-strain assumption while lower bound is calculated using the minimum complementary energy principle and iso-stress assumption. The direction vector of the CNT with respect to the global co-ordinate system (x, y, z) varies throughout the length of the RVE. We selected the principal (material) coordinate system at any location along the length of the RVE to be (x_0, y_0, z_0), where x axis is along the central line of the CNT. In the present case, co-ordinate transformation matrix is simply given by:

$$[\Omega] = \begin{bmatrix} \cos\gamma & -\sin\gamma & 0 \\ \sin\gamma & \cos\gamma & 0 \\ 0 & 0 & 1 \end{bmatrix} \tag{5}$$

The angle of rotation γ at any location along the length of RVE can be calculated as

$$\gamma = -\tan^{-1}\left(\frac{dy}{dx}\right) = -\left(-\frac{2\pi A}{\lambda}\sin(2\pi x/\lambda)\right) \tag{6}$$

The stress and strain vectors in the global and principal co-ordinate systems are dependent by the following equations:

$$\begin{aligned} \{\varepsilon'\} &= [T]\{\varepsilon\} \\ \{\sigma'\} &= [T]^{-T}\{\sigma\} \end{aligned} \tag{7}$$

where the strain transformation matrix $[T]$ is given by

$$[T] = \begin{bmatrix} p^2 & q^2 & 0 \\ p^2 & q^2 & 0 \\ 0 & 0 & p^2 - q^2 \end{bmatrix} \tag{8}$$

with $p=\cos\gamma; q=\sin\gamma$. The stress–strain relation in the principal material direction at any point in the RVE is given by

$$\{\varepsilon'\}=[S']\{\sigma'\}$$
$$\{\sigma'\}=[C']\{\varepsilon'\} \qquad (9)$$

where $[S']$ is the compliance matrix and $[C']$ is the stiffness matrix in a principal coordinate system. These matrices are same as the corresponding matrices for a unidirectional nanocomposite in which the CNT is straight and aligned along the principal axis. These are given by

$$[S']=\begin{bmatrix} 1/E_1 & -\nu_{12}/E_1 & 0 \\ -\nu_{12}/E_1 & 1/E_2 & 0 \\ 0 & 0 & 1/G_{12} \end{bmatrix} \qquad (10)$$

$$[C']=[S']^{-1} \qquad (11)$$

The engineering constants of unidirectional nanocomposite, appearing in the matrix can be determined either by different micromechanical models or by using finite element simulation. Rules of mixture provide the following relations:

$$E_1=\upsilon_{cnt}E_{cnt}+(1-\upsilon_{cnt})E_m$$
$$E_2=\left(\frac{\upsilon_{cnt}}{E_{cnt}}+\frac{1-\upsilon_{cnt}}{E_m}\right)^{-1} \qquad (12)$$
$$G_{12}=\left(\frac{\upsilon_{cnt}}{G_{cnt}}+\frac{1-\upsilon_{cnt}}{G_m}\right)^{-1}$$
$$\nu_{12}=\upsilon_{cnt}\nu_{cnt}+(1-\upsilon_{cnt})\nu_m$$

Utilizing the minimum potential energy principle with iso-strain assumption, the following form of average (effective) stiffness matrix of the RVE can be employed:

$$[C_{rve}]=\frac{1}{L_{rve}}\int_L [T]^T[C'][T]dx \qquad (13)$$

where L_{rve} is the length of the RVE.The average compliance matrix is given as $[S_{rve}]=[C_{rve}]^{-1}$ and from the average compliance matrix, the effective engineering constants of the RVE can be calculated. The values of the elastic constants thus calculated give the upper bound of the values. Similarly, from the principle of minimum complimentary energy with iso-stress assumption, effective compliance matrix of the RVE is obtained as

$$[S_{rve}]=\frac{1}{L_{rve}}\int_V [T]^T[S'][T]dx \qquad (14)$$

From the effective compliance matrix $[S_{rve}]$, the lower bounds of effective engineering constants of the RVE can be estimated.

2.2.2 Simulation of agglomerated CNT

To some extent, continuum mechanics is a useful tool to describe the relationship between nano-mechanics parameters and micromechanical properties of CNT based composites. An effective method to evaluate the macroscopic properties of nanocomposite is construction of RVE from a sample of the polymer matrix with embedded CNT. A square RVE is selected here (Figure 6 (a)) as the square and hexagonal RVEs are already accepted to yield the best outcomes due to their capacity to fill the space as contrasted with the cylindrical shaped RVE.

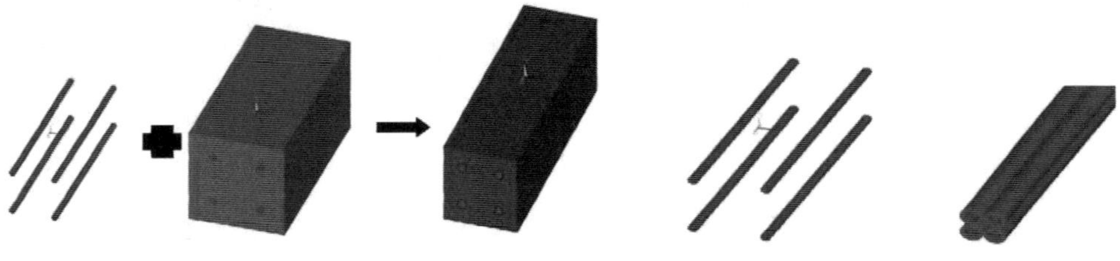

(a) A square RVE reinforced by 4 CNTs (b) Uniform and agglomerated CNTs

Fig. 6 A three dimensional RVE reinforced with four CNTs

Agglomeration of CNTs is represented using a four CNT based RVE (see Figure 6 (b)), which touch each other. Agglomerates are expected to result-in lower overall modulus than individual CNTs. Due to agglomeration, maximum shear stress is generated in RVE at inter-touching face of CNTs. A 4-CNT based finite element model is employed by considering uniform distribution and agglomeration respectively. The elastic properties of CNT/polymer composites is measured as a function of inter-tube spacing and compared with uniformly distributed CNT reinforced composite. A square RVE consisting of *4* agglomerated CNT and Polymer matrix is modeled using FEM. A three-dimensional (3D) RVE along with appropriate boundary conditions is explored under uniform extension to focus the effective elastic modulus of CNT/Polymer composites. Higher order brick structure element (quadratic element SOLID186-20-node with three degrees of freedom per node – ANSYS 15) is used for discretization. RVE is constrained at $z = 0$ in the displacement direction (z direction) and free to move in the transverse directions. The four free edges are also constrained to their respective perpendicular directions in order to permit contraction of the RVE because of extension. An extension with in elastic region, approximately 5% of the total experimental strain, is applied to all nodes on the end surface ($z = L_{rve}$). A very fine mesh is used for mesh the RVE and glue operation is performed at the interface of polymer and fiber, which gives better converged results. Firstly, uniform dispersion of four CNTs within the PP matrix is considered for the modeling and then distance between CNTs is varied to take account of agglomeration. The CNT content is varied by changing the cross-sectional area of RVE. The effective elastic modulus of the composite is computed as the ratio of the average stress to the applied strain. The average stress is computed taking the average of elemental stresses obtained from the FEA reaction forces generated at fixed end of RVE.

3. Results and discussion

In this section numerical result of effective elastic properties of CNTs and CNT reinforced polymer composite are presented. Carbon nanotube was considered as space frame structure and its analogous finite element structure was developed in ANSYS environment. A numerical study was performed to simulate the elastic behavior of different configuration of CNTs. In addition, the elastic behavior of DWCNTs and MWCNTs were also investigated using same approach. Further, an investigation was made to determine the effect of waviness and agglomeration of CNTs on elastic behavior composite material.

3.1 Properties of CNTs

Carbon nanotube is having a complex structure with hexagonal array. In present study a finite element analysis was carried out to evaluate the elastic and shear modulus of SWCNTs. The deformation of SWCNT and MWCNTs under axial stretch is shown in Figure 7. Maximum deformation can be seen at free end and its vary as function of length of CNT. In table 1, a comparison study is tabulated to validate the present analysis of (50, 50) armchair CNT with previous work outcomes and it shows the good agreement. Moreover, elastic modulus of DWCNT and MWCNT was also estimated using finite element approach. The predicted elastic modulus of DWCNT and MWCNT are 0.625 TPa and 0.520 TPa respectively, which are lower than corresponding SWCNT.

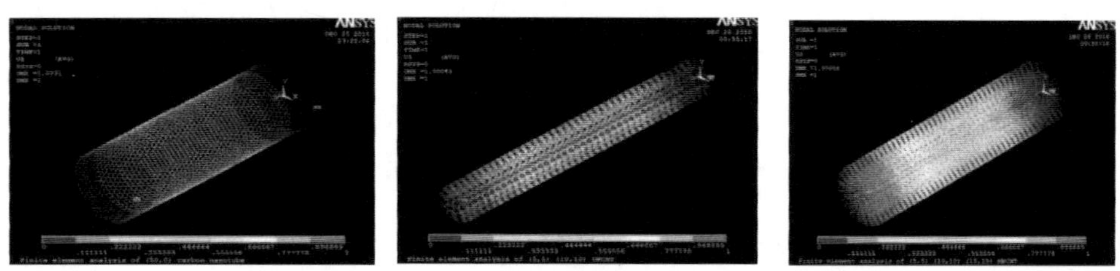

(a) (50,50) SWCNT (b) (20,20), (50,50) DWCNT (c) (10, 10), (20, 20), (50, 50) MWCNT

Fig.7 Axial deformation of different configuration of CNTs

Table 1 Comparison of elastic and shear modulus of single walled CNT

Study	Elastic modulus (E_{cnt}) TPa	Shear modulus (G_{cnt}) TPa
Present study	0.715	0.381
Natsuki et al (Ref. [33])	0.48-0.61	0.27-0.3
Zhou et al. (Ref.)	0.764	-
Guo et al (Ref. [34])	0.69	-
Chandraseker and Mukherjee (Ref.[35])	0.46-0.68	0.186-0.236
Agarwal et al. (Ref. [36])	0.73-0.82	-

The shear deformation of SWCNT under torsional loading and effect of vacancy defect in carbon nanotube are shown in Figure 8. It is observed that elastic modulus is continuously decreasing as more carbon atoms are missing in carbon nanotube. A comprehensive analysis was carried out to investigate the shear and elastic modulus of different configuration of SWCNTs. Table 2 provides the elastic and shear modulus values of zigzag and armchair CNTs.

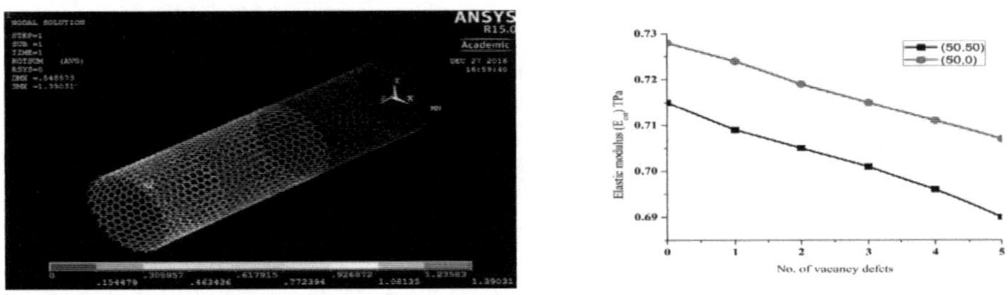

(a) Shear deformation of SWCNT (b) vacancy defects

Fig. 8 (a) Shear deformation and (b) effect of vacancy defects on elastic modulus of SWCNT

Table 2 Estimated elastic and shear modulus of different configuration of CNTs

Configuration	E_{cnt} (TPa)	G_{cnt} (TPa)	Configuration	E_{cnt} (TPa)	G_{cnt} (TPa)
(5,5)	0.711	0.364	(5,0)	0.676	0.354
(10,10)	0.712	0.375	(10,0)	0.708	0.366
(15,15)	0.713	0.377	(15,0)	0.716	0.371
(20,20)	0.714	0.378	(20,0)	0.720	0.374
(25,25)	0.714	0.379	(25,0)	0.723	0.380
(50,50)	0.715	0.381	(50,0)	0.728	0.384

3.2 Effect of CNTs waviness

Waviness of CNT was modelled sine wave form and 2 D RVE model was employed to estimate the effective elastic properties. Results are computed for small displacement along the CNT direction at a 1.2 % volume fraction. Figure 9 shows the dependency of longitudinal modulus of CNT reinforced polymer composite over the CNTs waviness ratio. It can be observed from figure that longitudinal elastic modulus is strongly dependent over the curvature and drastically decreases as the CNTs waviness increases. Therefore, it can be concluded that waviness of CNTs deceases the stiffening ability of CNTs inside the polymer resin. Figure 10 shows that transverse modulus variation with the CNTs waviness and it can be seen that transverse elastic modulus is almost unchanged. But it shows significant increment at higher waviness ratio (w=0.1). So, longitudinal modulus shows remarkable change as compared to transverse modulus with waviness ratio.

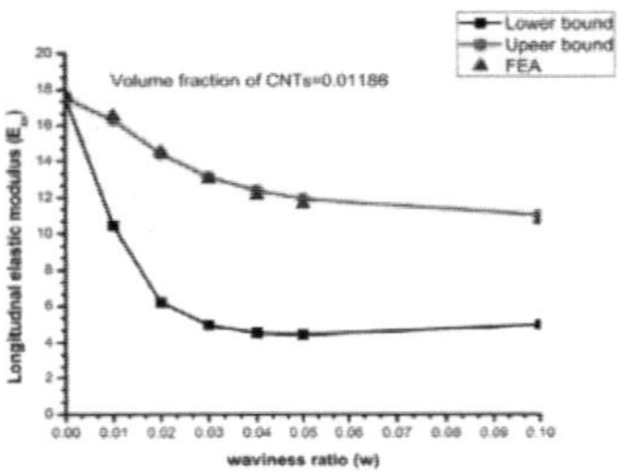

Fig.9 Effect of waviness on the longitudinal elastic modulus

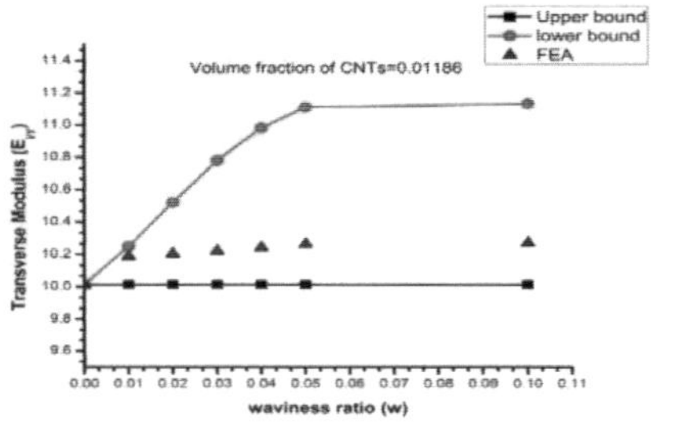

Fig.10 Effect of waviness on the transverse elastic modulus

Figure 11 shows the Poisson's ratio variation with the increase in waviness ratio of CNTs. It can be seen from results, Poisson's ratio of CNT reinforced polymer composite remains unchanged. So it can be concluded that waviness of CNTs have very less effect over the Poisson's ratio of composite.

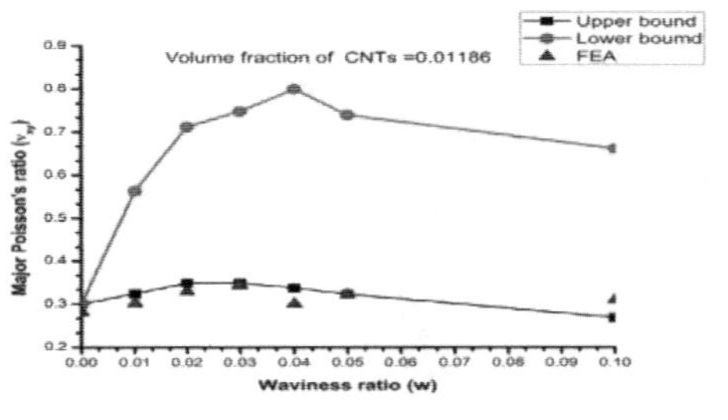

Fig.11 Effect of Waviness on the Poisson ratio

3.3 Effect of CNTs agglomeration

The approach presented here to give the fundamental idea of agglomeration (or clustering) of CNTs by a numerical technique. In this analysis square RVE with side edge dimension, a =24 nm and length L=50 nm was employed. As an initial step we have tried to investigate the effect of agglomeration on the longitudinal strain and stress distribution as shown in Figure 12. It is clear from figure that maximum longitudinal strain developed in a small region as compared to uniform distribution case of CNTs that may cause fracture or separation between CNT and polymer matrix. It can be seen from Fig. 13, that the effective elastic modulus varies with inter tube distance. There is no significant increment but it is having minimum effective elastic modulus when CNTs touch each other or inter tube distance is equal to zero.

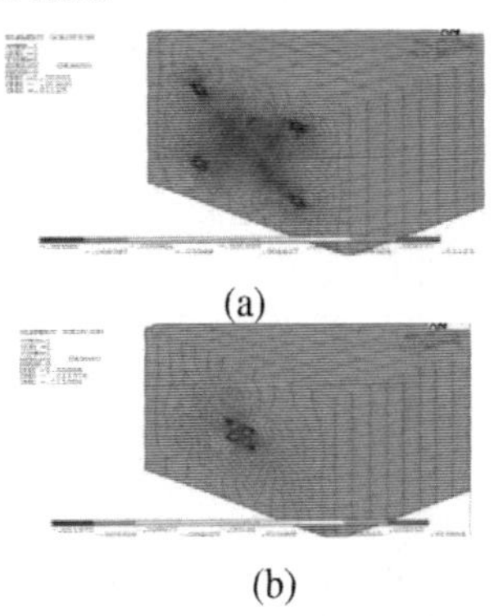

(a)

(b)

Fig.12 Axial strain distribution in RVE (a) without and (b) with agglomeration

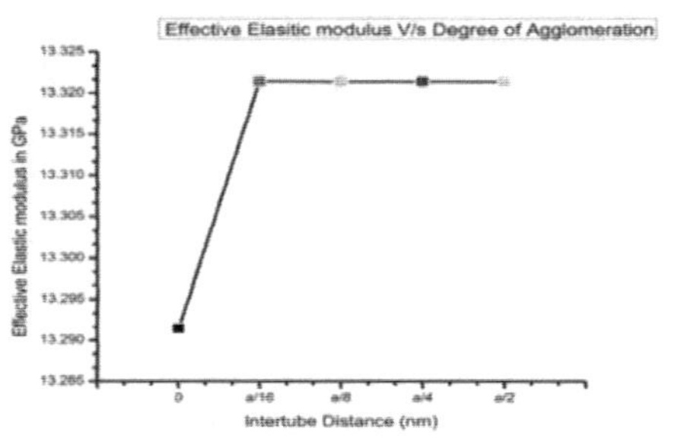

Fig.13 Effect of inter tube distance on elastic modulus of CNTs reinforced composite

The stress concentration region is clearly seen in agglomerated case. To investigate the effect of CNTs clusters on longitudinal elastic modulus we have varied the width of RVE and the elastic modulus values are listed in Table 3. These results give interesting information that the effective elastic modulus decreases as width of RVE increases, but RVE with agglomerated CNTs having low elastic modulus than RVE without agglomeration.

Table 3 Effective elastic modulus with and without agglomeration

Side of RVE (*a* in nm)	Volume fraction of CNT (V_{cnt}) in %	Effective Elastic Modulus without agglomeration	With agglomeration
24	1.11	13.3213	13.2914
22	1.31	15.2261	15.190
20	1.59	17.7367	17.6938
18	1.96	21.1411	21.0875
16	2.48	25.9216	25.8544
14	3.25	32.9411	32.8543

4. Conclusion

Effective elastic properties of CNTs and CNT reinforced polymer composite were investigated by using finite element Modelling. FE analysis of SWCNTs have shown that nanotube chirality and diameter influence the effective elastic property of SWCNTs. Additionally, effect of carbon atom vacancy in SWCNT was also investigated and it was observed that elastic modulus of CNT decreases with increase in number of carbon atom vacancy. It was also found that MWCNTs have lower modulus as compared to SWCNTs. Further, an attempt was made to predict the effect of waviness and agglomeration on over all elastic properties of CNT reinforced composite. By selecting optimal mesh size in simulation process, FEA results were compared with micromechanics wherever possible. Predicted results show that CNTs curvature was responsible for decrement in longitudinal modulus whereas transverse modulus was almost constant for small waviness. Transverse elastic modulus showed the sudden improvement at waviness ratio order of 0.1, while Poisson's ratio was not affected by the waviness. In addition, agglomerations of CNTs in polymer matrix was elaborated using 3-D RVE and it was observed that effective elastic modulus of 4 CNT based RVE with inter-tube spacing zero was less as compared to RVE with uniform inter-tube spacing. Maximum increase in elastic modulus of composite was observed when inter-tube spacing change zero to *a/16*. Present work can be extended further to include the effect of orientation, reinforcing effect of MWCNTs with interatomic Van der Waal forces and also a similar thermal analysis of CNT reinforced polymer composite material may be carried out.

Reference

[1] Iijima S and Ichihashi T 1993 Single-shell carbon nanotubes of 1-nm diameter *Nature* **363** 603–5
[2] Qian D, Dickey E C, Andrews R and Rantell T 2000 Load transfer and deformation mechanisms in carbon nanotube-polystyrene composites *Appl. Phys. Lett.* **76** 2868–70
[3] Xiao K Q and Zhang L C 2004 The stress transfer efficiency of a single-walled carbon nanotube in epoxy matrix *J. Mater. Sci.* **39** 4481–6
[4] Shokrieh M M and Rafiee R 2010 A review of the mechanical properties of isolated carbon nanotubes and carbon nanotube composites *Mech. Compos. Mater.* **46** 155–72
[5] Thostenson E T, Ren Z and Chou T-W 2001 Advances in the science and technology of carbon nanotubes and their composites: a review *Compos. Sci. Technol.* **61** 1899–912
[6] Lau A K-T and Hui D 2002 The revolutionary creation of new advanced materials—carbon nanotube composites *Compos. Part B Eng.* **33** 263–77
[7] Gupta S S and Batra R C 2008 Continuum structures equivalent in normal mode vibrations to single-walled carbon nanotubes *Comput. Mater. Sci.* **43** 715–23
[8] Arroyo M and Belytschko T Continuum Mechanics Modeling and Simulation of Carbon Nanotubes *Meccanica* **40** 455–69
[9] Tsai C, Zhang C, Jack D A, Liang R and Wang B 2011 The effect of inclusion waviness and waviness distribution on elastic properties of fiber-reinforced composites *Compos. Part B Eng.* **42** 62–70
[10] Nafar Dastgerdi J, Marquis G and Salimi M 2013 The effect of nanotubes waviness on mechanical properties of CNT/SMP composites *Compos. Sci. Technol.* **86** 164–9

[11] Motamedi M, Eskandari M and Yeganeh M 2012 Effect of straight and wavy carbon nanotube on the reinforcement modulus in nonlinear elastic matrix nanocomposites *Mater. Des.* **34** 603–8
[12] Shady E and Gowayed Y 2010 Effect of nanotube geometry on the elastic properties of nanocomposites *Compos. Sci. Technol.* **70** 1476–81
[13] Farsadi M, Öchsner A and Rahmandoust M 2013 Numerical investigation of composite materials reinforced with waved carbon nanotubes *J. Compos. Mater.* **47** 1425–34
[14] Shi D-L, Feng X-Q, Huang Y Y, Hwang K-C and Gao H 2004 The Effect of Nanotube Waviness and Agglomeration on the Elastic Property of Carbon Nanotube-Reinforced Composites *J. Eng. Mater. Technol.* **126** 250–7
[15] Jarali C S, Basavaraddi S R, Kiefer B, Pilli S C and Charles Lu Y 2013 Modeling of the Effective Elastic Properties of Multifunctional Carbon Nanocomposites Due to Agglomeration of Straight Circular Carbon Nanotubes in a Polymer Matrix *J. Appl. Mech.* **81** 021010–021010
[16] Paunikar S and Kumar S 2014 Effect of CNT waviness on the effective mechanical properties of long and short CNT reinforced composites *Comput. Mater. Sci.* **95** 21–8
[17] Herasati S and Zhang L 2014 A new method for characterizing and modeling the waviness and alignment of carbon nanotubes in composites *Compos. Sci. Technol.* **100** 136–42
[18] Zare Y 2016 Study of nanoparticles aggregation/agglomeration in polymer particulate nanocomposites by mechanical properties *Compos. Part Appl. Sci. Manuf.* **84** 158–64
[19] Alian A R, El-Borgi S and Meguid S A 2016 Multiscale modeling of the effect of waviness and agglomeration of CNTs on the elastic properties of nanocomposites *Comput. Mater. Sci.* **117** 195–204
[20] Bhuiyan M A, Pucha R V and Kalaitzidou K 2016 3D RVE Models Able to Capture and Quantify the Dispersion, Agglomeration, and Orientation State of CNT in CNT/PP Nanocomposites *Compos. Mater.* **3** 1–13
[21] Kumar P with J. Srinivas 2014 Analysis of elastic properties of transversely isotropic CNT-reinforced polymers *Nano Sci. Nano Technol. Indian J.* **8** 291–197
[22] Thakur A K, Kumar P and Srinivas J 2016 Studies on Effective Elastic Properties of CNT/Nano-Clay Reinforced Polymer Hybrid Composite *IOP Conf. Ser. Mater. Sci. Eng.* **115** 012007
[23] Rafiee R and Firouzbakht V 2014 Predicting Young's Modulus of Aggregated Carbon Nanotube Reinforced Polymer *Mech. Adv. Compos. Struct.* **1** 9–16
[24] Chanteli A and Tserpes K I 2015 Finite element modeling of carbon nanotube agglomerates in polymers *Compos. Struct.* **132** 1141–8
[25] Kumar P and Srinivas J 2014 Numerical evaluation of effective elastic properties of CNT-reinforced polymers for interphase effects *Comput. Mater. Sci.* **88** 139–44
[26] Tserpes K I and Papanikos P 2005 Finite element modeling of single-walled carbon nanotubes *Compos. Part B Eng.* **36** 468–77
[27] Shokrieh M M and Rafiee R 2010 Prediction of Young's modulus of graphene sheets and carbon nanotubes using nanoscale continuum mechanics approach *Mater. Des.* **31** 790–5
[28] Sakharova N A, Pereira A F G, Antunes J M, Brett C M A and Fernandes J V 2015 Mechanical characterization of single-walled carbon nanotubes: Numerical simulation study *Compos. Part B Eng.* **75** 73–85
[29] Fan C W, Liu Y Y and Hwu C 2009 Finite element simulation for estimating the mechanical properties of multi-walled carbon nanotubes *Appl. Phys. A* **95** 819–31
[30] Shokrieh M M and Rafiee R 2010 On the tensile behavior of an embedded carbon nanotube in polymer matrix with non-bonded interphase region *Compos. Struct.* **92** 647–52
[31] Pourasghar A and Kamarian S 2014 Mechanical material characterization of an embedded Carbon nanotube in polymer matrix by employing an equivalent fiber *Int. J. Nano Dimens.* **6** 167–75
[32] Oliva-Avilés A I, Avilés F, Seidel G D and Sosa V 2013 On the contribution of carbon nanotube deformation to piezoresistivity of carbon nanotube/polymer composites *Compos. Part B Eng.* **47** 200–6
[33] Natsuki T, Tantrakarn K and Endo M 2004 Prediction of elastic properties for single-walled carbon nanotubes *Carbon* **42** 39–45
[34] Guo X, Wang J B and Zhang H W 2006 Mechanical properties of single-walled carbon nanotubes based on higher order Cauchy–Born rule *Int. J. Solids Struct.* **43** 1276–90
[35] Chandraseker K and Mukherjee S 2007 Atomistic-continuum and ab initio estimation of the elastic moduli of single-walled carbon nanotubes *Comput. Mater. Sci.* **40** 147–58
[36] Agrawal P M, Sudalayandi B S, Raff L M and Komanduri R 2006 A comparison of different methods of Young's modulus determination for single-wall carbon nanotubes (SWCNT) using molecular dynamics (MD) simulations *Comput. Mater. Sci.* **38** 271–81

IOP Conf. Series: Materials Science and Engineering **178** (2017) 012017 doi:10.1088/1757-899X/178/1/012017

Impact of Annealing Temperature on the Phase of CZTS with the Variation in Surface Morphological Changes and Extraction of Optical Bandgap

J Raiguru[a], BVRS Subramanyam[b], K Sa[b], I Alam[b], S Das[b], J Mukherjee[b], P C Mahakul[b], B Subudhi[a], P Mahanandia[*]

[a]Department of Electrical Engineering, National Institute of Technology, Rourkela, India

[b,*]Department of Physics & Astronomy, National Institute of Technology, Rourkela, India

***Email: pitam@nitrkl.ac.in**

Abstract. Quaternary CZTS (Cu_2ZnSnS_4) is an emerging alternative semiconductor material for solar cell technologies due to its earth abundancy, low cost and non-toxic nature. In addition, CZTS has a direct band gap of ~1.5 eV which is the optimal value for converting the maximum amount of energy from the solar spectrum into electricity. The aim of the present study is to investigate the impacts of annealing temperature on the phase formation and morphological behavior of CZTS and material optical characteristics. X-ray Diffraction (XRD) reveals the single phase of kesterite CZTS that has been grown from the material precursors. Scanning electron microscope (SEM) shows the variation of morphological changes of CZTS with respect to annealing temperature. UV-Vis analysis confirms the optical band gap of 1.5 eV in the visible region which is suitable for photovoltaic applications.

Keywords: X-ray Diffraction (XRD), Scanning Electron Microscope (SEM), Ultraviolet- Visible absorption Spectroscopy (UV-Vis), Kesterite CZTS, Optical Bandgap

Introduction

Copper zinc tin sulfide (Cu_2ZnSnS_4 called CZTS) emerged as an alternative material which is abundant, environmentally benign, and inexpensive with a high absorption coefficient (10^4 cm^{-1}) with the suitable optical energy bandgap of 1.5eV which minimizes the quantity of material used in devices for thin film semiconductor solar cells [1].These advantages of CZTS make it a possible substitute of other thin film solar cells such as CdTe, CdSe, InP or GaAs which are having variety of concerns such as low abundance and high cost [2]. The quaternary kesterite CZTS, a competitive candidate for photovoltaic (PV) semiconductor material has its own challenge at its initial synthesis process because of simultaneous presence of secondary phases during its formation. The presence of secondary phases predicted to be the barriers for carriers, limits the exciton movement by formulating traps at interfacial layers that decreases the performance of PV cell which is not good for the prospective CZTS solar cell application [3-5].Therefore the elimination of ternary Cu_3SnS_4 is a challenging task without affecting the objective phase CZTS in an ambient condition. So it is utmost important to remove these secondary phases from the kesterite CZTS material which could harm the optoelectronic properties such as optical energy band gap that characterizes the transition process of electrons within visible-near infrared region for PV applications [8]. To avoid secondary phases various study have been reported about the widely used post annealing process to get proper CZTS phase in both vacuum and non-vacuum deposition technique[2]. Post annealing process plays an intensive major role to enhance the crystallinity as well the single kesterite phase of CZTS. But the secondary phases are also the inevitable during the annealing ambient condition and have the close tendency to temperature changes. Weber *et al.* have reported that the copper sulfides and zinc sulfides are formed due to the CZTS phase decomposes at temperatures above 350 °C [6]. Morphological evidence like nano flower shaped two dimensional secondary phases of Cu_3SnS_4 has been reported by Yang et al which is difficult to exclude from the objective CZTS phase [7]. So it needs to suppress the undesired secondary phase formation near to the quaternary CZTS semiconductor with the ramping up annealing temperature but with a careful and controlled manner. The degree of temperature variation in the ambient condition has proportionate effect for the elimination of secondary phases. This study explores that the kesterite CZTS could be grown in single phase with the impact of annealing temperature with the evidence of no secondary phase formation; which could enhance the performance of the cell. A generalized photovoltaic device structure of CZTS is shown in Fig 1.

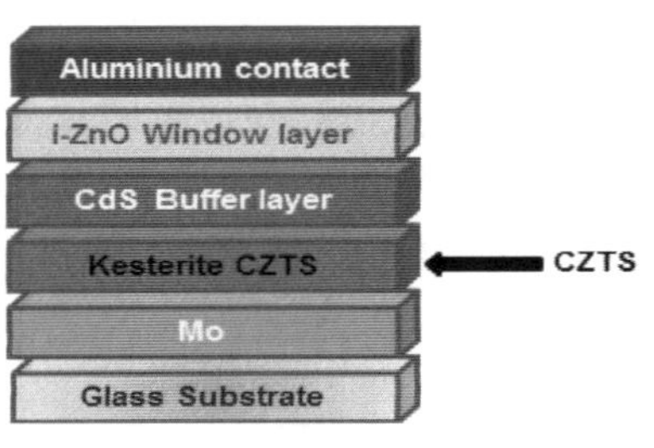

Figure 1. Generalized photovoltaic device structure of CZTS

IOP Conf. Series: Materials Science and Engineering **178** (2017) 012017 doi:10.1088/1757-899X/178/1/012017

Experimental Details

Synthesis of CZTS material

The proposed work delivers an alternative route for removal of secondary phases by the impact of annealing temperatures that are synthesized by the compatible non-toxic and low cost subsequent solution process which adds a beneficiary progress in the current state of art technology for single phase kesterite CZTS (Cu_2ZnSnS_4). A schematic representation of CZTS (Cu_2ZnSnS_4) preparation is shown in Figure 2. The solution is prepared using Cu, Zn, Sn precursors and Sulphur content solutions. The final solution is subjected for stirring with ethanol for 30 minutes at room temperature. The collected product then annealed at three different temperatures (i.e. 350 ^{0}C, 375 ^{0}C, 400 ^{0}C) which is performed by a tubular furnance at working pressure of 0.05 milibar with argon as carrier gas for 1 hour.

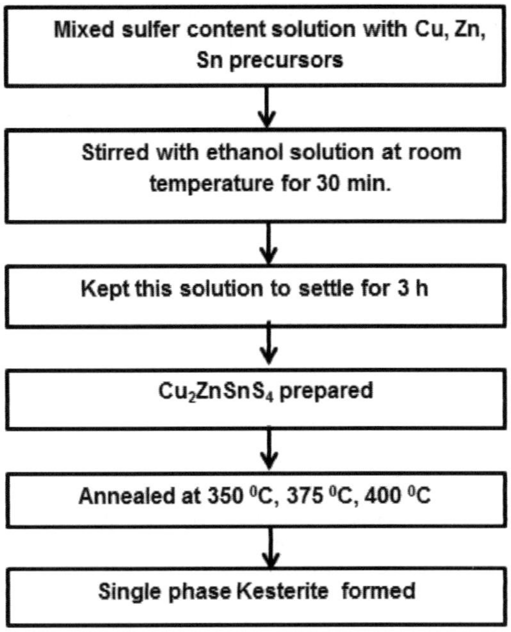

Figure 2. Synthesis of single phase of kesterite CZTS (Cu_2ZnSnS_4) semiconductor material

Characterization techniques

Phase analysis has been characterized by X-ray-diffraction (XRD) technique with CuKα radiation (λ=0.15406nm) by using **Rgakujapan/Ultima-IV** model X-ray diffractometer. Surface morphologies have been studied by using scanning electron microscopy (**JEOLJSM-6084LV**) resolution: 3.0nm (ACC V 30 Kv, working distance 8mm).Optical properties were investigated by studying the absorbance spectra recorded by UV-Vis-NIR spectrophotometer (**Shimadzu/model 3600**).

Results and Discussion

XRD (X-ray-Diffraction) analysis

The annealed CZTS at 350 ^{0}C, in argon atmosphere at working pressure of 0.05 milibar for 1 hour has been characterized by XRD and is shown in Figure 3(a). The peaks have been analyzed using Joint Committee on Power Diffraction Standards data (JCPDS 26-0575).The result displays the spectra from 2θ = 20 - 70^0 in which strong peaks of CZTS at 29.02^0, 33.01^0, 47.45^0, 56.83^0, 69.40^0 corresponds to the tetragonal CZTS crystal planes (112), (200), (220),(312), (008).

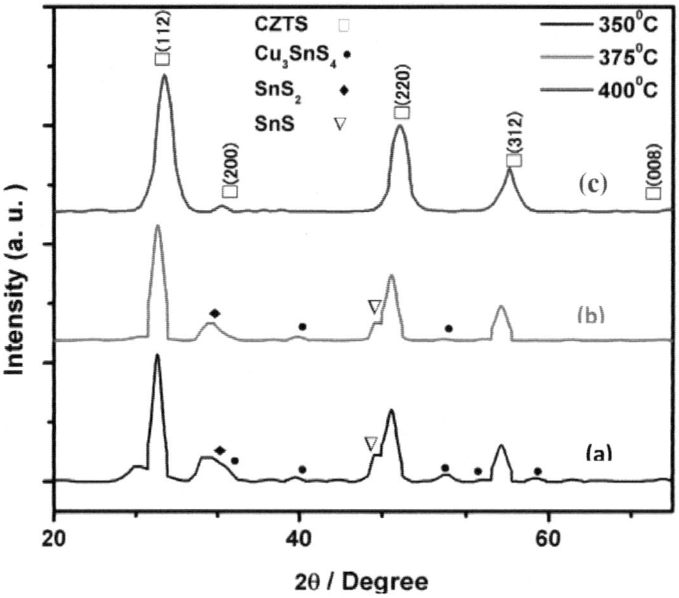

Figure 3. Powder XRD patterns of annealed CZTS (Cu2ZnSnS4) semiconductor material at (a) 350 ^{0}C, (b) 375 ^{0}C and (c) 400 ^{0}C annealing temperatures

In association with the main CZTS characteristics peaks, secondary phases Sn S_2 at 33.02^0 ;(011) (JCPDS-83-1707), SnS at 45.8^0 :(200) (JCPDS-831758) and Cu_3SnS_4 at 37.27^0, 39.81^0, 51.8^0, 56.36^0, 58.39^0 (212),(2012),(2018),(3212)(406) are also observed. Though the increased in annealing temperature to 375^{0}C keeping the duration of period constant resulted partial elimination of secondary phases, but still presence of a few secondary phases in CZTS is reflected from the XRD pattern as shown in Fig 3(b). However, complete elimination of secondary phases are observed in CZTs when it is annealed at 400 ^{o} C (Fig3 (c)). The peaks 29.04^0, 33.50^0, 47.97^0, 56.75^0, 69.42^0

correspond to the planes (1 1 2), (0 0 2), (2 2 0), (3 1 2), and (0 0 8) infer to a single phase of kesterite CZTS structure. Peak broadening in the XRD of CZTS as a result of annealing temperatures from 350 ^{0}C to 400 ^{0}C reveals decrease in crystalline size. The formation of single phase CZTS with increased annealing temperature could be attributed to the diffusion of binary sulfides to maintain appropriate distance for improved density [9]. A negligibly shifts of XRD peak of CZTS annealed at 400 ^{0} C towards the right side might be due to the compressive stress and the pronounced antisite defect complexes presence inside the CZTS material with a low level Cu/Zn disorder [10].

SEM (scanning electron microscope) analysis

Figure 4. SEM image of CZTS (Cu_2ZnSnS_4) annealed at 350 ^{0}C full of Nano flower like structure

SEM micrographs of CZTS at different annealing temperature are shown in the Figure 4 - 6. It is observed from the SEM images (Fig.4) that the formation of flower like nanostructures might be due to the presence of predominant secondary phase Cu_3SnS_4 along with CZTS annealed at 350 ^{0}C [7]. The observed less flower like structures in the SEM micrograph (Fig.5) of CZTS annealed at 375 ^{0}C might be due to the partial elimination of secondary phases Cu_3SnS_4 along with other phase such as SnS and/or SnS_2 which has been observed in XRD [7]. Disappearance of flower like structures is seen when the CZTS is annealed at 400 ^{0}C as shown in SEM micrograph (Fig.6) suggesting complete removal of secondary phases supported by XRD result.

Figure 5. SEM image of CZTS (Cu2ZnSnS4) annealed at 375 ^{0}C with less Nano flower like structure and formation of less amount of euhedral blocky shaped structure

Figure 6. SEM image of CZTS (Cu_2ZnSnS_4) annealed at 400 ^{0}C Nano flower structure completely disappears and all structures are changed to crystalline euhedral blocky shaped structure.

UV-Vis Absorption Spectroscopy

Due to the influence of particle composition at high annealing temperature the bandgap changes from 1.37 eV to 1.45 eV with CZTS morphology from, which further provides evidence that the annealing temperature strongly affect the CZTS properties [1]. Absorption spectrum of CZTS (Cu_2ZnSnS_4) annealed at 400 ^{0}C is measured using UV-Vis NIR range .The absorption spectrum appears from 300 nm to 1000 nm at room temperature. Since the optical absorption coefficient is related to the optical energy band gap E_g near the absorption edge that could be determined by the Tauc's formula [1]:

$$\alpha = \frac{A(hv - E_g)^n}{hv} \quad \ldots\ldots\ldots\ldots\ldots\ldots;\ (1)$$

where, A is an energy-independent constant, h is the Planck's constant, v is frequency, and n is an index that characterizes the transition process and is theoretically equal to 2 and ½ for indirect and direct allowed transitions for indirect band gap and direct bandgap semiconductors, respectively.

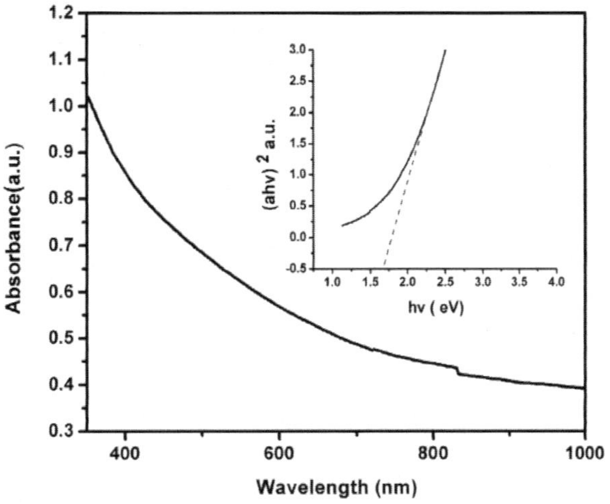

Figure 7. UV-vis absorption spectra of single phase kesterite CZTS annealed at 400 ^{0}C with good crystallinity without any secondary phases

Extrapolation of the straight line to zero absorption coefficient (α = 0) allows estimation of E_g. The band gap of single phase kesterite CZTS annealed at 400 ^{0}C are estimated to be around 1.52 eV by extrapolating the linear region of curve to intercept of photon energy as shown in the inset of the Figure 7.The straight line fit that yields E_g=1.5eV for the exponent n=2 in Tauc's formula reveals that the prepared single phase kesterite CZTS is a direct bandgap semiconductor [1].

Conclusion

The quaternary CZTS (Cu_2ZnSnS_4) compound semiconductor material powder has been synthesized at room temperature by solution method taking copper, zinc chloride, Tin and Thiourea at a required molar concentration ratio (Cu: Zn: Sn: S). XRD characterization resembled with JCPDS number (26-0575) confirmed the kesterite phase of CZTS (Cu_2ZnSnS_4). Our method of increased annealing temperature has significant positive impact on removal of secondary phases and achieved single phase kesterite CZTS at annealing temperature 400 ^{0}C. The surface morphology of CZTS material attributed to subjective changes with good crystallinity. In conclusion we have demonstrated a low cost promising state of art to yield single phase kesterite CZTS (Cu_2ZnSnS_4).

References

[1] Aldalbahi A, Mkawi EM, Ibrahim K. and Farrukh MA, 2016 *Sci. Rep* **6 (32431)**

[2] Suryawanshi MP, Agawane GL, Bhosale SM, Shin S W, Patil P. S., Kim J. H.and Moholkar AV 2013 *Materials Technology* **28** 98-109

[3] Polivtseva S, Acik IO, Katerski A, Mere A, Mikli V and Krunks M 2015 *P hys. Status Solidi C* **13** 18–23

[4] Chen H, Gu M, Pu X, Zhu J and Cheng L 2016 *Mater. Res. Express* ***3*** **(065002)** 1-6

[5] Deng M, Shen S, Zhang Y, Xuc H and Wang Q 2014 *NewJ.Chem* **53** 77—83

[6] Weber A, Mainz R and Schock H. W. 2010 *J. Appl. Phys* **107** 013516-6

[7] Yang F, Ma R, Zhao W, Zhang X and Li X 2016 *Journal of Alloys and Compounds* **689** 849-856

[8] Zhang X, Guo G, Ji C, Huang K, Zha C, Wang Y, Shen L, Gupta A and Bao N 2014 *Sci. Rep.* **4** (**5086**)

[9] Li Z, C. W. Ho J, Lee Kian K, Xin Z, Zhang T, Wongab L H and Lam Y M 2014 *RSC Adv.* **4** 26888–26894

[10] Paris M, Choubrac Léo, Lafond A, Catherine G D, and Jobic S 2014 *Inorg. Chem* **53** 8646−8653

IOP Conf. Series: Materials Science and Engineering **178** (2017) 012018 doi:10.1088/1757-899X/178/1/012018

Sea water corrosion behavior of plasma sprayed abradable coatings

M Parida*[1], S.P.Nanda[2] and S.C.Mishra[3]

[1]Dept. Of Chemistry, SCITM, Similiguda.
[2]Dept. Of Chemistry, CUTM, Paralakhemundi.
[3]Metallurgical & Materials Engineering Department, NIT, Rourkela.

E-mail: rinaa26@yahoo.com

Abstract. Aluminum based abradable coating is used for sealing purpose in compressor casing of aero engines to withstand up to a service temperature of 450^{0}C. Al-BNSiO$_2$ composite coating is deposited using thermal plasma spray technique. Coating thickness measured and porosity of the coating is evaluated.Coating morphology is observed and EDSanalysis is done with SEM (Jeol make). The effect of time on the sea water corrosion behavior of the coating is evaluated. It is observed that, there is a sharp increase in weight gain of the coating up to six weeks of immersion. This behavior is attributed to the adsorption /deposition of other elements/reactions taking place during interaction with sea water.

1. Introduction

Abradable aluminum base composite coating is used for clearance control and sealing in compressor of aero engines, thereby improving engine efficiency [1]. This abradable aluminum coating is used as top coat with a bond coat being nickel-aluminide etc. [2].Abradable aluminum coating is used in the area of cold end components like compressor casings, rings of gas generator, and internal rings etc. To deposit such coatings, flame or plasma spraying is considered as the most effective method. Since 1960, atmospheric plasma spraying (APS) is widely used in many fields of engineering applications. One of the main advantages of plasma spraying process is that, the substrateis heated very little during the coatingdeposition. During plasma spray process the plasma thermal energy melts the coating/feed material (usually in the form of powder/wire). Molten material is propelled by the carrier gas on a cleaned and prepared specimen surface, where it bonds to the surfacepredominantly by mechanical bonding. Coating characteristics are very much dependent on substratesurface and critical plasma spraying parameter (CPSP) [3-6].

Parameters those are important in choosing the abrasive for surface roughening are type, size, shape, purity and hardness of blasting powder. Specific CPSP is used for this aluminum based abradable coating for clearance control [7]. These parameters also affect the corrosion behavior of plasma spray ceramic coatings [8].This coating is intended for operation, when the operating temperature is below 450°C in the rotor sealing of gas turbine engine at general climatic condition. But however study has not been reported for this abradable aluminum coating related to corrosion testing in saline climatic condition. The main aim of the work is to study the sea water corrosion behavior of this abradable aluminum-based composite coating.

2. Experimental Procedure

Before carrying out Nickel plating of Aluminum powder, it was prepared by sheaving the powder with mesh size between 1-60. Granularity structure of the powder was maintained. Nickel chloride (30kg) was added to 50litersof solution made with (Boric acid + NaCl +Oxalic acid + HF).

All the chemicals were dissolved in distilled water at 70- 80 ^{0}C. Then sprinkling of already prepared aluminium powder was doneat a temperature of 22 – 25 ^{0}C within 5to 10minutes at a rate of 96gm /ltr. Then excess solution was taken out and flushedwith drinking water for 6 – 8 times. Finally the powder was taken in a stainless steel tray and heated at 180- 200 ^{0}C. The powder should have granular size between 16-56 mesh.

Aluminium powder is sieved through mesh size 016. Aluminium powder 74- 76 % and BoronNitride 24-26% were mixed with 40-60 rpm in a ball mill for 10-12 hrs. The dry mixture is mixed with liquid glass 140 -160 cc for 1 kg of dry mixturefor 30-40 minutes. The blend is dried at 140-160 ^{0}C for 8- 10 hrs and sieved through vibro-sieve no 315.

2.1 *Substrate Preparation*

Nickel based Nimonic(70% Ni,23%Cr) plate was selected as substrate for coating. The surface of the substrate prepared for coating was made free from moisture, oil, and other contaminates. The time gap between surface preparation and spraying process was lessthan 2 hours,to avoid disintegrationof coating. Substrate surface was roughened using Al_2O_3 powderof grit size 60.

2.2 *Spray Deposition*

Coating was made on the surface with a sub layer of nickel-aluminide, of 50 micron thickness by plasmaspray techniqueusing a specific critical plasma spraying parameters (CPSP). The plasmaspray parametersused for spraying are listedin table – 1. After coatingthe sampleswere heat treatedat 450^{0}C for 30 minutes.

Table 1.CPSP for deposition of coatings

Sl. No.	Parameter	Value
1	Nozzle Dia	6 mm
2	Primary plasma Gas	Argon (45 lpm)
3	Secondary Plasma-gen gas	Argon (7 lpm)
4	Current	570 Amp
5	Powder carrier gas	Argon (3 lpm)
6	Powder feed rate	90 % on machine scale
7	Spray distance	120 mm

2.3 *Specimen preparation for corrosion test*

Ten specimensof thickness 3 -4 mm and weight (ranging between 5-9 grams)were cutfrom preparedcoatings by wirecut EDM.The abradablealuminumcomposite coatings were subjected to corrosion test, deeping in sea water for different time periods. Sea water was collected from Bay of Bengal; the composition of sea water is shown in table 2.

Table 2. Composition of sea water.

Type of ion	Element	Concentration (Mg/ltr)	% by wt.
Cation	Na^{+1}	10500	30.42
	Mg^{+2}	1350	3.91
	Ca^{+2}	400	1.16
	K^{+1}	380	1.10
	Sr^{+2}	8	0.02
Anion	Cl^{-1}	19000	55.04
	So_4^{-2}	2655	7.04
	$C0_3^{-2}$	140	0.41
	Br^{-1}	65	0.19
	Bo_3^{-3}	20	0.06
	Sio_3^{-2}	8	0.02
	F^{-1}	1	0.003

3. Results and discussion

3.1 Hardness test

The heat-treated coatedsample surfaceswere polishedusing emery paperfor hardness measurement. The coatingswere checked for any defect using 7X magnifying glass. Blow holes, cracks, separation and unfused particles were not found. Hardness was measured (with Brinell hardness tester) on five areas of the sample after making the surface smooth. Ball indenter of diameter 5mm was used at a force of 62.5 kg/cm2., having indentation dia of 3.15 to 2.4mm. Ten readings were taken and the Hardness of the specimens was in the range of 7 – 13 HB.

3.2 Corrosion Test

Specimens were weighed in a balance of least count 0.01 gm. Prior to corrosion test the samples were washed with distilled water and dried at 110 – 120 ^{0}C for 1 hr. The specimenswere put in the sea water at different interval of time and kept under observation. After 10 weeks of observation samples were taken out and again dried at110 – 120 ^{0}C for 1 hr. Again weight of specimens were taken and listed at table – 3 and shown in Figure 1.

Table 3. Change in weight percent of the samples with treatment time.

Sl. No.	Time in days	Initial weight	Final weight	Change in weight	Percent weight gain
1	86	6.28	6.86	0.58	9.24
2	79	5.02	5.38	0.36	7.17
3	65	8.1	8.88	0.78	9.63
4	58	8.26	8.99	0.73	8.84
5	44	6.97	7.62	0.65	9.33
6	36	6.47	6.7	0.23	3.55
7	28	6.7	6.98	0.28	4.18
8	19	6.34	6.61	0.27	4.26

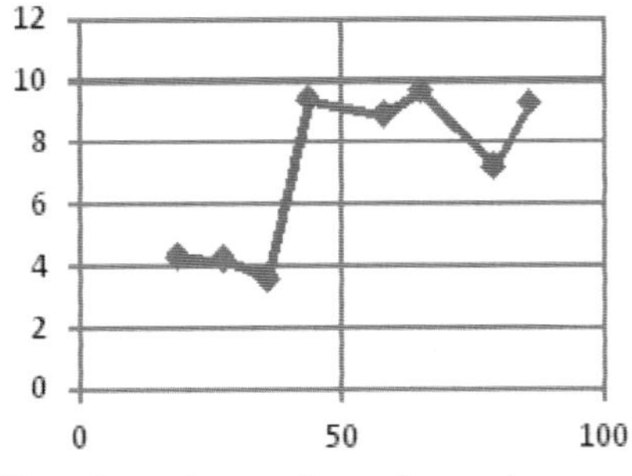

Figure 1. Time dependence of wt. change in sea water corrosion.

From the figure, it is evident that, the rate of increase in wt. is very rapid up to six weeks and then takes a linear trend.

3.3 Microstructural observations

SEM and EDS analysis is done with the raw and corroded samples, shown in figure 2 and figure 3 for these samples, respectively.

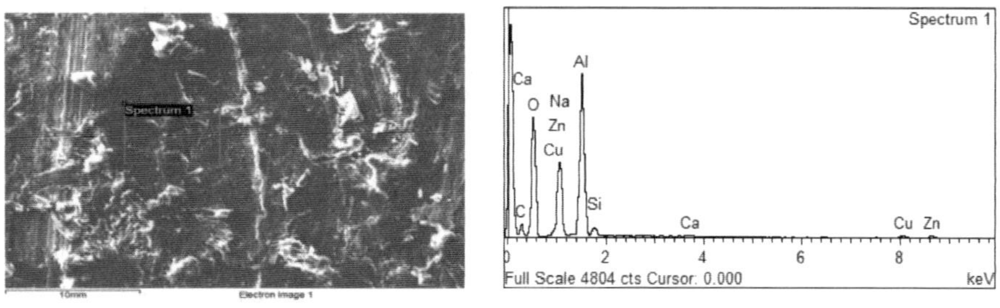

Figure 2. SEM and EDS spectrum of the surface of the coating.

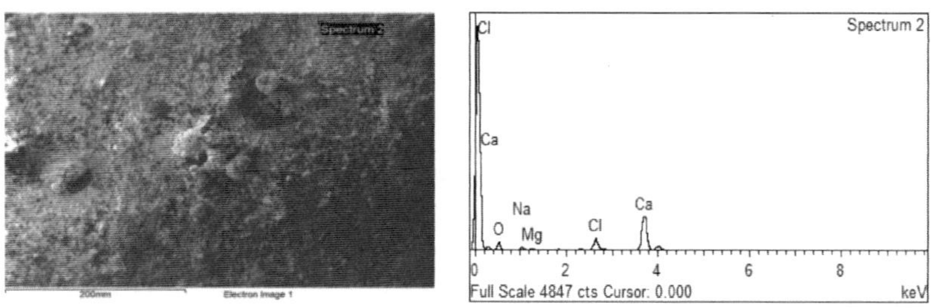

Figure 3. SEM and EDS spectrum of the surface of the untreated specimen.

From the above figures, it can be seen that, very good quality coatings could be made with using the above said plasma operating parameters. It is well known that, plasma spraying parameters plays an important role on deposition and the microstructure and the surface morphology is more important for optimization of plasma spray parameters for developing the best quality coatings [9-11]. Hence the selection of plasma spray parameters (in our case) is suitable to develop dense coatings. In case of the corroded specimen, fig.2, deposition of small particles and dimples are developed on the coating/surface. From the EDS composition analysis, table – 4 and 5, it can be seen that there is major increase in amount of Ca, Cl and a little increase in Mg also. This implies that, some kind/type of ionic change/reactions is taking place on the samples and investigations in this regard are on the way.

Table 4. EDS elemental analysis of the coating surface prior to treatment.

Element	Weight%	Atomic%
C K	11.40	17.71
O K	49.04	57.20
Na K	12.35	10.03
Al K	17.30	11.97
Si K	1.16	0.77
Ca K	0.27	0.13
Cu K	1.61	0.47
Zn K	2.19	0.63
Br L	4.67	1.09
Totals	100.00	

Table 5.EDS elemental analysis of the coating surface after dipped in sea water for 10 weeks.

Element	Weight%	Atomic%
O K	51.43	70.76
Na K	3.92	3.75
Mg K	1.06	0.96
Cl K	8.06	5.00
Ca K	35.54	19.52
Totals	100.00	

4. Conclusions

Aluminum based composite coating is deposited through plasma spraying bears the hardness of 7-13HB. With sea water treatment, weight gain of the samples is observed. The rate of increase in wt. is very rapid up to six weeks and then takes a linear trend. There a change in surface morphology of the coating after corroded in sea water. Deposition of Ca and Cl is observed on the surface of the sea water treatment.

References

[1] Mishra S 2012 *Thermal Plasma Application in Metallurgy* Germany Lap Lambert Academic Publishing ISBN: 978-3-659-23522-1.

[2] Das S 2007 *Study of Plasma Spray Alumina-Aluminide Composite Coating on Metals* (Doctoral dissertation).

[3] Parthasarathi N L and Duraiselvam M 2010 Effect of sliding velocity and plasma spraying parameters on wear resistance of NiCrBSiCFe plasma sprayed on austenitic steel at high temperature *Computer Modelling and New Technologies* **14** 17–29.

[4] Kulkarni A, Wang Z, Nakamura T, Sampath S, Goland A, Herman H, Allen J, Ilavsky J, Long G, Frahm J and Steinbrech R W 2003 Comprehensive microstructural characterization and predictive property modeling of plasma-sprayed zirconia coatings *ActaMaterialia* **51** 2457–75.

[5] Yang C Y, Lee T M, Lu Y Z, Yang C W, Lui T S, Kuo A and Huang B W 2010 The influence of plasma-spraying parameters on the characteristics of fluorapatite coatings *Journal of Medical and Biological Engineering* **30** 91-8.

[6] Lin Y B, Shen Y Z, Chen T F and Taoa J 2015 Optimization of Deposition Parameters for α-Al2O3 Coatings by Double Glow Plasma Technique *KEMIJA U INDUSTRIJI-JOURNAL OF CHEMISTS AND CHEMICAL ENGINEERS* **64** 457-66.

[7] Miller R A 1987 Current status of thermal barrier coatings-An overview, Surface Coatings and Technology **30** 1-11.

[8] Thirumalaikumarasamy D, Shanmugam K, Balasubramanian V 2012 Influences of atmospheric plasma spraying parameters on the porosity level of alumina coating on AZ31B magnesium alloy using response surface methodology *Progress in Natural Science: Materials International* **22** 468-79.

[9] Lamuta C, Girolamo G Di, Caliandro P and Pagnotta L *Influence of process parameters on the microstructural and mechanical properties of plasma sprayed nanostructured YSZ coatings;* Recent Advances in Energy, Environment and Materials Proc. of the International Conference on Energy, Environment and MaterialScience(EEMAS'14) (23-25September,2014) Saint Petersburg State Polytechnic University Saint Petersburg, Russia 58-65.

[10] Lamuta C, Girolamo G Di, Caliandro P and Pagnotta L *Influence of process parameters on the microstructural and mechanical properties of plasma sprayed nanostructured YSZ coatings;*

Recent Advances in Energy, Environment and Materials (3-5 October 2011) Thessaloniki, Greece 275-282.

[11] Racek O and Berndt C C 2007. Mechanical property variations within thermal barrier coatings *Surface and Coatings Technology* **202** 362-369.

Mechanical and wear characteristics of epoxy composites filled with industrial wastes: A comparative study

A Purohit and A Satapathy

Department of Mechanical Engineering, National Institute of Technology, Rourkela-769008, Odisha, India

E-mail: mr.abhilash.purohit@gmail.com, alok@nitrkl.ac.in

Abstract: Use of industrial wastes, such as slag and sludge particles, as filler in polymers is not very common in the field of composite research. Therefore in this paper, a comparison of mechanical characteristics of epoxy based composites filled with LD sludge, BF slag and LD slag (wastes generated in iron and steel industries) were presented. A comparative study among these composites in regard to their dry sliding wear characteristics under similar test conditions was also included. Composites with different weight proportions (0, 5, 10, 15 and 20 wt.%) of LD sludge were fabricated by solution casting technique. Mechanical properties were evaluated as per ASTM test standards and sliding wear test was performed following a design of experiment approach based on Taguchi's orthogonal array. The test results for epoxy-LD sludge composites were compared with those of epoxy-BF slag and epoxy-LD slag composites reported by previous investigators. The comparison reveals that epoxy filled with LD sludge exhibits superior mechanical and wear characteristics among the three types of composites considered in this study.

Keywords: Epoxy composites; Characterization; Sliding wear; Industrial wastes

1. Introduction

Significant quantities of slag and sludge are generated as byproduct or waste material every day from steel industries. During the production of steel, 2–4 tonnes of wastes are being generated per tonne of steel produced. They generally have significant amounts of valuable metals and materials. It is normally possible to regain some values by chemical or physical processing techniques, such as magnetic separation, crushing, flotation, grinding, leaching or roasting [1]. Transforming these solid wastes from one form to another, to be reused either by the same production unit or by different industrial installation, is very much essential not only for conserving metals and mineral resources, but also for protecting the environment.

The different solid wastes in the form of sludge and slags that are generated in huge quantity from steel plants are Linz-Donawitz sludge (LDS), blast furnace slag (BF slag) and Linz-Donawitz slag (LD slag). The composition of these solid wastes differ extensively depending on the source of generation but generally consists of some reusable materials, such as carbon, iron, calcium, phosphorous, sulfur, etc., which can be recovered and recycled in a judicious manner. Various efforts have been made on the utilization of these steel industry wastes in the past. The BF slag is generated in a blast furnace, where combustion is intensified by a blast of air, especially a furnace for smelting iron by blowing air through a hot mixture of ore, coke and flux. It contains mainly inorganic constituents such as SiO_2, CaO, Fe_2O_3, MgO, Al_2O_3 etc. The fine solid particles recovered after wet cleaning of the gas emerging from LD converters in the sludge form are termed as LD sludge. It mainly contains FeO, Fe_2O_3 and CaO along with small quantities of MgO, SiO_2, Al_2O_3, P and MnO. Huge quantities of slags from steel plants are produced through LD converter or basic oxygen furnace route of steel making. The main components present in LD slag are CaO, Fe, and SiO_2.

As an advanced engineering material, polymer composites are used in many low load bearing applications, where high wear resistance is required. The wear resistance of neat epoxy can generally be enhanced by introducing a secondary phase(s) into the matrix material [2–6]. There have been various reports of applications of polymers and their composites in dry sliding wear situations in the literature [7-8]. Although available references suggest a large number of materials being used as fillers in polymers [9], utilization of solid industrial wastes for this purpose has been rare. Only very few studies on the use of industrial wastes like red mud, fly ash etc. in polymer composites have been reported so far [10-15]. Due to the presence of hard phases, the solid wastes generated from steel industry can also be used as filler material in polymer matrix to enhance hardness and tribological properties of the composite [16-18]. In view of this, in the present work a comparative study is presented for dry sliding wear response of epoxy-based composites filled with LDS, BF slag and LD slag under similar test conditions.

2. Materials and Methods

2.1 Composite fabrication

In the present work, epoxy LY-556 which belongs to the epoxide family was used as the matrix material. Epoxy LY-556 is a low-temperature curing resin, and it was mixed with hardener HY-951 in the ratio of 10:1. LD sludge collected from Rourkela Steel Plant (Rourkela, India), was used as the filler material in this investigation. These sludge particles were sieved to obtain an average particle size in the range of 70–100 μm. This particulate filler was thoroughly mixed with the epoxy resin (LY556) in different weight proportions (0, 5, 10, 15 and 20 wt. %). The dough was then slowly poured into the mould of different shapes, coated beforehand with silicone-releasing agent. These were left to cure for 24 h after which the moulds were broken and specimens were released. Then samples of desired shapes were cut to conduct tensile, flexural, micro-hardness and sliding wear tests.

2.2 Characterization

The tensile tests were carried out on flat specimens as per ASTM D 638 M91. The dimensions of the specimens used to carry out the test were 150 mm × 20 mm × 3 mm and a uniaxial load was applied through both ends of the specimen. In the present research work, tensile and flexural tests were carried out in the universal testing machine Instron 1195 at a crosshead speed of 10 mm/min and the results were used to calculate the strength of the composite. The tests were repeated five times and the mean value was reported as the tensile and flexural strength of that specimen respectively. A Vaiseshika micro-hardness tester is used to measure the micro-hardness of the composite samples. A right pyramid form of diamond indenter with a square base having an angle of 136° between the opposite faces was forced into the material under a load of 50g. The two diagonals X and Y of the indentation left on the surface of the material after removal of the load were measured and their arithmetic mean L was calculated. Vickers hardness was then calculated by using the relation:

$$H_v = 0.1889\frac{F}{L^2} \tag{1}$$

2.3 Test apparatus

Wear tests were carried out as per ASTM G 99 using a pin-on-disc type friction and wear monitoring test rig. The counter body is a disc made of hardened ground steel (hardness 72 HRC, surface roughness 0.6 μmR_a). The specimen was held stationary while the disc rotates. A series of tests were conducted with five sliding velocities of 63, 125,190, 250 and 315 cm/s each under five different normal loads of 5, 10, 15, 20 and 25 N. The material loss from the composite surface was measured using a precision electronic balance with accuracy ±0.01 mg and the specific wear rate (mm^3/N-m) is expressed on a volume loss basis as:

$$W_s = \frac{\Delta m}{\rho t} \times V_s F_n \qquad (2)$$

where, Δ m is the mass loss of the composite in the test duration (g), ρ is the density of the composite (g/mm^3), t is the test duration (s), V_s is the sliding velocity (cm/s), and F_n is the average normal load (N).

To conduct sliding wear test, four major factors sliding velocity, normal load, sliding distance and filler content have been considered, each at five levels according to Taguchi's L_{25} orthogonal array as shown in Table 1.

Table 1. Control factors and their selected levels for dry sliding wear tests.

Control factors	Levels					Units
	I	II	III	IV	V	
Sliding velocity (A)	63	125	190	250	315	cm/sec
Normal load (B)	5	10	15	20	25	N
Sliding distance (C)	500	1000	1500	2000	2500	m
LDS content (D)	0	5	10	15	20	wt. %

3. Results and discussion

3.1 Tensile, Flexural and Micro-Hardness Test Results

The tensile strength of the epoxy-LD sludge composites was evaluated and the test results for various compositions of composites are presented in figure 1. It was found out that with increase in the filler content, tensile strength of the composite gradually decreased. Figure 1 also represents the variation of tensile strength of epoxy-BF slag and epoxy-LD slag composites with filler content as reported previously by many authors [16-17]. It has been observed that out of all the three fillers, epoxy-LD sludge composites have higher strength in tension.

In the present work, the flexural strength values of the epoxy-LD sludge composites were evaluated and presented in figure 2 along with the flexural strength values of epoxy-BF slag and epoxy-LD slag composites [16-17]. A gradual decrease in flexural strength was recorded for all the three classes of composites and this decrement was found to be proportional to the filler content. From figure 2 it was also observed that flexural strength of the epoxy-LD sludge composites showed better result than the other two types of composites.

The micro-hardness values of the epoxy matrix composites filled with different wt. % of LD sludge, were obtained and compared with BF slag and LD slag [16-17] and are presented in figure 3. It was evident that with addition of filler, micro-hardness of the composite had improved. With an addition of 5 wt.% of LD sludge in the neat epoxy, the micro-hardness of the neat epoxy increased from 0.085 to 0.12 GPa. The micro-hardness value further increased to 0.181, 0.232 and 0.273 GPa with the addition of 10, 15 and 20 wt. % of LD sludge to the neat epoxy, respectively. The micro-hardness values of epoxy-LD sludge composites were found to be higher than the other two classes of composites, filled with BF slag and LD slag, at same filler content. There can be two reasons for the decline in the strength of these particulate-filled composites. One possibility is that the chemical reaction at the interface between LD sludge particles and the matrix may be too weak to transfer the tensile stress; the other is that the irregular-shaped LD sludge particles act as stress raisers in the polymer matrix.

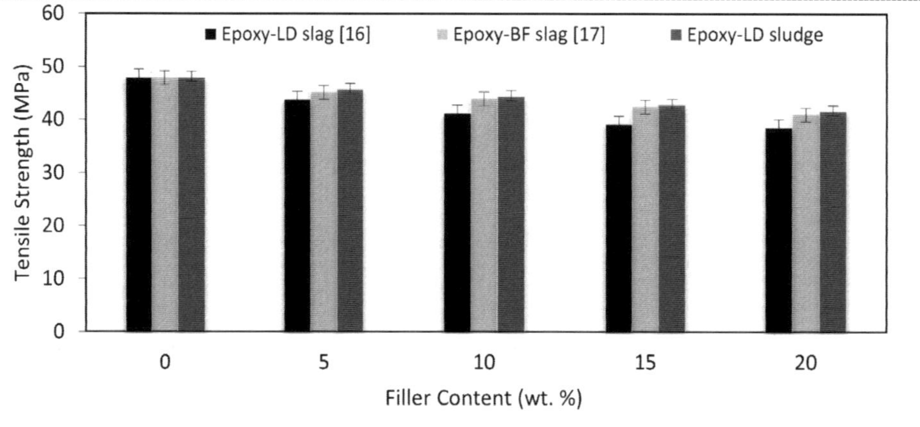

Figure 1. Comparison of tensile strength of LD sludge, BF slag and LD slag filled composites.

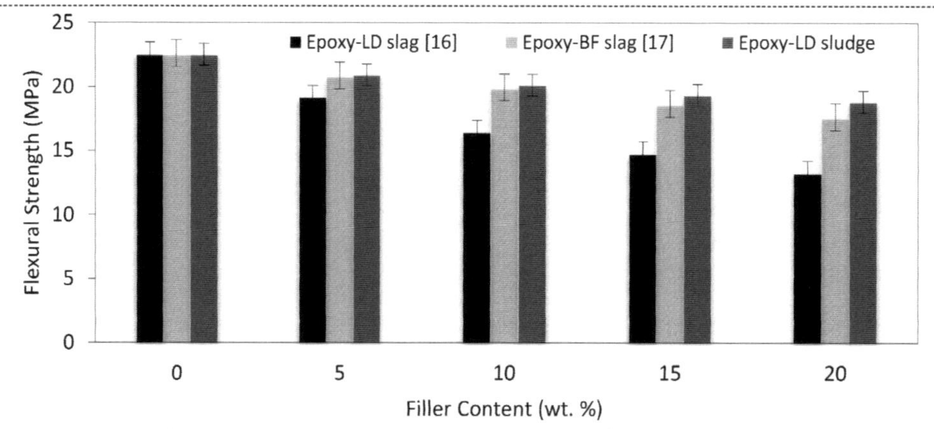

Figure 2. Comparison of flexural strength of LD sludge, BF slag and LD slag filled composites.

3.2 Sliding Wear Test Results

The specific wear rates for composites filled with LD sludge, BF slag and LD slag for all the 25 experiments are presented in table 2. It has been observed that LD sludge filled epoxy composite showed better sliding wear resistance than epoxy-BF slag and epoxy-LD slag composites. A graphical representation of variation in specific wear rate for different test runs are shown in figure 4. It is observed from the figure 4 that for a particular sliding velocity, the specific wear rate decreased with increase in filler addition, irrespective of other two control factors. Hence filler content was found to be the most affecting control factor that controls the specific wear rate of the composites. This decrease in specific wear rate is due to the presence of hard solid particles in the composite, which is also responsible for higher values hardness with increase in filler content.

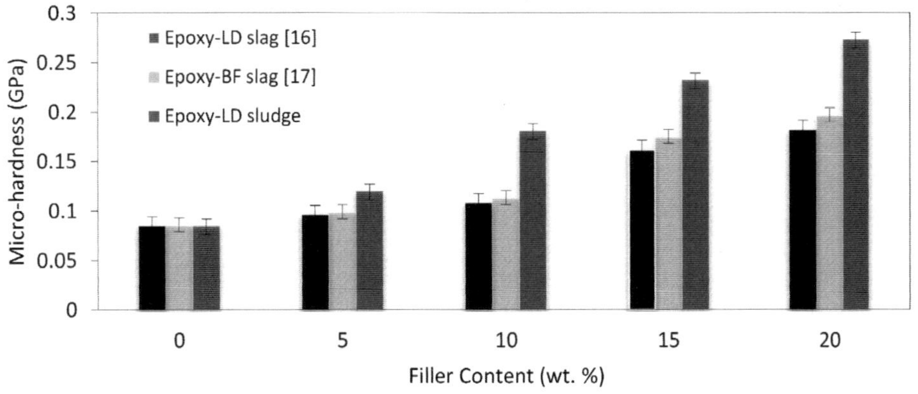

Figure 3. Comparison of micro-hardness of LD sludge, BF slag and LD slag filled composites.

Table 2. The sliding wear test results for LD sludge, BF slag and LD slag filled composites.

Test Run	Sliding Velocity A(cm/sec)	Normal Load B (N)	Sliding Distance C (m)	LDS Content D(wt.%)	Specific Wear Rate ($10^{-9}mm^3/N\text{-}m$)		
					LD sludge	BF slag	LD slag
1	63	5	500	0	1.612	1.894	2.098
2	63	10	1000	5	1.325	1.584	1.711
3	63	15	1500	10	0.873	1.609	1.147
4	63	20	2000	15	0.786	0.879	0.919
5	63	25	2500	20	0.688	0.835	0.927
6	125	5	1000	10	1.053	1.167	1.252
7	125	10	1500	15	0.982	1.195	1.293
8	125	15	2000	20	0.788	0.851	0.931
9	125	20	2500	0	1.736	2.107	2.123
10	125	25	500	5	1.442	1.617	1.866
11	190	5	1500	20	0.864	0.949	1.116
12	190	10	2000	0	1.823	2.188	2.221
13	190	15	2500	5	1.518	1.695	1.841
14	190	20	500	10	1.192	1.178	1.565
15	190	25	1000	15	1.118	1.217	1.473
16	250	5	2000	5	1.615	1.827	2.052
17	250	10	2500	10	1.316	1.591	1.559
18	250	15	500	15	1.214	1.483	1.592
19	250	20	1000	20	0.986	1.086	1.288
20	250	25	1500	0	2.012	2.256	2.607
21	315	5	2500	15	1.396	1.664	1.719
22	315	10	500	20	1.117	1.381	1.331
23	315	15	1000	0	2.226	2.514	2.843
24	315	20	1500	5	1.834	2.211	2.385
25	315	25	2500	10	1.528	1.827	1.954

Figure 4 clearly revealed that for similar test conditions, the epoxy-LD sludge composites exhibited minimum wear rate as compared to those obtained for epoxy-BF slag and epoxy-LD slag composites. This may be attributed to the relatively higher value of micro-hardness of epoxy-LD sludge composites.

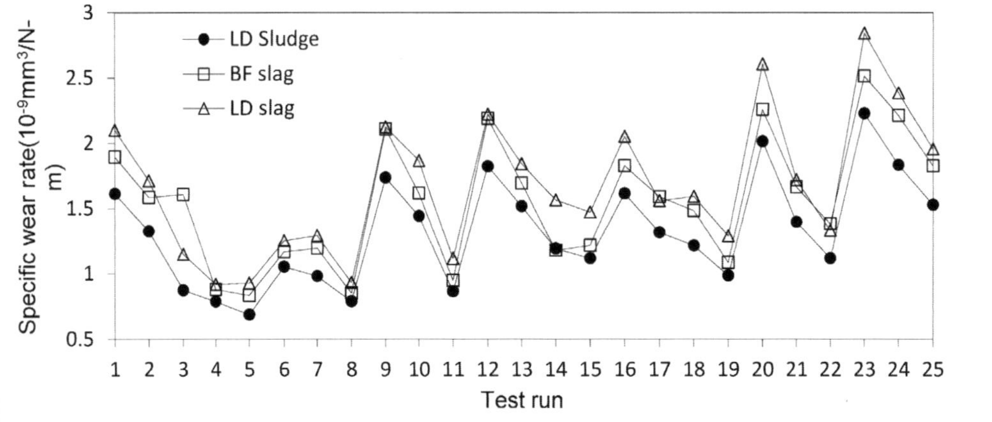

Figure 4. Comparison of specific wear rates of EP-LD sludge, EP-BF slag and EP-LD slag composites for all test runs.

4. Conclusions

i. Successful fabrication of industrial waste filled epoxy matrix composites is possible through solution casting technique.
ii. LD sludge filled composite possesses superior mechanical properties as compared to BF slag and LD slag filled composites.
iii. Dry sliding wear test result revealed that wear resistance of epoxy-LDS composite was higher than other two classes of composite and it increased with increase in filler content.
iv. The analysis also revealed that filler content is the most predominant controlling factor affecting the wear rates of the composites.

References

[1] Das B Prakash S Reddy P S R and V N Misra 2007 *Resources, Conservation and Recycling* **50** 40.
[2] Khruschov M M 1974 *Wear* **28** 69.
[3] Lee G Y, Dharan C K H and Ritchie R O 2002 *Wear* **252** 322.
[4] Harsha A P Tewari U S and Venkatraman B 2003 *Wear* **254** 693.
[5] Ahmed S and Jones F R 1990 *J. Mater. Sci.* **25** 4933.
[6] Basavarajappa S and Ellangovan S 2012 *wear* **296** 491.
[7] Carvalho O Buciumeanu M Madeira S Soares D Silva F S and Miranda G 2015 *Tribology International* **90** 148.
[8] Srinivasa K and Bhagyashekarb M S 2014 *Procedia Engineering* **97** 488.
[9] Katz H S and Mileski J V 1987 "Handbook of fillers for plastics", A Von Nostrand Reinhold Book International Journal of Plastics Technology.

[10] Patnaik A Satapathy A Mahapatra S S and Dash R R 2009 *J. Reinf. Plast. Compos.* **28** 513.
[11] Mahapatra S S 2010 *Part J: Journal of Engineering Tribology* **224** 157.
[12] Siddhartha V Patnaik A and Bhatt A D 2010 *Part J: Journal of Engineering Tribology* **224** 1103.
[13] Biswas S and Satapathy A 2010 *Waste Manage. Res* **28** 615.
[14] Biswas S and Satapathy A 2009 *Mater. Des* **30** 2841.
[15] Zhang Y Zhang A Zhen Z Lv F Chu P K and Ji J 2011 *J. Compos. Mater* **45** 2811.
[16] Pati P R and Satapathy A 2013 *International Journal on Mechanical Engineering and Robotics* **1** 2321.
[17] Padhi P K and Satapathy A 2013 *Tribology Transactions* **56** 789.
[18] Purohit A and Alok Satapathy 2016 *J. Compos. Mater* **0** 1-13 DOI: 10.1177/0021998316656478.

IOP Conf. Series: Materials Science and Engineering **178** (2017) 012020 doi:10.1088/1757-899X/178/1/012020

Fabrication of flexible piezoelectric PMN-PT based composite films for energy harvesting

Satyabati Das *, Asutya Kumar Biswal, Amritendu Roy
School of Minerals, Metallurgical and Materials Engineering, Indian Institute of Technology, Bhubaneswar, India

Email: das.satyabati@gmail.com

Abstract

Flexible piezoelectric composite films of lead magnesium niobate and lead titanate (PMN-PT) ceramic and multiwalled carbon nanotube (MW-CNT) in the matrix of polyvinyldene fluoride (PVDF) were fabricated for green energy harvesting and self-powered sensing. Compositions of PMN-PT was varied from 10-50 volume (vol. %) in PVDF matrix while a constant concentration of MW-CNT was maintained (1vol. %). Phase purity of the synthesized composites was confirm by X-ray diffraction (XRD) analysis. PMN-PT powder was found to be in single phase without the presence of any additional peak, which generally arises due to crystallization of another pyrochlore phase. Surface morphology study by scanning electron microscopy (SEM) revealed a well dispersed PMN-PT/CNT in PVDF matrix. The maximum measured output voltage and current during mechanical pressing and releasing conditions were found to be ~ 3 volt and 30 nA, respectively in 30 vol. % PMN-PT composite.

1. Introduction

Extraction and utilization of ubiquitous energy from the environment have been a challenging issue in energy harvesting technology. The most common and accessible renewable energy sources such as solar, wind, thermal and wave can provide the large scale of power. However, for driving the microelectronics devices, wireless sensor networks in healthcare and implantable biomedical devices, etc., milliwatt-scale power should be scavenge from the environment. The harvesting of one such low power energy from mechanical vibrations viz., vibrations from machines, movements in human beings, flow of water, etc., can therefore be a simple and innovative approach. The most common way to generate electrical energy from mechanical vibrations is the use of ceramic piezoelectric actuators made from lead zirconate titanate (PZT), or barium titanate (BTO). However, those monolithic piezoelectric ceramic materials are brittle and hence difficult to integrate in devices. Polymer-ceramic composite materials can combine the flexibility of polymer and the high piezoelectric coupling coefficient of piezoelectric ceramic materials and hence can avoid the breakdown and cracking of the piezoelectric material under stress. Piezoelectric semiconductor like ZnO[1], GaN[2], CdS[3], InN[4], etc., based devices are extensively studied to achieve high performance by fabricating in different processes such as dense nanowire growth, electrospinning and transfer techniques. In order to increase the efficiency and piezo sensitivity, higher piezoelectric symmetry with high mechanical compliant are indeed for energy harvesting. Recently, perovskite based $PbZr_xTi_{1-x}O_3$ (PZT)[5], $BaTiO_3$[6] and $NaNbO_3$[7] piezoelectric composites are fabricated on flexible substrates to generate high level of power density. Park *et al.*[5] have reported carbon nanotube based composites used in PZT-5H and have demonstrated its applications to energize the RGB (Red, Green, Blue) LED (Light emitting diode) arrays and the commercially available seven segment decoder.

Flexible Composites are mostly intend to study for energy harvesting and self-powered sensing due to its improvisation in energy production. Among the commonly used piezoelectric ceramic materials, the lead magnesium niobate and lead titanate ($PbMg_{1/3}Nb_{2/3}O_3 - PbTiO_3$), abbreviated as PMT-PT have drawn much attention due its highest piezoelectric coupling coefficient (d_{33}) up to 2500 pm/V which is almost 4 times higher than conventionally used PZT bulk and 30 times higher than $BaTiO_3$. It has other specific advantages such as negligible hysteresis, lower creep and lack of high voltage as compared to PZT[8]. One can also manipulate the piezoelectric properties by varying the compositional ratio of PMN and PT around the morphotropic phase boundary i.e., PMN-65% and PT-35%. Flexible piezoelectric PMN-PT nanowires based device has produced very large output voltage and current as compared to $BaTiO_3$ nanoparticle and $NaNbO_3$ nanowires based devices [9]. So by reviewing the different kind of fabrication techniques and procedures, we are focused on fabrication of flexible composite which is made up of PMN-PT ceramic particles and commercially available multiwalled carbon nanotube (MW-CNT) as organic nanofiller in a polymeric matrix such as Polyvinyldene fluoride (PVDF). MW-CNT organic filler is commonly used in polymeric nanocomposites in order to achieve good mechanical, thermal and electroactive properties because of its low mass density and large aspect ratio and hence chosen as an energy improviser in composites. The important role of the MW-CNTs in these composites is to enhance the stress transfer from the polymer to the PMN-PT particle.

2. Experimental

a) Synthesis of PMN-PT

The PMN-PT powder was synthesized by solid state route via columbite method [10], Fig.1(a). The commercially available powders MgO and Nb_2O_3 were first stoichiometrically mixed in a roller point mill (Model.No-Secor142(B)) with high purity Yttria-stabilized zirconia (YSZ) balls in Isopropanol medium for 24 h. The ball to powder ratio was 4:1. After that it was dried and calcined at 950°C for 4 h. The calcined powders were then crushed and again ball milled with high purity Yttria-stabilized zirconia (YSZ) balls in Isopropanol medium for 24 h. Then the powder was compacted using a hydraulic press at a pressure of 2 MPa and pellets of 10 mm diameter was made. The pellets were then sintered at 1250°C for 4 h to form $MgNb_2O_3$. The pellet was then crushed and mixed with TiO_2 and PbO in stoichiometric ratio for synthesis of 0.65 (PMN) and 0.35 (PT). Then the above procedure, i.e, ball milling-calcination-ball milling-sintering, was repeated to form the final PMN-PT product.

b) Synthesis of PMN-PT/CNT/PVDF composite

The composites were prepared via solution casting route, Fig.1(b). The synthesized PMN-PT powder (10, 20, 30, 40, 50 vol. %) was mixed with commercially available MW-CNT powder (1 vol.%). The mixture was stirred in ethanol and water for ~1 h by using magnetic stirrer and then dried up at 80 °C. Again it was stirred in Acetone and a surfactant (span 80) for 24 h and dried at 80 °C. The PVDF solution was made by adding calculated amount of Polyvinyldene fluoride (PVDF) powder in dimethylformamide (DMF) solvent (in the ratio of 1:10 w/w) with continuous stirring in a magnetic stirrer and heating at 85 °C. The dried mixture of PMN-PT/CNT powder was then mixed with PVDF solution by magnetic stirrer. The final PMN-PT, CNT and PVDF well mixed slurry was casted on to a Petri dish and allowed to harden at 85 °C. The dried PMN-PT/CNT/PVDF with approximate dimension of 10 cm diameter was peeled off from the Petridish. This flexi layer was taken as the final composite.

c) Characterizations

X-Ray diffraction (XRD) patterns of the synthesized ceramic powders as well as polymer composites were recorded using Bruker Advance D8 by employing CuK_α radiation. The microstructure was studied by Scanning electron microscope (Zeiss Gemini SEM) and energy dispersive X-ray (EDX) analysis. The composites were cut into 1.5 cm x 1.5 cm and were painted with silver paste on both sides that served as electrodes. The connecting wires were connected to the top and bottom electrodes and the whole assembly was sandwiched between laminate pouch and then hot rolled to measure the electrical outputs. The film was repeatedly pressed and released to measure the generated output voltage and current using Keitheley Source meter 2410.

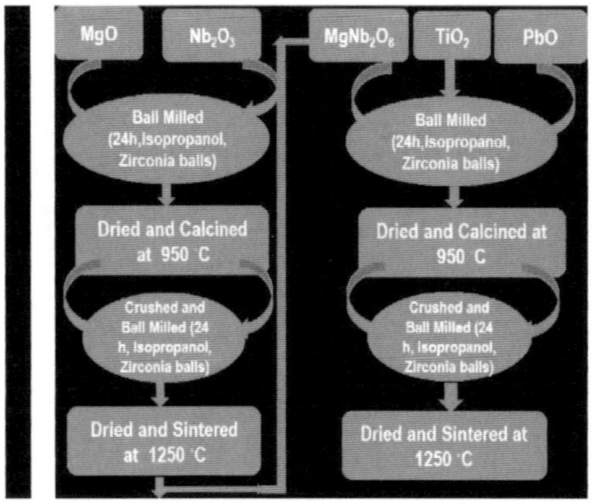

Figure 1. (a) Schematic for synthesis of PMN-PT.

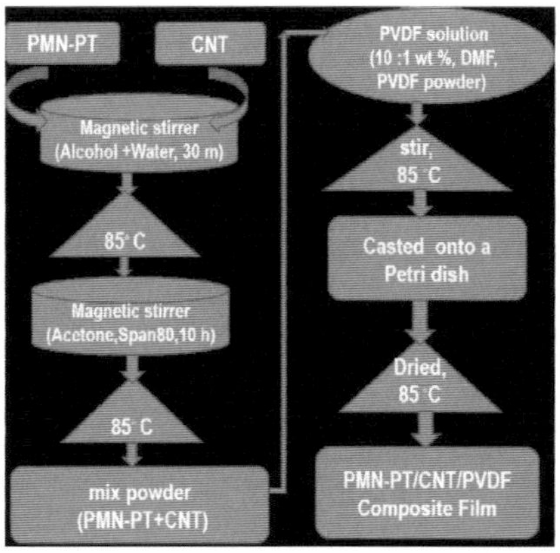

Figure 1. (b) Schematic for synthesis of PMN- PT/CNT/PVDF.

3. Results and Discussion

3.1. Structural and Microstructural Analysis

The X-Ray Diffraction pattern of synthesized PMN-PT/ CNT/PVDF composites with varying (10-50) vol. % of PMN-PT are shown in Fig. 2. The peaks marked with star correspond to PVDF and is observed in 10 vol.% PMN-PT/CNT/PVDF composite due to higher concentration of polymer. However, with the increase in vol. % of PMN-PT only PMN-PT peaks are observed and the PVDF peaks have disappeared due to the dominance of ceramic content.

The SEM micrographs of the composites as shown in Fig. 3 reveal the presence of both the polymer phase as well as the ceramic phase. The composites with 10 vol.% and 30 vol.% PMN-PT content, Fig. 3. (a) and (b), show uniform dispersion of ceramic in the polymer matrix. However, for 50 vol.% PMN-PT, Fig. 3 (c), particle-particle agglomeration takes place due to higher volume fraction of ceramics. Fig. 3(d) shows the crossectional view of 30 vol.% PMN-PT composite from which the average thickness is found to be ~70 μm.

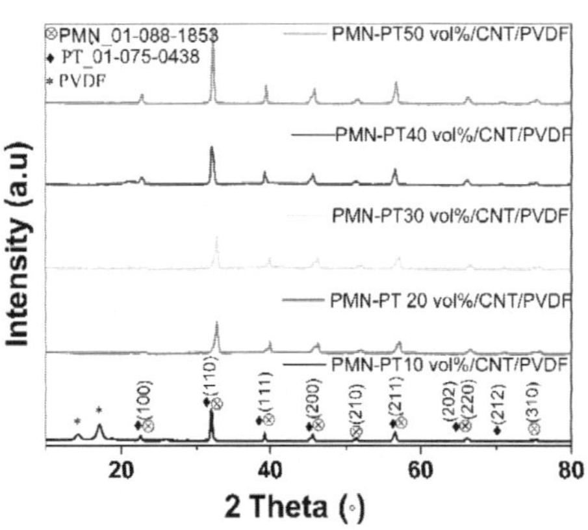

Figure 2. XRD patterns of PMN-PT/CNT/ PVDF composites with varying vol.% of PMN-PT.

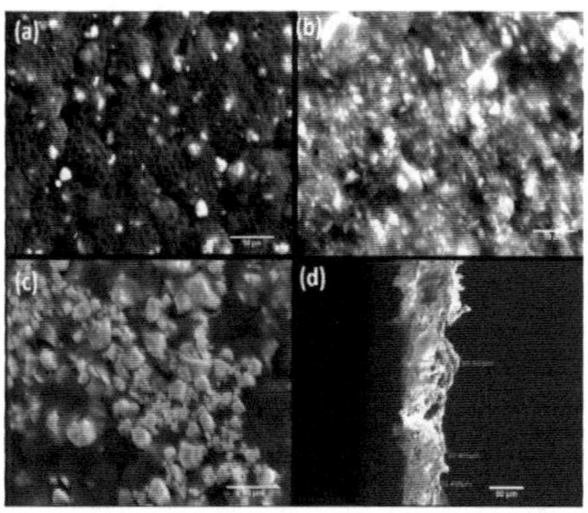

Figure 3. SEM images of (a) 10 vol.% (b) 30 vol.% (c) 50 vol.% PMN-PT /CNT /PVDF (d) crossectional view of film.

3.2. Output Voltage and Output Current

The typical voltage generation graphs, shown in Fig. 4 are obtained on a 3 cm × 3 cm sample area, by manually finger tapping which depicts the real life energy harvesting. The 10 vol.% PMN-PT content composite generates an output voltage of around 1V, Fig.4 (b). With the increase in PMN-PT content to 20 vol.%, the composite generates an output voltage of around 2 V and is shown Fig. 4 (c). The 30 vol.% PMN-PT content composite, Fig.4 (d), shows the maximum output voltage of ~ 4 V, which is sufficient to light a commercial LED. The generated output voltages decrease with further increase in PMN-PT content and is found to be in the range of 1-2 V for 40 and 50 vol.% PMN-PT, Fig.4 e-f. This may be due to agglomeration, as revealed from the SEM micrographs, of ceramics in the polymer composite.

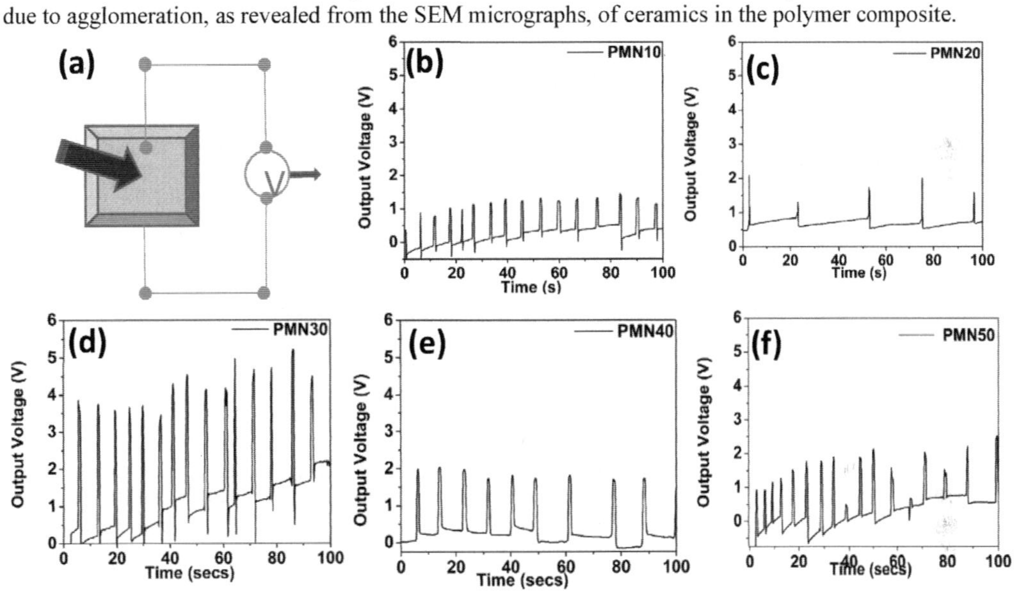

Figure 4. (a) Schematic representation of tapping or applying pressure on sample leading to output voltage generation. Output voltage signal under periodic tapping generated in (b) 10 vol.% (c) 20 vol.% (d) 30 vol.% (e) 40 vol.% (f) 50 vol.% PMN-PT composites.

The output current of all the synthesized composites are shown in Fig.5 which demonstrates an increases in output current with the increase in ceramic content upto 30 vol.% PMN-PT and then decreases with further addition of ceramics. The 10 vol.% PMN-PT composite generates an output current around 5 nA. The slight increase in magnitude of current is observed in 20 vol.% PMN-PT composite with a maximum magnitude of ~10 nA. However, an improvement in output current is clearly seen in 30 vol.% PMN-PT composite, Fig.4(d), with a maximum value of ~30 nA. With further increase in reinforcement of ceramic particles to 40 and 50 vol.% PMN-PT in composites, the maximum output current shows a decrement and the magnitudes are found to be around 15-20 nA as shown in Fig. 5.e-f. This is ascribed to the fact that agglomeration of ceramic particles, as revealed from SEM micrograph in Fig. 3(c), might have contributed to poor distribution of ceramic particles in the polymer matrix.

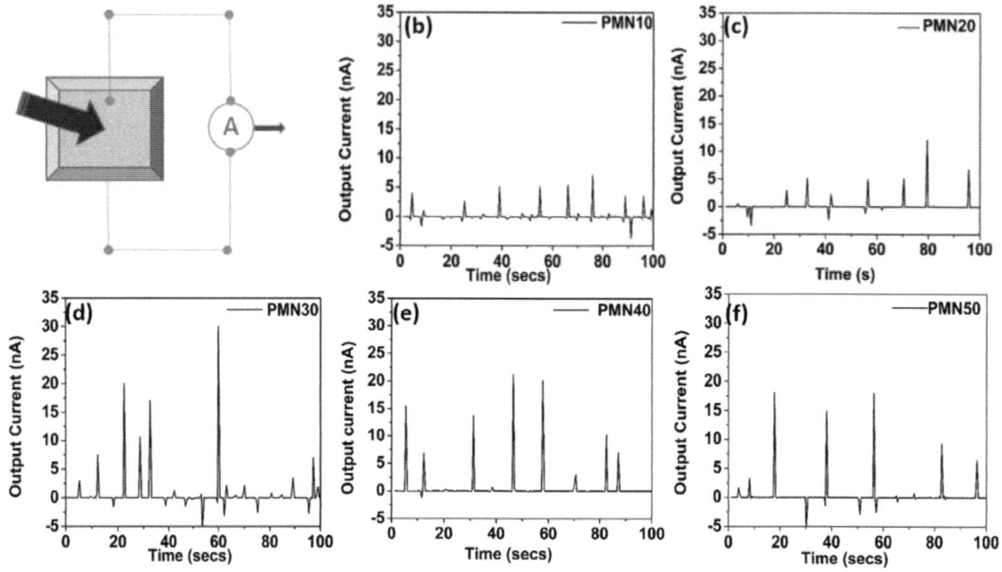

Figure 5. (a) Schematic representation of tapping or applying pressure on sample leading to output voltage generation. Output voltage signal under periodic tapping generated in (b) 10 vol.% (c) 20 vol.% (d) 30 vol.% (e) 40 vol.% (f) 50 vol.% PMN-PT composites.

4. Conclusions

A pyrochlore free PMN-PT and CNT based piezoelectric flexible composite films for energy harvesting has been fabricated. The investigation of the output voltage and current reveals that the 30 vol.% PMN-PT/CNT/PVDF composite has the highest output voltage and current with a maximum magnitude ~4 V and 30 nA, respectively. This output voltage and current is comparable with other reported flexible piezoelectric composites. Hence, this flexible composite fabricated with easy technique can be a promising candidate for energy harvesting.

Acknowledgement

This work is supported by Science and Engineering Research Board (SERB) under National Post Doc Fellowship scheme with project code SERB/PDF/2015/000721.

References

[1] Wang Z L and Song J 2006 *Science* **312** 242-246.

[2] Huang C-T, Song J, Lee W-F, Ding Y, Gao Z, Hao Y, Chen L-J and Wang Z L, 2010 *Journal of the American Chemical Society* **132** 4766- 4771.

[3] Zhou Y S, Wang K, Han W, Rai S C, Zhang Y, Ding Y, Pan C, Zhang F, Zhou W and Wang Z L, 2012 *ACS Nano* **6** 6478-6482.

[4] Huang C-T, Song J, Tsai C-M, Lee W-F, Lien D-H, Gao Z, Hao Y, Chen L-J and Wang Z L, 2010 *Advanced Materials* **22** 4008-4013.

[5] Park K-I, Jeong C K, Ryu J, Hwang G-T and Lee K J, 2013 *Advanced Energy Materials* **3** 1539-1544.

[6] Park K-I, Lee M, Liu Y, Moon S, Hwang G-T, Zhu G, Kim J E, Kim S O, Kim D K, Wang Z L and Lee K J, 2012 *Advanced Materials* **24** 2999-3004.

[7] Jung J H, Lee M, Hong J-I, Ding Y, Chen C-Y, Chou L-J and Wang Z L, 2011 *ACS Nano* **5** 10041-10046.

[8] Ye Z G, Noheda B, Dong M, Cox D and Shirane G, 2001 *Physical Review B* **64** 184114-184118.
[9] Xu S, Yeh Y-w, Poirier G, McAlpine M C, Register R A and Yao N, 2013 *Nano Letters* **13** 2393-2398.
[10] Augustine P and Ramachandra Rao M S, 2015 *Ceramics International* **41** 11984-11991.

IOP Conf. Series: Materials Science and Engineering **178** (2017) 012021 doi:10.1088/1757-899X/178/1/012021

Synthesis of copper doped Zinc oxide nanowires with enhanced ultraviolet photoresponse behavior

S P Ghosh[1], K C Das[1], N Tripathy[1], A Moharana[1], A Adhikari[1], G Bose[2], D H Kim[3], T I Lee[4], J M Myoung[3], J P Kar[1]*
[1]Department of Physics and Astronomy, National Institute of Technology, Rourkela, India 769008
[2]FST, IFHE University, Hyderabad, India 501203
[3]Department of Materials Science and Engineering, Yonsei University, Seoul, Republic of Korea, 120749
[4]Department of Bio-Nanotechnology, Gachon University, Gyeonggi-do, Republic of Korea, 461701

E-mail: karjp@nitrkl.ac.in

Abstract: This work mainly focused on the synthesis of copper doped zinc oxide nanowires by hydrothermal method, their characterization and their ultraviolet (UV) light sensing response at room temperature. The structural and morphological characterization was carried out using X-ray diffraction and scanning electron microscopy. Enhanced ultraviolet (UV) photodetection property has been achieved for a dopant concentration of 2.5wt% and 5wt% showing UV photosensitivity values (photo-to-dark current ratio) of around 800 and 1800, respectively. This is an order of magnitude higher than the undoped ZnO nanowires. High photocurrent gain is attributed to the trapping of carriers under dark and detrapping them under UV illumination by Cu-related defects. Enhancement in photosensitivity indicates that Cu-doped ZnO nanowires are promising candidates for UV photodetection applications.

Keywords: Copper; Doped; Nanowire; Ultraviolet; Photodetection; Zinc Oxide

1. Introduction

Low dimensional ZnO nanostructures such as nanowires (NW) become an inevitable material for ultraviolet (UV) photodetection due its excellent physical and chemical properties such as large surface area, radiation hardness [1], superior spatial resolution [2] and high electron mobility which provides a platform for versatile applications such as high temperature flame detection, fire alarm, space research, environmental monitoring, and missile warnings, where an installation of UV detector is highly required [3-5]. However, pure ZnO materials typically exhibit a comparatively poor UV sensing performance due to the large n-type carrier concentration and fast recombination rate of photoexcited electron–hole pairs [6]. Therefore, in order to succeed as an efficient photodetector material very high photocurrent gain is necessary. Till date, various types of modifications including creation of a junction, doping with metals, and fabrication of composites have been associated with nanostructured ZnO in order to enhance the photoconductive gain. Among them doping is an extensively used method to improve the electrical and optical properties of semiconductors. Various group III metals such as Al and Ga, transition metals such as Mn and Fe, and rare earth elements such as Eu and Er have been doped with ZnO nanostructures for different applications [7-16]. Conversely, most of the recent reports comprise either multi-step growth process or require a costly synthesis set-up for doping. Therefore, operationally simple and an efficient method to have enhanced UV detection property in ZnO is needed to be explored. However, the role of Cu as a luminescence activator and as a compensator of n-type materials is of substantial significance. The emission spectra of Cu-doped ZnO extend from the ultraviolet to the infrared region depending on the concentration of copper, defects in ZnO, and excitation conditions [17–19]. In this work, an attempt is being made to enhance the UV photoconductivity behaviour through fabrication of Cu-doped ZnO nanowires by a simple low cost hydrothermal method.

2. Materials and Methods

ZnO nanowires were synthesized by hydrothermal method on glass substrate with a ZnO seed layer deposited on to it by RF magnetron sputtering technique. The details of seed layer deposition are described elsewhere [22]. Here, an equimolar solution of Zinc nitrate and hexamethytetramine was taken and for copper (Cu) doping, required amounts copper nitrate have been added to the growth solution to obtain 2.5 and 5wt% of Cu doped samples. The investigations on the crystalline phase and the morphology were carried by X-ray diffraction (XRD; Rigaku/UltimaIV) with a CuKα (λ=0.154nm) radiation. The morphology of the ZnO NW has been investigated using scanning electron microscopy (SEM; JEOL JSM-6480LV). Top electrodes were deposited by thermal evaporation of aluminum. Ultraviolet photodetection properties were investigated by adopting current-voltage (I-V) and current-time (I-t) measurement using a Keithley picoammeter (model 6487). All the measurements were performed at room temperature. The nature of electrical contact between metal (aluminum) and semiconductor (ZnO) samples has been observed by taking the current-voltage measurement before studying the photodetection behavior for all the samples.

3. Results and discussion

Figure 1 shows the SEM micrographs of the undoped as well as the doped ZnO NW hydrothermally grown at 90°C for 4hrs. It can be seen that the substrate is covered with small ZnO nanowires of about 40–80nm in diameter and ~2μm in length. The overall morphology indicates uniformity in the shape and size, with no considerable change in length and diameter with various Cu doping concentration. The X-ray diffraction patterns of all the samples are shown in Figure 2, which show the presence of strong (002) peak, thereby confirms the formation of crystalline ZnO phase. The presence of copper oxide and metal copper has not been identified, which indicates that there is no phase transformation of the grown ZnO NW has been occurred.

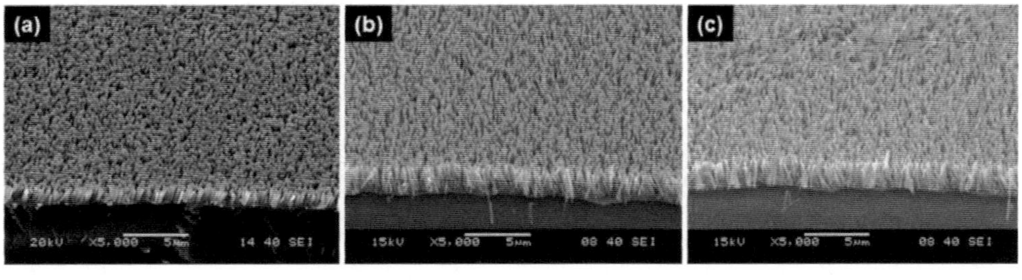

Figure 1. SEM images of ZnO nanowires grown with 0.01M at 90°C for (a) undoped ZnO NW (b) 2.5wt% Cu-doped ZnO NW (c) 5wt% Cu-doped ZnO NW. The scale bar represents 5μm.

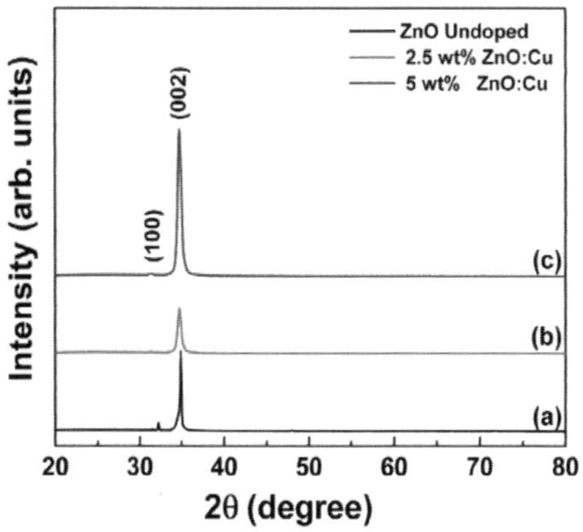

Figure 2. X-ray diffraction patterns of ZnO nanowires grown at 90°C with 0.01M concentration (a) undoped ZnO NW (b) 2.5wt% Cu-doped ZnO NW (c) 5wt% Cu-doped ZnO NW.

The I–V characteristics of the hydrothermally grown ZnO NW and the Cu-doped ZnO NW under dark conditions are shown in Figure 3(a). Linear enhancement of current with sweep in voltage of ±6V for both doped and undoped ZnO NW samples depicted good Ohmic nature of the electrical contact. Under dark condition, as-grown ZnO NW shows higher (I-V) current value of 2.0×10^{-6}A as compared to the dark current (1.05×10^{-8}A) of Cu-doped ZnO NW. This enhancement of dark current in the as-grown ZnO NW is attributed to Copper dopants which are good acceptor to trapped electrons in undoped ZnO NW. Figure 3(b), (c) & (d) show the growth and decay of the photocurrent against time. When UV illumination is 'on', an enhancement in current result and when the illumination is 'off', the current drops, indicating that the NW are highly UV sensitive. In undoped ZnO NW, the photo-to-dark current ratio is about 60 while the value increases upto 800 when ZnO is doped with 2.5wt% Cu. The value is enhanced further to ~1800 when 5wt% Cu is doped in the ZnO NW. From experimental results, the decay time constants for 5wt% Cu-doped ZnO are estimated to be around 4sec by fitting the decay curve using the exponential function. Thus, a considerable photocurrent gain and faster relaxation time is observed in Cu-doped ZnO NW. The UV light treatment generates electron–hole pairs, which are trapped by the surface states and thus the number of electrons in the conduction band of ZnO increases, which causes a sharp increase in the photocurrent. In addition, the photocurrent was not well saturated, which may be due to the presence of surface states in the aqueous grown nanowires. Furthermore, three repeated cycles are displayed in Figure 3(c), steady and reiterated photocurrent behavior indicates no rapid change during the UV exposure with time. Thereby, the hydrothermally grown Cu doped ZnO NW significantly improves the UV sensitivity behavior.

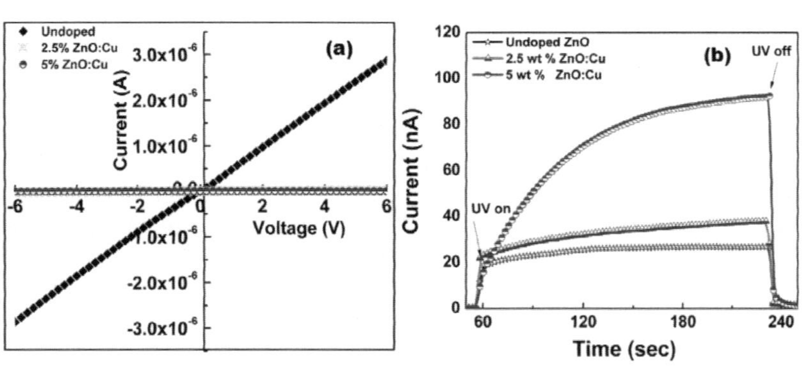

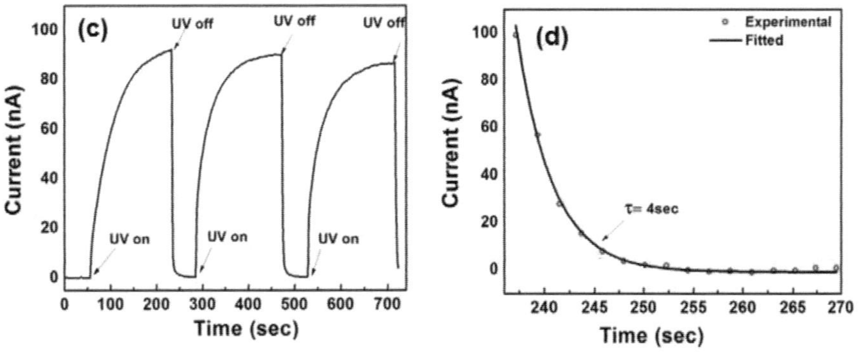

Figure 3.(a) I–V characteristics curves of ZnO NW and Cu-doped ZnO NW array film under dark conditions. (b) The growth and decay of the photocurrent of ZnO NW and Cu-doped ZnO NW under UV illumination.(c) Multi-pulse observation of 5wt% Cu-doped ZnO and (d) Fitted decay curve of the 5wt% Cu-doped ZnO NW under UV illumination.

4. UV detection mechanism of copper doped ZnO nanowires

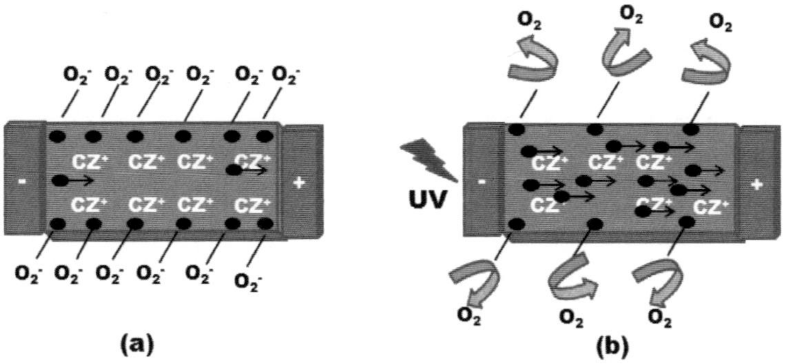

Figure 4. Schematic model to illustrate the mechanism of the hydrothermally grown Cu-doped ZnO NW based UV photodetector (a) During dark condition (b) Under UV illumination

Figure 4 schematic diagram of UV detection mechanism of Cu-doped ZnO nanowires. In the absence of UV light, adsorption of oxygen on ZnO NW surface results in the formation of depletion layer by capturing free conduction electrons. Under UV illumination, photogenerated holes recombine with the adsorbed oxygen molecules near to the surface of ZnO NW. Subsequently, oxygen molecules are desorbed from the surface of ZnO NW leads to the enhancement in photocurrent by generating more number of free carrier concentration. During dark condition Cu^{2+} ions acts as deep acceptor trap centre and interacts with the ZnO native donors to form $[Cu_{Zn} + Zn_i]^0$ complex [20-21]. Thus, the dark current is further reduced in the Cu-doped ZnO NW array, as supported from the I–V curves in Figure 3(a). However, when the ZnO NW are exposed to UV light, ionisation of Cu-Zn complex results in great enhancement in photocurrent by releasing the electrons back to the conduction band. Thus, due to copper doping the trapping and de-trapping rate of electrons increases as compared to the adsorption and desorption of oxygen molecules which results in faster relaxation time of hydrothermally grown copper doped ZnO NW.

5. Conclusions

In this work, Cu-doped ZnO NW was hydrothermally grown in order to improve the UV sensitivity response. UV photodetection behavior has been obtained by adopting current-voltage (I-V) and current-time (I-t) measurement. It has been observed that addition of Cu dopant to the hydrothermally grown ZnO NW, results in higher UV sensitivity and faster relaxation time due to faster trapping and detrapping of electrons by Cu dopants in ZnO NW. From experimental results, faster decay time constants for 5wt% Cu-doped ZnO is estimated by fitting the decay curve using the exponential function and found to be around 4sec. Through the introduction of Cu dopants into ZnO NW, a simple and efficient way is proposed to improve the UV sensing properties of ZnO NW which can be deliver a platform in order to achieve room temperature operated next generation of metal oxide UV detectors and sensors.

Acknowledgement

This work was supported by the Department of Science and Technology (DST), India sponsored Indo-Korea project (INT/Korea/P-16/2013). This research was also supported by the International Research & Development Program of the National Research Foundation of Korea (NRF) funded by the Ministry of Science, ICT & Future Planning (Grant number: 2012K1A3A1A19038371).

References

[1] Liu A, (2008) *Biosensors and Bioelectronics* **24** 167-177.
[2] Chen Y J, Nie L, Xue X Y, Wang Y G and Wang T H (2006) *Appl. Phys. Lett.* **88** 083105.
[3] Lin W Z, (2004) *J. Phys. Condens. Matter* **16** R829.
[4] Ramirez A, Subramanian M, Gardel M, Blumberg G, Li D, Vogt T (2000) *Solid State Commun.* **115** 217-220.
[5] Schreiber P, Dang T, Smith G, Pickenpaugh T, Gehred P and Litton C (1999) *Proc. SPIE* **3629** 230-48.
[6] Goldberg Y A (1999) *Semicond. Sci. Technol.* **14** R41.
[7] Schuler T and Aegerter M A (1999) *Thin Solid Films* **125** 351.
[8] Yamamoto T and Yoshida H K (2001) *Physica B* **155** 302/303.
[9] Han J, Senos A M R and Mantas P Q (2002) *Mater. Chem. Phys.* **75** 117.
[10] Han S J, Song J W, Yang C H, Park S H, Park J H, Jeong Y H and Rhie K W (2002) *Appl. Phys. Lett.* **81** 4212.
[11] Park Y K, Han J I, Kwak M G, Yang H, Ju S H and Cho W S (1998) *J. Lumin.* **78** 87.
[12] Zhao X, Komuro S, Isshiki H, Aoyagi Y and Sugano T (2000) *J. Lumin.* **1254** 87-89.
[13] Kryshtab T G, Khomchenko V S, Papusha V P, Mazin M O and Tzykunov Y A (2002) *Thin Solid Films* **76** 403/404.
[14] Kutty T R N and Raghu N (1989) *Appl. Phys. Lett.* **54** 1796.
[15] Lee J B, Lee H J, Seo S H and Park J S (2001) *Thin Solid Films* **61** 398/399.
[16] Dingle R (1969) *Phys. Rev.Lett.* **23** 679.
[17] Robbins D J, Herbert D C and Dean P J (1981) *J. Phys. C:Solid State Phys.* **14** 2859.
[18] Dahan P and Fleurov V (1998) *Phys. Rev. B* **57** 9690.
[19] Razeghi M and Rogalski A (1996) *J. Appl. Phys.* **79** 7433.
[20] Liu J, Smith RW, Mei WN (2007) *Chem. Mater.* **19** 6020-6024.
[21] Hill W, Coleman C (1980) Solid-State Electron **23** 987-993.
[22] Ghosh S P, Das K C, Tripathy N, Bose G, Kim D H, Lee T I, Myoung J M and Kar J P (2015) J. Mater. Sci.: Mater. Electron **26** 7860-7866.

Biochar, a substitute for fossil fuels

Akankshya Supriya*[1], R Samantray[1] and S C Mishra[1]
[1]National Institute of Technology, Rourkela, Odisha, India.
E-mail: akankshya.igit@gmail.com

Abstract. The present piece of research is envisaged for the possibilities and usefulness of coco peat (coir dust) which is generally not considered for usefulness. Pyrolysis of coco peat was done at two different temperatures to obtain biochars and their properties were investigated. Proximate analysis gave the fixed carbon content which gave a good value for the required purpose. Characterization techniques like Scanning Electron Microscopy, X-Ray Diffraction and Fourier Transform Infra-Red Spectroscopy were employed. It was observed that depending on the pyrolysis temperature and conditions, new substances formed, the diffraction peaks changed and also there were major changes in functional groups. The cross-links were found to disintegrate with increasing temperature. The peaks unveiled the presence of silicon carbide in the biochar pyrolysed at 1000°C. The change in the functional groups with increase in temperature manifested higher aromaticity which is an important attribute of fuels viz. petroleum, etc. Thus the present investigation can open up a new era for processing biochar from unutilized biowaste for its use as an alternative to fossil fuels.

Keywords: Coco peat (coir dust), biowaste, biochar, pyrolysis

1. Introduction

Biomass, the organic subject matter which is plentily available on earth has huge potential as feedstock for producing fuel, and generating thermal as well as electrical power. Lignocellulosic biomass is becoming the cynosure in the present era because of limited reserves of current energy resources. With the potential to append the depleting fossil fuel resources, biomass is the only renewable energy source that can be used without any energy storage systems. Although biomass primarily consists of carbohydrates comprising of carbon, hydrogen and oxygen; some biomass types may also contain a significant amount of inorganic species. The various thermal conversion or thermo-chemical processes viz. carbonization, pyrolysis, etc. may help to extenuate the growing environmental concerns, while providing greater amounts of a suitable, inexhaustible fuel and chemical resource. The thermal conversion of biomass can potentially lead to various applications leading to environmental as well as economic benefits. Pyrolysis is an example of thermo-chemical conversion which is performed in absence of air to generate solid, liquid or gaseous products. The solid black carbon or biochar obtained is usually rich in carbon and also pollution free potential solid bio-fuel. Such conversions have a shorter processing span and higher product yield. Also the pyrolysis gas can be used as a fuel in industries for combustion processes keeping in view the amount of carbon dioxide and methane it bears. The rising prices of petroleum, demand for energy, serious concerns about security of renewable sources and their supply and environmental problems are some critical reasons due to which there is much emphasis on alternative renewable energy sources **[1-5]**.

However, biomass as a fuel has certain disadvantages viz. bulkiness, high moisture content, low energy-density and high transportation cost in many cases **[6]** due to which they cannot be always used in various commercial purposes. But the biodegradability, environmental friendliness and sustainability of these organic materials have made them primary candidate in production of biofuels and bioenergy generation. Biochar which has a completely different set of properties in comparison to the respective feedstock, can bring about substantial as well as sustainable changes in securing a future supply of green energy and as well turn the bioenergy into carbon-negative industry **[7]**. Biofuels are referred to fuels which may be solid, liquid or gaseous in nature but are prevalently produced from biomass. These include energy security reasons, environmental concerns, etc. Biofuels are non-polluting and reliable, and also contain oxygen levels of 10-45%which make them absolutely different from petroleum feed stocks **[8]**. Pyrolysis produces high energy-density fuels, such as biochar. Biochars have heterogeneous char structure, contain abundant inorganic species that can be catalytically active during any thermal conversion process and also are low-rank fuels **[9]**.

In view of using biomass precursors for production of biochar fuels, several research studies have been performed. Production of biochar from rapeseed cake **[1]**, feasibility of using biomass residues like apricot stone, hazelnut shell, grapeseed and chestnut shell **[2]**, synthetic fuel production from tea waste **[5]**, and fuel properties of coffee residue biochars**[10],** have been already studied contributing some very positive results. Some other notable research studies regarding biochar and biofuels are: analysis of pyrolysis products of wheat straw **[11]**, algae **[12]** and lignocellulosic biomass viz. pinewood, perennial grass, etc. **[13]**. An interesting perspective regarding biomass is its substitution for coal and coke in ironmaking process. Biomass has favourable metallurgical properties, which can be used as heating as well as reducing agent to partially replace the raw feed materials in ironmaking process which in turn would reduce the consumption of underground carbon resources and fossil fuels **[14]**.Thus introducing new renewable fuels would be considered as a major alternative to conventional fossil fuels.

Thus the present research study emphasizes to establish the usefulness of the biomass and highlight the effects of biochar as a sustainable replacement for fossil fuels. In this study coco peat is the primary raw material which was subjected to pyrolysis to obtain the solid residue called biochar. The biochars obtained at two different temperatures were studied to reveal several unique characteristics. Therefore the present work focuses to evaluate biochar obtained from coco peat as a substitute for fossil fuels.

2. Materials and methods

2.1 Raw materials:
In the present study, the biomass feedstock used was coco peat or coir dust. It was collected from local area in Rourkela (India) and sun-dried followed by drying in a hot air oven for 2 h at 100(±5) °C in order to remove the moisture. Then it was ground using a Fritsch planetary ball mill (Pulverisette 5) with rotating speed of 300 rpm for 30 min. After a 30 min pause in the ball mill, the finely ground powder was removed and stored in an air-tight container.

2.2 Proximate analysis of raw material:
Proximate analysis was done for raw coco peat. 10 g sample was taken in a crucible and then kept in hot air oven maintained at a temperature of 110°C for 1 h to know the moisture content. Then, 10 g sample was taken in a closed crucible and was placed in a muffle furnace maintained at a temperature of 900°C for 10 min to know the content of volatile matter. Thereafter, same amount of sample was taken in an open crucible and kept in the muffle furnace at 775°C for 1 h. The crucible was taken out and the sample was stirred and again kept back inside the hot furnace at 800°C. After 10 min when the crucible was taken out, ash content was measured. So, the fixed carbon was calculated based on proximate analysis as described.

2.3 Carbonization or pyrolysis experiment:
Carbonization/pyrolytic treatment of the coco peat were done at two different temperatures in a tubular furnace (OKAY electric furnace Model 70 T7) in inert atmosphere. The inert gas used was argon and its flow was maintained at 100 mL/min. A crucible with 15-20 g of coco peat was placed inside the tubular furnace. After starting the furnace, the temperature was raised to the desired value at a heating

rate of 5°/min. Pyrolysis was done at temperatures of 400°C and 1000°C. For each of this procedure, when the desired temperature was achieved inside the furnace, isothermal holding was done for 2 h. Also the argon gas flow was maintained at 100 mL/min. After holding for 2 h, the furnace was switched off and the sample was cooled inside the furnace followed by sample removal on the next day. Fine powdery mass of coco peat biochar was obtained and kept in a zip-lock pouch to avoid moisture contamination. Including the raw sample, a total of three samples were analyzed.

2.4 Characterization of raw material and biochars:
Particle size distribution of all the three samples was done and the value of the particle at 50% in the cumulative distribution (medium value of particle size distribution/average particle size) or the D-50 value was calculated for each sample. The microstructure and elemental composition of all samples was studied using a JEOL JSM-6480LV Scanning Electron Microscope with an INCA analyzer (Penta FET Precision- Oxford Instruments) Energy dispersive X-ray spectroscopy (EDS) system. Sample after being coated with platinum was placed in a sample holder and images were captured at different magnifications. Wide angle X-Ray diffraction (WAXRD) was done using D8 Advance Bruker equipment in the range of 30 – 80° with a scan rate of 2 °/min and step size of 0.02°. Pellets were made taking sample and KBr in the ratio of 1:9 and then were subjected to IR radiations so as to determine chemical functional groups using Fourier transform infrared spectra (FTIR) using Perkin Elmer Spectrum Two.

3. Results and discussions

3.1 Proximate analysis of raw material:
After the proximate analysis of coco peat, very low percentage of moisture could be detected i.e 3%. While the volatile matter was 31%, the ash content after proximate analysis was found to be 36%. Thus the fixed carbon was 30% which showed positive potential of the biomass to be used for the purpose of alternative bio-fuel production.

3.2 Characterization and analysis of raw material and biochars:
*3.2.1. Particle size distribution.*The particle size distribution and D-50 value of all the three samples is shown in table 1.

Table 1. Particle size distribution of raw sample and biochars.

Material sample	2.16 μm	2.16 – 9.76 μm	9.76 – 30.22 μm	> 30.22 μm	D-50 value
	(Distribution in %)				(in μm)
Raw coir dust	5.21	22.57	51.31	20.91	15.54
Biochar at 400°C	3.06	28.13	68.18	0.63	12.27
Biochar at 1000°C	4.3	28.85	52.46	14.39	14.29

The major concentration of particles was in the range of 9.76 – 30.22 μm where the percentage increased for the biochar treated at 400°C which might be due to initial lump formation of the internal constituents as soon as degradation and collapsing of the lignocellulosic network begins. However the distribution of biochar at 1000°C was almost similar to that of the raw sample in the range of 9.76-30.22 μm. The D-50 value decreased for both biochar samples in comparison to the raw sample indicating reduction in the overall particle size after carbonization.

*3.2.2. Micro-structural analysis.*SEM analysis was done for raw material and both the biochar samples, and micrographs were obtained as shown in figure 1.

Figure 1. SEM micrograph of a) raw coco peat, b) biochar pyrolysed at 400°C and c) biochar pyrolysed at 1000°C.

Figure 1 a) revealed the compact and rigid raw material with cross-linked structure (highlighted in red) which defines the existence of cellulose, hemicellulose and lignin components. Also a very uniform surface was evident from the micrograph. Figure 1 b) showed broken cross-linked structure with large voids and terrains (highlighted in red) which confirmed the dehydration and fragmentation of the inherent biomass constituents. With increase in carbonization temperature, the particle distribution changed and disintegration could be seen in the microstructure. Figure 1 c) showed presence of elongated and flaky particles (the bigger circle highlighted in red) in the biochar. The smaller circle highlighted in red, depicted small whisker like bright deposits which may be attributed to the inorganic mineral deposits.

The weight percentage (wt %) of oxygen varied in the raw sample and the biochars as per the EDS data analysis. A decrease in oxygen wt% from 46.34 in raw coir dust to 42.93 in biochar at 1000°C was found indicating more carbonaceous percentage within the biochars. There was increase in the wt% of silicon in the biochars i.e. 0.02% in raw sample to 2.73% in biochar at 1000°C.

3.2.3. X-Ray diffraction analysis. The XRD diffractogram of all the samples is shown in figure 2. All the peaks for raw coco peat were listed and are shown in the plot. Only those peaks which shifted or were generated at higher temperatures have been indicated for the biochar samples (400 and 1000), rest of the peaks is same as that of coco peat.

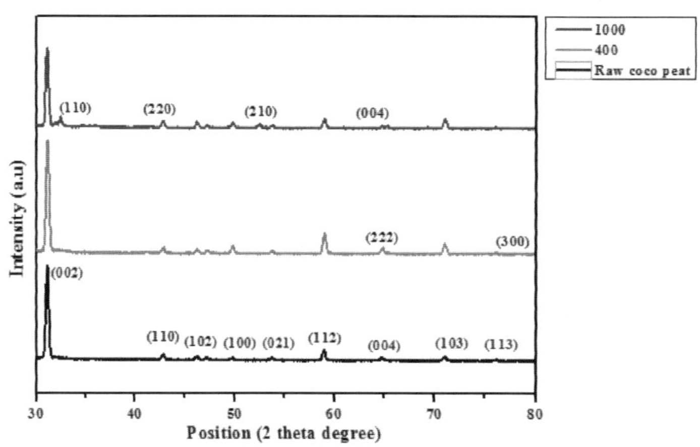

Figure 2. XRD pattern for raw coco peat and biochars pyrolysed at two different temperatures (in °C).

The intensity was highest for the peak at 2θ = 31.088° and was assigned as (002) reflection of graphite in case of all samples. The crystal structure of graphite indicated by (004) plane was hexagonal in raw coco peat. With increase in carbonization temperature, the crystal structure as well as hkl plane for carbon (graphite) was transformed. Rhombohedral structure was observed for the carbon present in biochar at 400°C as indicated by (222) in figure 2. The structure of carbon in biochar at 1000°C again changed back to hexagonal indicated by the plane (004). The presence of some inorganic species in the biochars viz. silica, silicon carbide and stable oxides of silicon could also be analyzed from the XRD data. The peak at 2θ = 52.476° in the biochar pyrolysed at 1000°C depicted the presence of silicon carbide as indicated by (210) plane.

3.2.4. FTIR analysis. FTIR pattern for raw sample and the biochars is demonstrated in figure 3. In case of the raw sample, the peak at 3416.02cm^{-1} was assigned to OH stretch from alcoholic groups. The band 2887.9 cm^{-1} was attributed to the aliphatic compounds **[15, 16]** or may have been contributed by aromatic CH_3 **[5]**.Hemicellulosic signal was observed at 1618.83 cm^{-1} **[17]** or the peak may be present due to the C=C groups **[5]**.The peak at 1384.39 cm^{-1} was assigned to C-H in-plane bending **[15]**. Bands between 1000 and 1300 cm^{-1} represent C-C groups. The band at 1078.83 cm^{-1} could also be either C-O stretch in the C-O-H bonds of glycosidic linkages **[15]**, or C-O-C asymmetric deformation of cellulose and hemicellulose **[17]** and according to a research study it may also be due the Si-O-Si stretch from the inorganic species present in the biomass feedstock **[18]**. The bands in fingerprint region were assigned as follows: 796.82 cm^{-1} - Si-O-Si bond vibrations **[18]**, 694.5 cm^{-1} - CH_2 rocking or out-of-plane ring deformation **[16]** and 460.3 cm^{-1} - Si-O-Si rocking motion **[18]**. Bands between 700 and 900 cm^{-1} may also represent the aromatic CH wagging **[5]**.

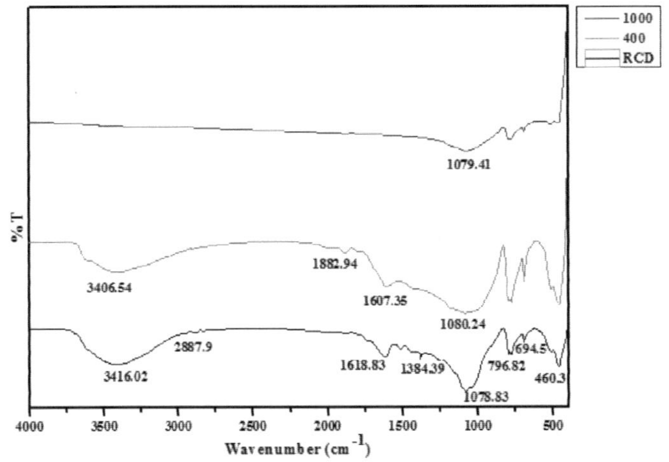

Figure 3. FTIR pattern for raw coir dust and biochar samples pyrolysed at two different temperatures (in °C).

In case of biochars the number of peaks lessened. The broad peak of OH group faded in biochar at 400°C as shown by the band at 3406.54 cm^{-1} and completely disappeared in biochar at 1000°C in comparison to raw coco peat. The ceasing of peaks at higher pyrolysis/carbonization temperature indicated loss of functionalities **[5]**. The peaks in between 1000 – 1100 cm^{-1} could be constantly observed in all three samples with changed transmittance percentage which was due to the C-C groups of the ethers, esters, carboxylic acids or anhydrides. The peaks in the fingerprint region for all samples were nearly same. It is known that as the number of low energy H-C and O-C bonds palliate and high energy C-C bonds increase, the energy density of biomass feedstock improves. With increase in the carbonization temperature, the degree of aromaticity increased. Dehydration and deoxygenation reactions led to the increase in carbon content as well as hydrophobicity. The increased hydrophobicity is an important parameter for fuel storage and handling as it has higher humidity resistance **[4]**. Thus the biochars experienced effective increase in the aromaticity which might be due to increased number of benzene-derivatives. Also the stability increased and carbon enrichment took place after the pyrolysis treatment of raw feedstock. The results were better for biochar treated at 1000 ^{o}C as all the OH groups were completely eliminated and stronger carbon-carbon bonds (C-C) existed in the residue indicating improved energy density.

4. Conclusions

The present study revealed various attributes of the raw biomass feedstock and biochar obtained from it. It was known from the study that the particle size of raw coco peat decreases when pyrolysed at higher temperatures. The morphology, crystal structure and phases within the raw material and biochars were evaluated along with the functional groups which varied in all the three samples. The increase in carbon content of the biomass feedstock with increased pyrolysis temperature made the biochar carbon-rich. Moreover, the biochar treated at 1000 ^{o}C showed effective results in comparison to the biochar treated at 400 ^{o}C owing to complete elimination of moisture and OH groups which can be helpful in generation of a large amount of heat. Also these biochars have a broad spectrum of concentration carrying numerous derivatives of benzene and other hydrocarbons making them rich in carbon. The increased aromaticity leads to better stability which is an important attribute in producing fuels.The analysis results obtained after characterizing the raw sample and biochars showed that biochars have the potential to replace the conventional fossil fuels owing to their high fixed carbon content and physico-chemical properties. In addition, biochars offer extremely high surface area thus providing sites where molecules of other substances can be held and thereby can be used as adsorbent materials too. Biochars may also help in mitigating some negative environmental effects by increasing the CO_2 uptake in the form of glomalin or by addressing the climate changes which in-turn may lead to economic as well as environmental benefits. Thus the unique characteristics of biochar in carbon sequestration as well as their biodegradability make them vital in this field. Thus the usage of biochars in various applications will prove absolutely beneficial in the near future.

References:

[1] Özçimen D and Karaosmanoğlu F 2004 Production and characterization of bio-oil and biochar from rapeseed cake *Renewable Energy* **29** 779-87.

[2] Özçimen D and Ersoy-Meriçboyu A 2010 Characterization of biochar and bio-oil samples obtained from carbonization of various biomass materials *Renewable Energy* **35** 1319-24.

[3] Winsley P 2007 Biochar and bioenergy production for climate change mitigation *New Zealand Science Review* **64** 5-10.

[4] Liu Z, Quek A, Hoekman SK and Balasubramanian R 2013 Production of solid biochar fuel from waste biomass by hydrothermal carbonization *Fuel* **103** 943-49.

[5] Uzun BB, Apaydin-Varol E, Ateş F, Özbay N and Pütün AE 2010 Synthetic fuel production from tea waste: characterisation of bio-oil and bio-char *Fuel* **89** 176-84.

[6] Abdullah H and Wu H 2009 Biochar as a fuel: 1. Properties and grindability of biochars produced from the pyrolysis of mallee wood under slow-heating conditions *Energy & Fuels* **23** 4174-81.

[7] Kwapinski W, Byrne CM, Kryachko E, Wolfram P, Adley C, Leahy JJ, Novotny EH and Hayes MH 2010 Biochar from biomass and waste *Waste and Biomass Valorization* **1** 177-89.

[8] Demirbas A 2008 Biofuels sources, biofuel policy, biofuel economy and global biofuel projections *Energy conversion and management* **49** 2106-16.

[9] Yip K, Xu M, Li CZ, Jiang SP and Wu H 2010 Biochar as a fuel: 3. Mechanistic understanding on biochar thermal annealing at mild temperatures and its effect on biochar reactivity *Energy & Fuels* **25** 406-14.

[10] Tsai WT, Liu SC and Hsieh CH 2012 Preparation and fuel properties of biochars from the pyrolysis of exhausted coffee residue *Journal of Analytical and Applied Pyrolysis* **93** 63-7.

[11] Mahinpey N, Murugan P, Mani T and Raina R 2009 Analysis of bio-oil, biogas, and biochar from pressurized pyrolysis of wheat straw using a tubular reactor *Energy & Fuels* **23** 2736-42.

[12] Chaiwong K, Kiatsiriroat T, Vorayos N and Thararax C 2013 Study of bio-oil and bio-char production from algae by slow pyrolysis *Biomass and bioenergy* **56** 600-6.

[13] Nanda S, Mohanty P, Pant KK, Naik S, Kozinski JA and Dalai AK 2013 Characterization of North American lignocellulosic biomass and biochars in terms of their candidacy for alternate renewable fuels *Bioenergy Research* **6** 663-77.

[14] Hu ZW, Zhang JL, Zuo HB, Tian M, Liu ZJ and Yang TJ 2011 Substitution of biomass for coal and coke in ironmaking process *Advanced Materials Research* **236** 77-82.

[15] Ballesteros LF, Cerqueira MA, Teixeira JA and Mussatto SI 2015 Characterization of polysaccharides extracted from spent coffee grounds by alkali pretreatment *Carbohydrate polymers* **127** 347-54.

[16] Namasivayam C and Kavitha D 2006 IR, XRD and SEM studies on the mechanism of adsorption of dyes and phenols by coir pith carbon from aqueous phase *Microchemical Journal* **82** 43-8.

[17] Mothé CG and de Miranda IC 2009 Characterization of sugarcane and coconut fibers by thermal analysis and FTIR *Journal of Thermal Analysis and Calorimetry* **97** 661-5.

[18] Xiao X, Chen B and Zhu L 2014 Transformation, morphology, and dissolution of silicon and carbon in rice straw-derived biochars under different pyrolytic temperatures *Environmental science & technology* **48** 3411-9.

Evidence of Magnetoelectric coupling in $Bi_{2(1-x)}Ho_{2x}Fe_4O_9$ (x=0, 0.01) multiferroic ceramics

S R Mohapatra[1], A K Singh[1*], S D Kaushik[2]

[1]Department of Physics and Astronomy, National Institute of Technology, Rourkela-769008, Odisha, India.

[2]UGC-DAE Consortium for Scientific Research Mumbai Centre, R-5 Shed, BARC, Mumbai-400085, India.

Corresponding author Email : singhanil@nitrkl.ac.in

Abstract: We report structural, magnetic, dielectric and magnetoelectric properties of orthorhombic $Bi_{2(1-x)}Ho_{2x}Fe_4O_9$ (x=0, 0.01) ceramics (space group *'Pbam'*) synthesized by conventional solid state reaction route. Rietveld refined X-ray diffraction (XRD) shows phase formation of the samples along with reduction in lattice parameters and cell volume due to lower ionic radii substitution of Ho^{3+} at Bi^{3+}-site. DC magnetization study revealed antiferromagnetic transition (T_N) of the parent (BFO = 250 K) as well as 1% Ho substituted sample (BHFO1 = 253 K) along with increase in magnetization. Opening of hysteresis was observed for BHFO1 sample implying onset of weak ferromagnetism with dominating antiferromagnetic behaviour. Moreover, temperature dependent dielectric (ε') study and its derivative (dε'/dT) for both the samples show significant anomaly around the antiferromagnetic transition (T_N) thus, implying a plausible magnetoelectric coupling between the magnetic as well as the electric order parameters in the studied samples. Magnetoelectric coupling was also confirmed from temperature and magnetic field (1.3 T) dependant tan loss plots.

Keywords: Rietveld refinement, Dielectric, Magnetization, Magnetodielectric.

1. Introduction

The quest for developing near room temperature magnetoelectric materials draws a considerable attention in the research community because of its rich technological applicability in developing pioneering devices [1-4]. These magnetoelectric materials exhibits magnetoelectric (ME) coupling usually defined as the coupling between the (ferro) magnetic and electrical ordering [4]. However, near room temperature ME materials are very rare in nature due to the incompatibility between these order parameters [3].The compound of our interest is $Bi_2Fe_4O_9$ (BFO), a well-known near room temperature ME material has gained substantial interest in the past few years owing to its rich potential applications [5-7]. Ressouche *et al.* [8] has reported that a unit cell of BFO contains tetrahedrally and octahedrally coordinated Fe^{3+} ions where the competing exchange interaction among various Fe^{3+} ions generates rarely observed peculiar pentagon spin frustration. Also, BFO shows antiferromagnetic (AFM) transition at T_N ~260 K and displays ferroelectric hysteresis loop along with novel ME coupling in the vicinity of T_N. Further, urge for improving its multiferroicity and tune its ME coupling, several chemical substitution has been done at A and B-site of BFO [9].

In the present work, for the first time we introduce Ho^{3+} substituted BFO sample using solid state route. Here, we report the effect of Ho^{3+} (1.01 Å) substitution at Bi^{3+} (1.17 Å) site of BFO in context to its crystal structure, magnetic and dielectric properties. Moreover, enhanced magnetization along with evidence of possible ME coupling that extends close to room temperature has been explored due to 1% Ho^{3+} substitution in BFO.

2. Materials and methods

The polycrystalline $Bi_{2(1-x)}Ho_{2x}Fe_4O_9$ (x = 0, 0.01) samples were prepared using high purity chemicals namely Bi_2O_3, Fe_2O_3 and Ho_2O_3 using solid state reaction route (SSR). The raw materials were calcined at 800 ^{0}C for 12 h followed by sintering at 850 ^{0}C for 10 h. The phase formation was confirmed by room temperature X-ray diffraction (XRD) study, RIGAKU, JAPAN using Cu-Kα radiation (λ = 1.5405 Å). Temperature dependant dielectric measurement was done using High precession impedance analyzer (Wayne Kerr 6500B) coupled with closed cycle refrigerator (Cryo-Industries of America, USA). Finally, magnetization measurement was carried out using 9T re-liquefier based Physical Property Measurement System (PPMS)-Vibrating Sample Magnetometer (Quantum Design).

3. Result and Discussion

Rietveld refined XRD pattern of BFO (parent) and 1% Ho doped BFO (BHFO1) is shown in figure 1. All the major diffracted peaks are found to crystallize in orthorhombic structure with space group 'Pbam'. The diffracted peaks are also consistent with the standard data (JCPDS no. 25-0090). The lattice parameters obtained using FULLPROF refinement are as follows: a = 7.9735(2)Å, b = 8.4414(4)Å, c = 6.0028(5) Å, volume = 404.033(4) Å^{3}for BFO and a = 7.9733(5)Å, b = 8.4412(2)Å, c = 6.0027(2) Å, volume = 404.007(4) Å^3 for BHFO1. However, we observe decrease in lattice parameters due to 1% Ho doping along with the shifting of (002) and (220) peaks to higher angles (not shown here). This feature can be explained on the basis of lower ionic radii Ho^{3+} (1.01 Å) substitution at the higher ionic radii Bi^{3+} ions (1.17 Å).

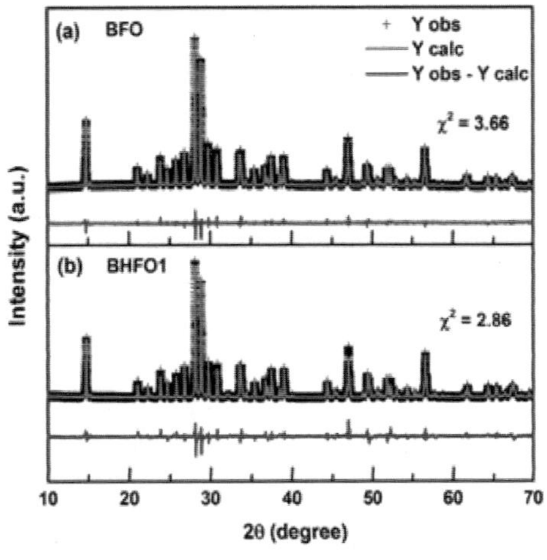

Figure 1. Room temperature Rietveld refined XRD pattern of (a) BFO and (b) BHFO1

Figure 2 depicts the temperature dependant DC magnetic susceptibility of BFO and BHFO1 samples under an external field of 1 kOe. It is observed that both the samples show no signature of irreversibility in ZFC-FCW plot clearly signifying no signature of cluster or spin glass in the samples. A broad cusp around 250 K is noticed in the BFO typically revealing the antiferromagnetic (AFM) behaviour and is similar to our previous reported data [10]. Contrastingly, a much broader shoulder is observed in case of BHFO1 with a very small hump around 253 K (inset of figure 2 (b)). Further, on studying the isothermal magnetization (M-H) plots of parent and substituted sample, the M-H plot shows contrasting behaviour as illustrated in figure 2 (c, d). The M-H plot for both the samples were carried out at 240 K, 280 K and 320 K up to a maximum field of ±20 kOe. BFO shows a linear behaviour of magnetization relating its AFM characteristics as discussed above. Instead, BHFO1 shows opening of a small hysteresis curve at low field region which later follows a linear path on increasing the field. This is a typical characteristic feature of superposition of AFM and weak ferromagnetism (FM) where the later has a substantial effect on the enhanced magnetic properties. At 320 K, the values of remnant magnetization (Mr) and coercive field (Hc) are 0.0117 emu/gm and 90 Oe respectively for BHFO1, subsequently confirming onset of weak FM even at room temperature.

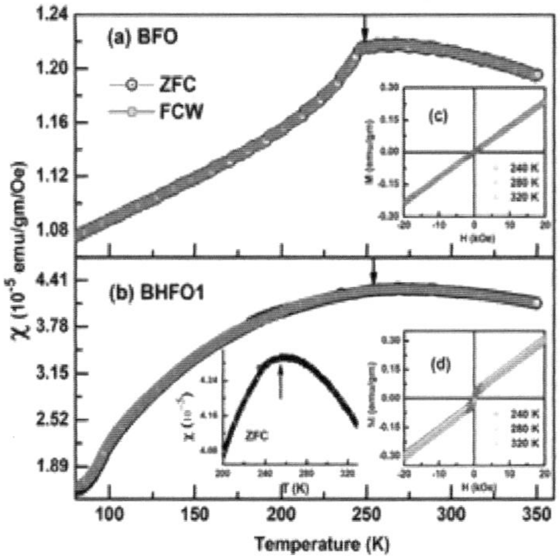

Figure 2. ZFC-FCW magnetization of (a) BFO and (b) BHFO1 at 1 kOe. Inset shows M-H plot of (c) BFO and (d) BHFO1 at selected temperatures

Now, we try to explore the existence of any coupling between magnetic and electric order parameters. This fact can be proved by observing any anomalies in the dielectric data in the vicinity of magnetic transitions (T_N), often defined as the magnetoelectric (ME) coupling [1]. Figure 3 shows the temperature dependant real part of dielectric constant (ε') of BFO and BHFO1 along with the derivative plot of the same at 12 kHz. Here, we observe a change in slope or a concomitant anomaly in ε' for both the samples which exactly matches with the peak obtained in the $d\varepsilon'/dT$ plot of the same. Interestingly, this anomaly happens to be at same temperature matching with T_N. This demonstrates a plausible coupling of both magnetic as well as (di)electric order parameters. The peak for BFO is obtained at 250 K whereas BHFO1 shows the peak at 253 K (shown in figure 3 with dotted circle). Thus, slight Ho substitution (1%)has a substantial effect of enhancing magnetization as well as magnetic transition temperature (T_N) along with a clear evidence of ME coupling.

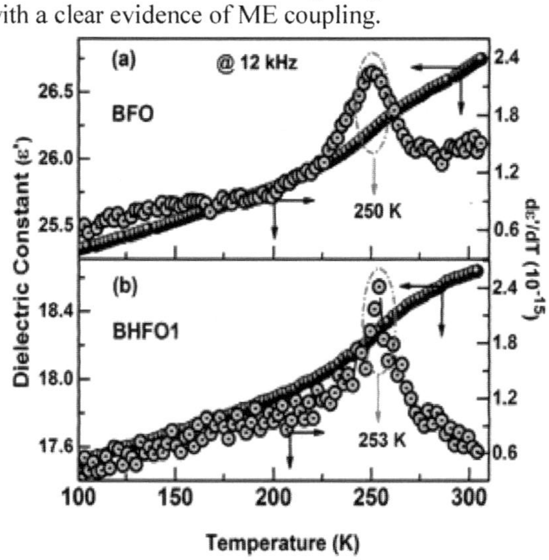

Figure 3. Temperature dependant ε' study of (a) BFO and (b) BHFO1 sample at 12 kHz and dε'/dT versus T (K) plot of (c) BFO and (d) BHFO1

Furthermore, the existence of ME coupling in the material is confirmed from the temperature (150 K-310 K) and magnetic field (1.3 T) dependant tan loss plot as shown in figure 4. An S-shape anomaly is seen in both the cases along with distinguished separation of peaks (broad shoulder) due to application of 1.3 T field which clearly reveals the evidence of strong coupling in the sample. Also, the change in slope (marked by a black arrow) due to 1.3 T field is seen near the vicinity of magnetic transition as well as the peaks obtained in dε'/dT plot. For comprehensive understanding of ME coupling or magnetodielectric (MD) effect and their variation due to substitution, one needs to probe field and temperature dependant ME/MD study, which we shall study in near future.

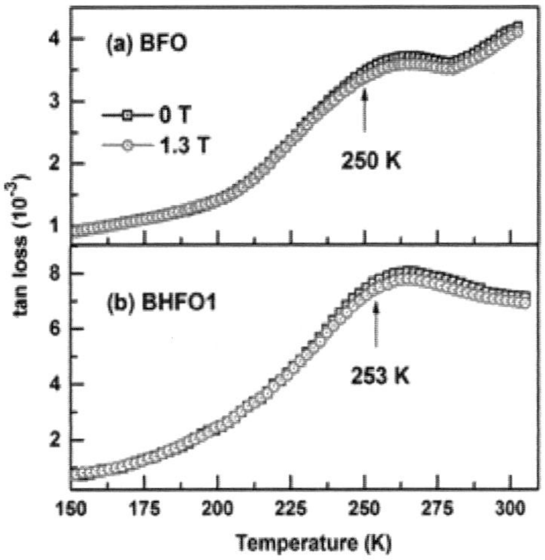

Figure 4. Variation of temperature dependant tan loss at zero external field and 1.3 T for (a) BFO and (b) BHFO1

4. Conclusion

In summary, we have synthesized polycrystalline $Bi_{2(1-x)}Ho_{2x}Fe_4O_9$ (x=0, 0.01) using SSR and subsequently investigated its crystal structure, magnetic and dielectric properties. Enhanced magnetization along with a weak FM is observed in the modified sample from M-H study. Slight substitution (1%) induced T_N to increase from 250 K (BFO) to 253 K (BHFO1). Significant anomalies in ε' and tan loss around T_N unambiguously establishes the existence of ME coupling between the electric and magnetic order parameters in this material. Future study of magnetic field dependant dielectric behavior and temperature dependent neutron diffraction will further enable us to understand the microscopic origin of magnetoelectric coupling in a much better way.

5. Acknowledgement

We acknowledge Board of Research in Nuclear Science (BRNS), Mumbai (Sanction No: 2012/37P/40/BRNS/2145), Department of Science and Technology (DST), New Delhi (Sanction No: SR/FTP/PS-187/2011) and UGC-DAE-CSR Mumbai (Sanction No: CRS-M-187,225) for funding. SRM is thankful to BRNS for financial assistance.

Reference

[1] ErensteinW, Mathur ND, Scott JF 2006 *Nature***442** 759.
[2] Khomskii D 2009*Physics***2** 20.
[3] Hill NA 2000*J. Phys. Chem. B* **104** 6694.
[4] Fiebig M 2005*J. Phys. D: Appl. Phy.***38** R123.

[5] Mohapatra SR, Sahu B, Badapanda T, Pattanaik MS, Kaushik SD, Singh AK 2016 *J Mater Sci: Mater Electron***27**3645.
[6] Singh AK, Kaushik SD, Kumar Brijesh, Mishra PK, Venimadhav A, Siruguri V, Patnaik S 2008 *Appl. Phys. Lett.***92** 132910.
[7] Shamir N, Gurewitz E 1978*ActaCrystallogr. Sect.***A34** 662.
[8] Ressouche E, Simonet V, Canals B, Gospodinov M, Skumryev V 2009*Phys. Rev. Lett.* **103**267204.
[9] Mohapatra SR,Sahu B, Chandrasekhar M, Kumar P, Kaushik SD, Rath S, Singh AK 2016 *Ceram. Int.***42** 12352.
[10] Mohapatra SR, Swain A, Yadav CS, Kaushik SD, Singh AK 2016 *RSC Adv.* **6** 112282.

Structural and electrical characteristics of solution processed P3HT-carbon nanotube composite

Prakash Chandra Mahakul[1] and Pitamber Mahanandia[1,*]

[1]Department of physics and astronomy, National Institute of Technology, Rourkela, Odisha, 769008-India

Corresponding Author Address:

Department of Physics and Astronomy, National Institute of Technology, India-769008

Corresponding author E-mail: pitam@nitrkl.ac.in

Abstract

Organic semiconductors have been identified as a fascinating class of low cost and flexible novel semiconductor materials that have the electrical and optical properties which can be easily processed. Due to their interesting physical properties, organic semiconductors have attracted tremendous research attention for next generation electronics and optoelectronics. Multiwalled carbon nanotubes (MWCNT) incorporated Poly[3-hexylthiophene-2,5-diyl] (P3HT) hybrid nano-composite film have been fabricated by solution processing technique followed by spin coating method using 1,2-dichlorobenzene as an intermediate solvent. Structural and morphological characteristics of the composite film have been studied by x-ray diffraction (XRD) and scanning electron microscope (SEM). The MWCNTs were observed to be well dispersed in the polymer matrix. Crystallites were found to be more ordered barely affecting the lamellar structure of P3HT in the nano-composite film. Structural and functional characteristics of P3HT and its hybrid nano-composite have been studied by UV-Visible (UV-Vis), Fourier transform infrared (FTIR) and Raman spectroscopic characterization. Excellent electrical properties have been observed from I-V and cyclic-voltammetric characterization of the well dispersed MWCNT in the P3HT composite. Improvement in electrical properties can be attributed to the higher carrier mobility of MWCNTs in the composites.

1. Introduction: Potentially low cost processing, simple manufacturing process, flexibility and high throughput have made organic semiconductors promising for future optoelectronic devices [1-3]. Relatively high absorption coefficient of organic semiconductors allows light to be absorbed very efficiently within a very thin layer leading to advanced optoelectronics device applications [4-6]. Poly[3-hexylthiophene-2,5-diyl] (P3HT) in combination with some n-type semiconductors have been intensively investigated during the last several years as active layer in most of the organic devices due to its easy solution processability [7-10]. However, the conductivity, stability and carrier diffusion length of spin coated P3HT have been a serious concern for their applicability in electronic devices. Carrier mobility of spin coated is comparatively low due to loss of crystallinity during evaporation of solvent [11]. Conductivity and optical properties in π-conjugated organic semiconductors mainly depends on the arrangement of chains in the polymer matrix. Incorporation of inorganic filler materials which have higher carrier mobility compared to their organic counterparts, have been widely researched for enhancing the physical properties of organic semiconductors. Incorporation of carbon based nanomaterials like multiwalled carbon nanotubes (MWCNT), graphene and fullerene [12] have been widely studied as filler material with improved physical properties [13-15].Due to excellent conductivity, high surface areas, low mass density, and large charge carrier mobility carbon nanotubes (CNTs) are being used in different optoelectronic devices as electrodes, semiconductors, transparent electrodes, gas sensors, etc. [16,17]. However, depositing an uniform thin film out of MWCNT is extremely difficult due to their chemical inertness and insolubility in most of the solvents. The improvement in the field effect mobility by incorporation of carbon based nanomaterials has already been reported [11]. P3HT based nanocomposites with better performance have been reported for solar cells, Light emitting diodes (LEDs), supercapacitor, thermoelectric devices, thin film transistors (TFTs) applications[18,19]. Only, solar cell device performance has been studied by of P3HT-single walled carbon nanotubes composite reported by Kyamkis *et al.*[18]. Photovoltaic device performance of P3HT-rGO composite has been investigated by Dingshan *et al.* [19]. In their work the used rGO was obtained during preparation of P3HT-rGO composites following many steps. Kuila *et al.* [20] have investigated photovoltaic properties of the composite using functionalized P3HT as well as MWCNT. They have focused mostly on the synthesis and device performance in which the structural properties have not been investigated in the composite. In this work, we report the structural and electrical properties of the solution processed P3HT-MWCNT composite without functionalizing either P3HT or MWCNT which is different from others [18-20].

2. Experimental: P3HT (poly 3-hexyl thiophene-2, 5-diyl) and 1, 2-dichlorobenzene were purchased from TCI chemicals, India and used as it was without further purification or functionalization.

2.1 Material synthesis: Carbon nanotubes used for our studies have been synthesized by pyrolysis of benzene using ferrocene as catalyst for CNT growth using a cylindrical reactor tube at 900^0C [21] and purified by oxidation and acid bath treatment [22].The purified CNTs were subjected for washing in distilled water many times and recovered followed by drying.

2.2 Composite preparation: For composite preparation, around 40 mg of P3HT and 0.1 mg of CNT were dispersed in 10 ml of 1, 2-dichlorobenzene to prepare the P3HT-CNT solution mixture. The solution mixture has been stirred overnight to prepare the uniform dispersion.

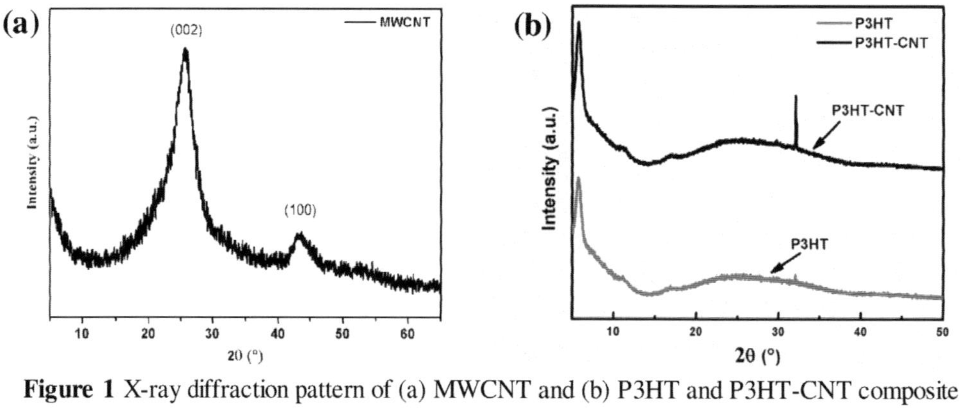

Figure 1 X-ray diffraction pattern of (a) MWCNT and (b) P3HT and P3HT-CNT composite

2.3 Fabrication of composite film: For the deposition of thin film, glass substrate was washed in detergent, rinsed in acetone and isopropanol for 10 minutes each and dried under a lamp. P3HT-CNT film have been

IOP Conf. Series: Materials Science and Engineering **178** (2017) 012024 doi:10.1088/1757-899X/178/1/012024

prepared by spin coating the precursor dispersion over ITO substrate at 600 RPM for 1 min. 40 μl of the above dispersion was spun onto the cleaned glass substrate at 600 RPM for 1 min and dried over a hot plate at 70^0C. All the chemical processing, deposition and measurements have been carried at room temperature.

3. Characterization:

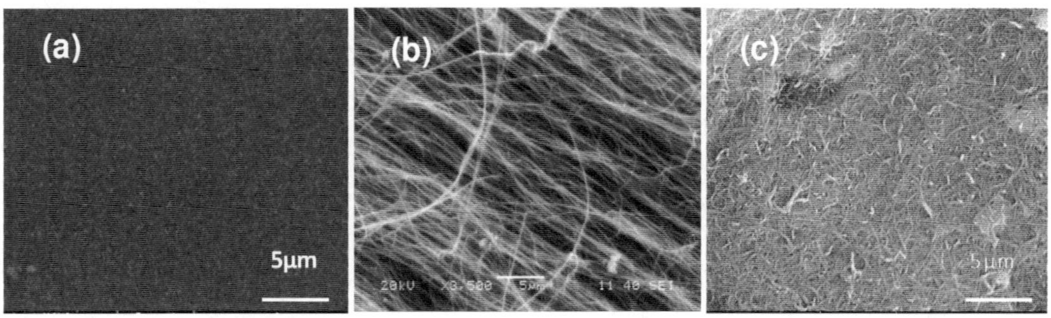

Figure 2 SEM Micrographs of (a) P3HT, (b) MWCNT and (c) P3HT-CNT composite

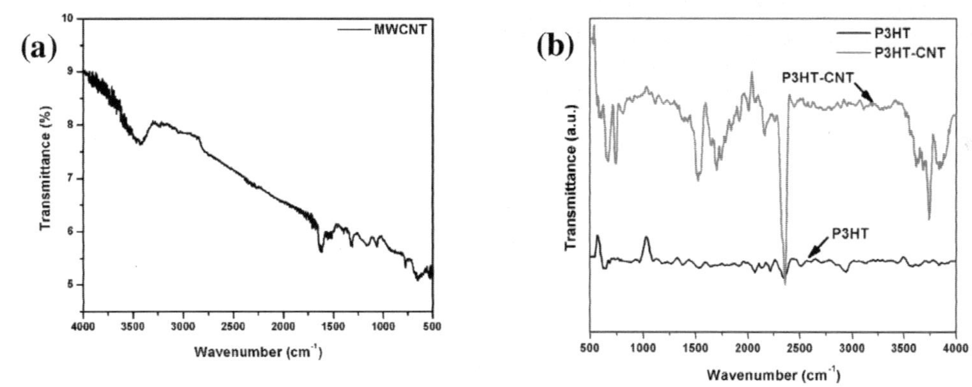

Figure 3 FTIR spectra of (a) MWCNT (b) P3HT (black line) and P3HT-CNT composite (red line)

X-ray diffraction data of MWCNT, P3HT and its MWCNT based composite recorded from a x-ray diffractometer (Ultima IV, RIGAKU) using a copper Kα-source (λ=1.54 Å) in the range of 4° to 60° are presented in Figure 1. SEM images obtained using a Nova NanoSEM 450/ FEI are presented in figure 2. Figure 2(a) shows the FESEM micrograph of P3HT film over a glass substrate. FESEM micrographs of MWCNTs and P3HT-CNT film have been shown in figures 2(b) and 2(c). FTIR spectra of the prepared materials is shown in figure 3. Raman spectra(Figure-4a) of all the samples have been collected at an excitation wavelength of 532 nm. The obtained UV-Visible spectra of P3HT and P3HT-CNT dispersed in 1,2-dichlorobenzene is shown in figure 4 b. The current-voltage (I-V) characteristics of P3HT and P3HT-CNT composite film studied by 2-probe technqiue using a keithley 2400 sourcemeter is shown in figure 5(a). Cyclo-voltagram of P3HT and P3HT-CNT has been carried out using glassy carbon as the working

IOP Conf. Series: Materials Science and Engineering **178** (2017) 012024 doi:10.1088/1757-899X/178/1/012024

electrode with tetrabutyl ammonium hexafluorophosphate (Bu_4NPF_6) in acetonitrile as the supporting electrolyte as shown in figure 5(b).

4. Discussion:

The XRD pattern of MWCNT presented in figure 1 (a) reveals a broad peak at 26.6° corresponding to (200) planes of the hexagonal lattice. P3HT and its MWCNT composite reveal that scattering peaks corresponding to 2θ value 5.18° with d-spacing 17.13Å. The peak at 23.8° arising from the higher order reflection from large d-spacing represents the stacking distance of thiophene ring in P3HT [23-27]. With increase in MWCNT content in the composites, the intensity corresponding to (100) planes gradually increases and gradual decreases in (010) planes are observed. Hence the in-plane lamellar structure of P3HT is expected to be distorted while the out of plane inter-chains ordering is improved due to the incorporation of the fillers. FTIR spectra of MWCNT records small O-H bands that correspond to the unwanted residual intercalated water. The peak at 1634 cm^{-1} due to the C=C stretching vibration reveals hexagonal structure of MWCNT. The peak about 3400 cm^{-1} in FTIR is due to the adsorbed water or OH functional group [28]. No major peak shift in observed FTIR spectra of P3HT-CNT composites,

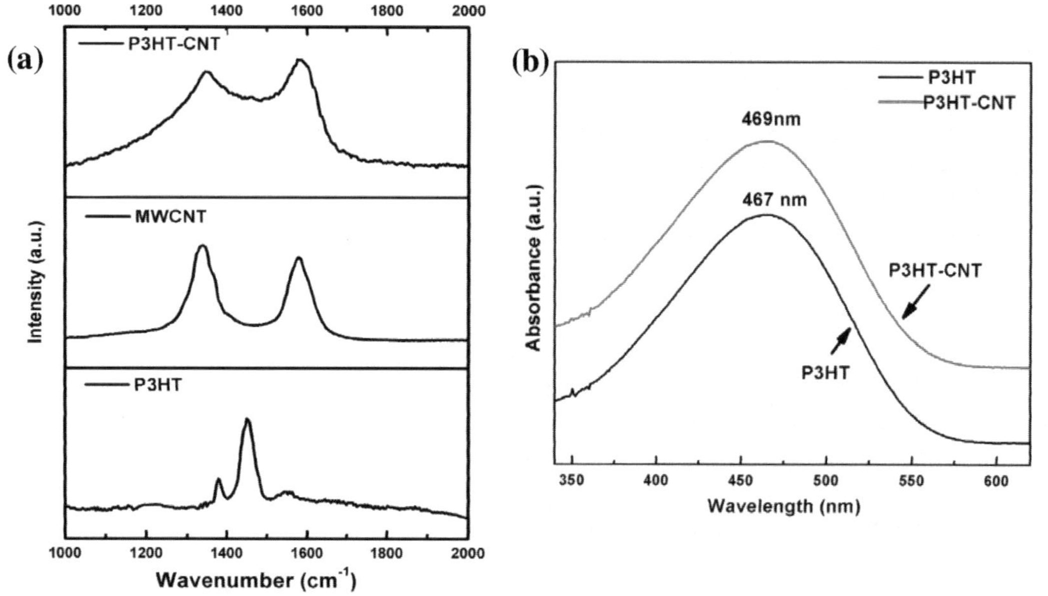

Figure 4 (a) Raman spectra of P3HT_CNT composite, MWCNT and P3HT (b) UV-Vis spectra of P3HT (black line) and P3HT-CNT composite (red line)

suggests absence of strong interaction between CNT and P3HT except π-π stacking interaction [27].Employing FTIR spectra intensity ratio conjugation length of P3HT could be evaluated and has been reported by Furukawa *et al.* [29]. The absorption bands at 2855 cm^{-1}, 2927 cm^{-1} and 2950 cm^{-1} could be ascribed to the $-CH_2$ out-of-plane mode, $-CH_2$ in-plane mode and $-CH_3$ asymmetry mode of P3HT, respectively. The peaks at 1410 cm^{-1} and 821 cm^{-1} are the symmetric ring stretching and C–H out-of-plane mode of P3HT [30]. Blue shift in the absorption band at 821 cm^{-1} in the composite suggests the presence of weaker interaction between MWCNT with P3HT in the composite. The intensity ratio between the antisymmetric C=C stretching peak (1510 cm^{-1}) and the symmetric stretching peak (1456 cm^{-1}) directly affects the conjugation length of P3HT chain, as reported earlier [31,32]. The increase in absorption peak height at 1510 cm^{-1} signifies the increase in the conjugation length of the P3HT chain with increase in MWCNT concentration in the composite [31].

The Raman spectra of P3HT, MWCNT and P3HT-CNT [Figure 4(a)] composite recorded at an excitation wavelength of 532 nm have been presented in figure 4(a). P3HT shows a strong peak at 1449 cm^{-1} and another at 1378 cm^{-1} representing the C-C skeletal deformation and C=C ring deformation of P3HT matrix [33]. The Raman active D-band and G-band of MWCNT has been recorded at 1347 cm^{-1} and 1580 cm^{-1}, respectively. Incorporation of MWCNT into the P3HT matrix has revealed widening of the D-band and G-band in the composite. From the UV-Vis spectra of P3HT and MWCNT-P3HT composites in 1,2-dichlorobenzene shown in figure 4(b), it can be observed that an absorption peak at 467 nm is observed. Ovsyannikova *et al.* [34] have reported no significant contribution of MWCNT addition to the absorption spectra of the composite. However, in the present investigation slight red shift of the band has been observed in the nano-composites upon addition of MWCNT to the P3HT suggesting the ground state interaction and appreciable charge transfer between MWCNT and the polymer. This phenomenon can be attributed to the increase in the conjugation length of the polymer chain [31, 35].The observed increase in absorbance by incorporation of MWCNTs results in decrease in the torsional deformation in P3HT and thereby increase in the polymer ordering [36].

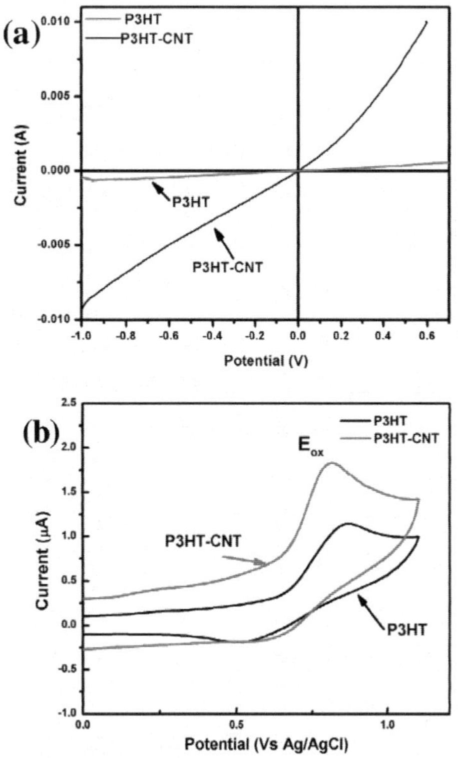

Figure 5 (a) I-V characteristics of P3HT (red line) and P3HT-CNT composite (black line) (b) cyclic-voltammetric characteristics of P3HT (red line) and P3HT-CNT composite (black line)

I-V characteristics of P3HT and its MWCNT composite [Figure 5(a)] reveals the increase in conductivity by the incorporation of the highly conductive MWCNTs. Enhanced conductivity in the nano-composite can be attributed to the conductive network created by carbon nanotubes that have high carrier mobility. The wrapping of MWCNT by P3HT in the composite eases the carrier transformation. To demonstrate the potential application of the prepared composite material, cyclic voltammetry measurement has been carried out for pristine P3HT and solution processed P3HT-CNT composite using glassy carbon electrode as working electrode, Platinum as counter electrode and Ag/AgCl as supporting electrode [Figure 5(b)]. The HOMO level of P3HT was calculated to be 5.197 and that of P3HT-CNT was found to be 5.082. Incorporation of MWCNT in the P3HT matrix decreases the HOMO level of P3HT that could be ascribed to the MWCNT network which acts like a membrane for the passage of charge carriers in the composite.

5. Conclusions:-

P3HT-CNT composite has been prepared successfully by solution processing method and improve in physical properties have been observed compared to pristine P3HT. Structural changes such as change in conjugation length of polymer chain and ordering of the stacked structure of P3HT has been observed due to incorporation of MWCNT into the P3HT matrix .Wrapping of P3HT around MWCNT decreases the agglomeration that accounts for better dispersion of fillers. Weaker π-π interaction between MWCNT and P3HT ease the flow of electrons from P3HT to MWCNT that leads to enhanced electrical properties of the composite. Smaller decrease in the HOMO level of P3HT-CNT composite compared to pristine P3HT might be useful for the improved performance of optoelectronic devices.

6. Acknowledgement: The above research work has been supported by SERB, DST, Government of India through grant Number SB/S2/CMP-109/2013.

References:

[1] Mutolo K L, Mayo E I, Rand B P, Forrest S R and Thompson M E 2006 *J. Am. Chem. Soc.*, **128**, 8108–8109.
[2] Rogers J A, Bao Z, Baldwin K, Dodabalapur A, Crone B, V Raju R, Kuck V, Katz H, Amundson K, Ewing J and Drzaic P 2001 *Proc. Natl. Acad. Sci. U. S. A.* **98** 4835–4840.
[3] Dimitrakopoulos C D and Malenfant P R L 2002 *Adv. Mater.* **14** 99–117.
[4] Uladzimir Z, Tobias E, Gerhard G, Ibrahim M A and Oliver A 2006 *Chem. Phys. Lett.* **418** 347–350.
[5] Hoppe H and Sariciftci N S 2004 *J. Mater. Res.* **19** 1924–1945.
[6] Kietzke T 2007 *Adv. Optoelectron.* **2007** 40285.
[7] Murphy A R and Frechet J M J 2007 *Chem. Rev* **107** 1066–1096
[8] Chou C C, Wu H C, Lin C J, Ghelichkhani E, and Chen W C 2013 *Macromolecular Chemistry and Physics* **214**(7) 751–760.
[9] Li L, Jacobs D L, Che Y, Huang H, Bunes B R, Yang X and Zang L 2013 *Organic Electronics* **14** (5) 1383–1390.
[10] Nam S, Seo J, Park S, Lee S, Jeong J, Lee H, Kim H and Kim Y 2013 *ACS Applied Materials & Interfaces* **5**(4) 1385–1392.
[11] Park Y D, Lim J A, Jang Y, Hwang M, Lee H S, Lee D H, Lee H J, Baek J B and Cho K 2008 *Organic Electronics* 9 317–322.
[12] Yu G, Gao J, Hummelen J C, Wudl F, and Heeger A J 1995 *Science* **270** (5243) 1789–1791,.
[13] Ajayan P M and Tour J M 2007 *Nature* **447** 1066–8.
[14] Kymakis E and Amaratunga G A J 2002 *Appl Phys Lett* **80**(1) 112.
[15] Zou J, Liu L, Chen H, Khondaker S I, McCullough R D and Huo Q 2008 *Adv Mater* **20**(11) 2055–60
[16] Skakalova V, Dettlaff-Weglikowska U and Roth S 2005 *Synth Met* **152** 349.
[17] Landi B J, Raffaelle R P, Castro S L and Bailey S G 2005 *Progress in Photovoltaics* **13**(2) 165–172.
[18] Kymakis E, Alexandrou I. and Amaratunga G. A. J., 2003 *Journal of Applied Physics* **93**(3) 1764–1768.

[19] Dingshan Yu, Yan Yang, Michael Durstock, Jong-Beom Baek,and Liming Dai, 2010 *ACS Nano* **4**(10) 5633–5640.
[20] Kuila B K., Park K and Liming Dai 2010*Macromolecules* **43** 6699.
[21] Mahanandia P and Nanda K K 2008 *Nanotechnology* **19** 155602.
[22] Mahanandia P, Vishwakarma P N, Nanda K K, Prasad V, Barai K, Mondal A K, Sarangi S, Dey G K and Subramanyam S V 2008 *Solid State Commun.* **145** 143.
[23] Zhu L, Chang D W, Dai L and Hong Y 2007 *Nano Lett.* **7** 3592.
[24] Jin H D, Zheng F, Xu W L, Yuan W H, Zhu M Q and Hao X T 2014 *J. Phys. D: Appl. Phys.* **47** 505502
[25] Bolognesi A, Porzio W, Provasoli A, Botta C, Comotti A, Szanni P and Simonutti R 2001 *Macromolecular Chemistry and Physics*, **202**(12) 2586–2591.
[26] Chen S A and Ni J M 1992 *Macromolecules* **25**(23) 6081–6089.
[27] Shiga T and Okada A 1996 Journal of *Applied Polymer Science*, **62**(6) 903–910.
[28] Wen-Cheng Ming-Hui Yang, Shiang-Bin Jong, Po-Chiao Lin, Shyng-Shiou Yuan, Chi-Yu Lu, Pei-Wen Chiang, Tze-Wen Chung, Wan-Chi Tsai, and Yu-Chang Tyan 2014 *BioMed Research International*, , 209469
[29] Furukawa Y, Akimoto M and Harada I 1987 *Synthetic Metals* **18**(1–3) 151–156.
[30] Motaunga D E, Malgasa G F, Christopher J. Arendseb, Sipho E. Mavundlaa,c and D. Knoesen, 2009 *Materials Chemistry and Physics* **116** 279–283.
[31] Trznadel M, Pron A, Zagorska M, Chrzaszcz R and Pielichowski J 1998 *Macromolecules* **31** 5051-5058.
[32] Musumeci A W, Silva G G, Liu J W, Martens W N and Waclawik E R 2007 *Polymer* **48**(6) 1667–1678.
[33] Mohammad Rezaul Karim 2012 Journal of Nanomaterials, 2012 174353.
[34] Ovsyannikov E V, Efimov O N, Moravsky A P, Loutfy R O, Krinichnaya E P and Alpatova N M 2005 *Russian Journal of Electrochemistry* **41**(4) 439–446.
[35] Hotta S, Rughooputh S D D V, Heeger A J and Wudl F 2002*Macromolecules* **20** 212-215.
[36] Giulianini M , Waclawik Eric R , Bell John M , Scarselli Manuela, Castrucci Paola , Crescenzi M D and Motta N 2011 *Polymers* **3** 1433-1446.

Wear behavior of Al-Si alloy based metal matrix composite reinforced with TiB2

J.K.Sahoo[1]*, S.K.Sahoo[1], H.Sutar[2] and B.Sarangi[1]
[1]Department of Metallurgical and Materials Engineering,
Indira Gandhi Institute of Technology, Sarang, Dhenkanal (ODISHA), INDIA-759146.
[2]Department of Chemical Engineering,
Indira Gandhi Institute of Technology, Sarang, Dhenkanal (ODISHA), INDIA-759146

Email: jayantasahoo91@gmail.com

Abstract: Al-Si alloy based composites are widely used in automotive, aerospace and for structural application due to improved strength to weight ratio, low density, and better wear resistance. In the present work, Al-xSi-5TiB2 (x=7, 11, 12.6) in-situ composite was synthesized successfully by stir casting method. Here the composites were prepared by the exothermic reaction of $K2TiF_6$ and KBF_4 salts with the molten Al-x Si alloy. The dry sliding wear behavior of Al-Si matrix composites reinforced with 5 % TiB_2 was studied using a pin-on-disc wear testing machine to study the effect of % Si, load (10, 20, 30 N), sliding speed (1.36, 1.82, 2.27 $m.s^{-1}$) and sliding distance on stir cast Al–xSi-5TiB_2 composites. The Al-Si alloy and the reinforcement mixers were confirmed by the X-ray Diffraction analysis. The microstructure of Al-xSi-5TiB_2 composite was investigated by using Optical Microscope to determine the phases present in the prepared composites. The prepared AMC composites were tested for hardness using Vickers Hardness tester with the variation of Si. Wear rate (mm^3/m), Wear resistance (m/mm^3), Specific Wear rate ($m^3/N.m$) and were analyzed with various conditions. The worn surfaces of the specimens were analyzed before and after wear testing by Scanning Electron Microscope (SEM) to determine the governing wear mechanisms in the composites. Wear rate and specific wear rate decreases at all the operating condition with increase in wt% Si. Wear resistance all most increases with increase in wt% Si. Hardness values are increased with increase in amount of Si.

1. Introduction

The most rapidly growing area of composites is that of Aluminium matrix composites as a result of their improved physical, chemical and mechanical properties. Amongst different Aluminium alloys as matrix phase Al-Si alloys are widely used as engineering materials due to their light weight, good castability, high specific strength, enhanced wear resistance and reasonable cost. Addition of ceramic particles like TiB_2, SiC or Al_2O_3 to the aluminium based matrix phase does not increase the density considerably but ensures significant rise in the specific strength, modulus and wear resistance [1-3]. Al-TiB_2 composite exhibits some useful and unique characteristics with respect to its properties and processing route. [4-5]
Synthesizing the ceramic particles within molten Aluminium alloys is called the in-situ method of preparing the AMCs. In the in-situ synthesis of salt-metal reaction the exothermic reaction between Potassium Hexa Fluoro Titanate (K_2TiF_6) and Potassium Tetra Fluoro Borate (KBF_4) halide salts yields TiB_2 particles within the matrix phase. [6-8]

In the present work an attempt is to be made in order to study, synthesize and characterize the Al-varying amount of Si alloys dispersed with fixed amount of TiB_2 for different engineering applications.

2. Experimental Details

The overall experimental procedure can be divided into two stages. In the first stage, Al-xSi alloy (x=7, 11, 12.6) and Al-xSi-5TiB_2 composite samples are synthesized and casted. In the second stage, the samples were characterized for their microstructure, and wear behavior.

2.1 Synthesis of in-situ Al-axis alloy and Al- xSi-5TiB_2 composite

Initially Al-50Si master alloy and commercially pure aluminum (99 %) was melted in a graphite crucible in a non-ferrous melting furnace at 720°C to prepare Al-axis alloy. The Al-xSi-5TiB_2 composite was synthesized by Flux Assisted Synthesis (FAS) technique which consists of addition of halide salts, K_2TiF_6 and KBF_4, to molten Al-Si alloy at 800° C and a reaction time of one hour. The exothermic reaction within the melt results in formation of titanium diboride (TiB_2) particles. The melt was stirred intermittently every 10 minutes to ensure complete reaction of salts with molten aluminium and homogeneous distribution of TiB_2 particles. After completion of reaction, the lighter dross was decanted. The molten composite was then poured into a preheated (450°C) cast iron mould for preparation of different specimen.

2.2 Characterization

2.2.1 Optical Microscopy

Microstructures of the composite samples were observed under computerized optical microscope. The samples were polished using emery papers followed by disc polishing and etching with Keller's reagent (1 % HF-1.5 % HCl-2.5 % HNO_3-95 % H_2O). The micrographs of the samples were obtained.

2.2.2 X-Ray Diffraction Analysis

X-ray diffraction analysis of the composite samples was performed with Co-Kα target to identify different phases. Sample of size 10mm×10 mm×2mm were taken for x-ray diffraction. The diffraction pattern was then analyzed by matching the peaks with JCPDS data files to identify all the phases present in the sample.

2.2.3 Hardness test

The microvicker's hardness test was conducted for measuring the hardness of the AMC in-situ composite with varying amounts of Si in the matrix. The samples to be tested were machined to require dimension and both their horizontal faces were made parallel by polishing .The square based diamond pyramid was used as indenter and 1kgf load applied for 15 seconds. The hardness values were measured in three locations over the sample and the average length of the diagonals of the indentation were calculated. Then the Vickers Hardness Number (VHN) of the samples was obtained.

2.2.4 Wear Testing

Wear tests were conducted on a pin-on-disc wear testing machine. The cylindrical pins (10mm diameter and 30mm height) of the composite sample were cut by specimen cutter ,machined and then polished metallographically .Wear test are conducted by varying loads ranging from 10 to 30N ,and sliding speed 1.36,1.82,2.27 $m.s^{-1}$ for 300 sec and 600 sec. at room temperature without any lubricants . During the test the pin is pressed to the steel disc by applying normal load. The wear testing machine is microprocessor controlled where height loss and frictional force was monitored simultaneously. After the test completed the specimen are removed, cleaned through acetone, dried and weighed to determine weight loss due to wear. Difference in weight before and after testing was measured to determine weight loss (i.e. wear) in the composite specimen. The wear of the composite specimen were studied as a function of weight percentage of silicon in the alloy matrix, sliding speed, applied load, and sliding distance.

2.2.5 Scanning electron microscopy

Worn surfaces of the composite were examined with Field emission electron microscope (FESEM).

3. Results and discussion

3.1 Microstructure

The optical microstructure of Al-7Si-5TiB_2, Al-11Si-5TiB_2 and Al-12.6Si-5TiB_2 insitu composite are respectively shown the fig-1.it is observed from the figure that the Si particles are distributed with the α-Al in the Al-7Si-5TiB_2. Al-12.6Si is the eutectic composition of Al-Si binary alloy which exhibit niddle shaped Si well dispersed throughout aluminium matrix but in the formation of TiB_2 particles in the melt , the space available for the growth of Si decreases.TiB_2 particles thus restrict the growth of Si during the solidification.

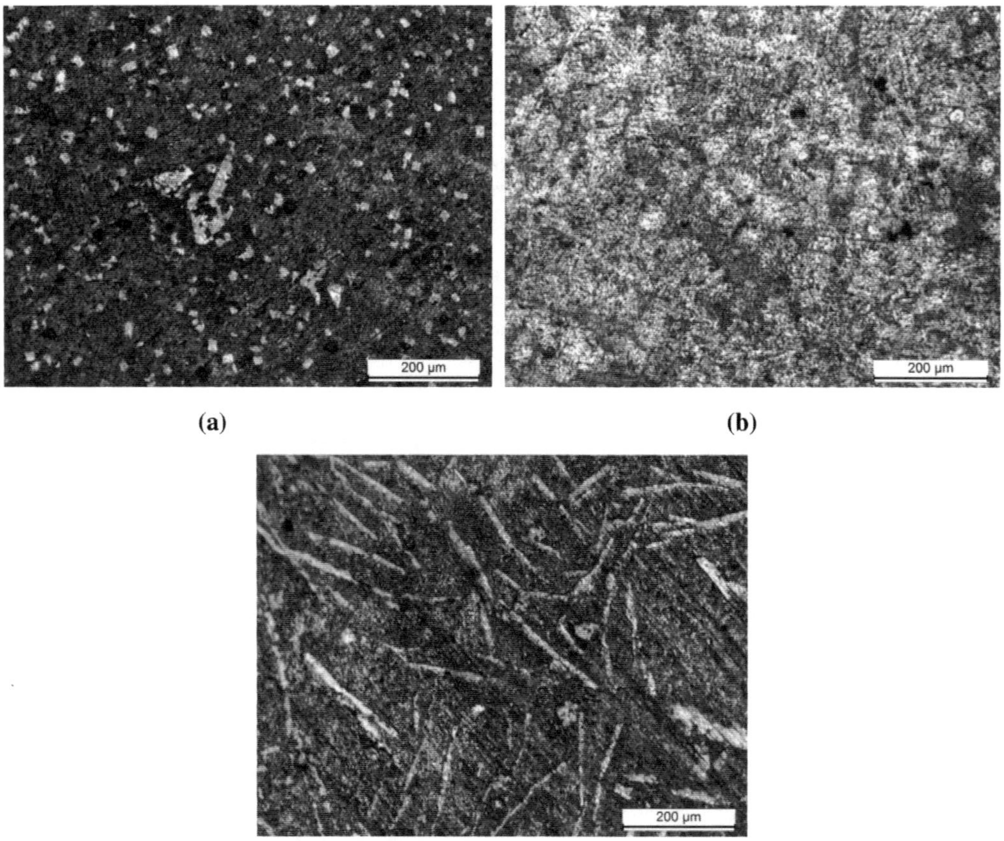

(a) (b) (c)

Figure1.Optical micrograph at (100X) of (a) Al-7Si-5TiB_2 (b) Al-11Si-5TiB_2(c) Al-12.6Si- 5TiB_2 in-situ AMC composite

3.2 XRD Analysis of in-situ Composite

In XRD result that has been obtained reveals that the presence of Aluminium is in the form of the largest peaks.Minor peaks indicate the presence of Silicon and Titanium diboride particles.It is evident that there is an increase in the weight percentage of the Silicon the peaks of Si also increases.The peaks of TiB_2 in the XRD pattern confirms the formation of TiB_2 in the Al-Si alloy matrix.

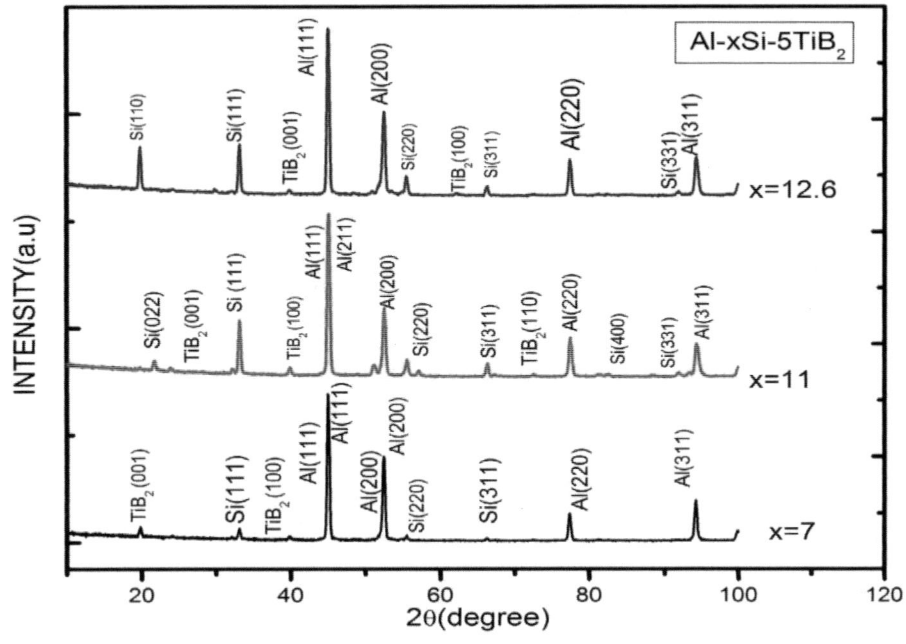

Figure.2 XRD results for the prepared in-situ composites

3.3 Hardness test results

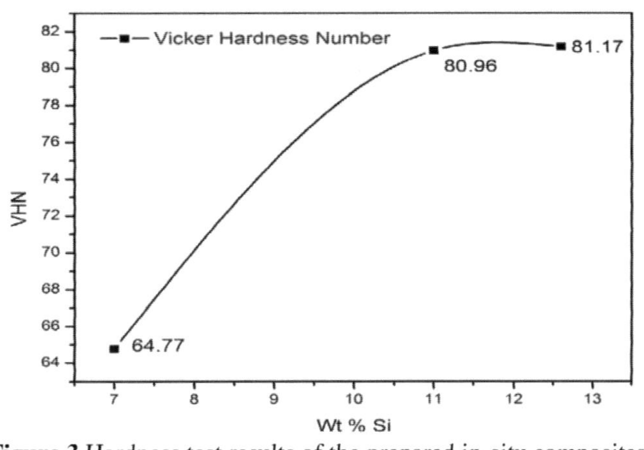

Figure.3 Hardness test results of the prepared in-situ composites

From the above graph it can be observed that AMC material having 12.6 % Si is showing maximum hardness and AMC having 7% Si is showing the least hardness. Increase in amount of silicon increases the eutectic Al-Si phase which resists the plastic deformation which leads to in the hardness of the composites.

3.4 Analysis of wear test results

3.4.1 Wear rate of the composite as a function of wt% of Si and load,speed,sliding distance

Figures 4-6 indicates the wear rate of the prepared composite as a function of amount of Si and load, sliding distance, sliding speed. The wear rate of three varying wt% of Si of the in-situ composites werecompared with different controlling condition. The wear rate was decreased along with the increasing wt% of Si. The higher wt% of Si has the minimum wear rate in all the conditions of the experiment. Figure. 4 shows that the wear rate decreases with increase in load. Fig 4.2 confirms the minimum wear rate achieved to the various sliding distances.

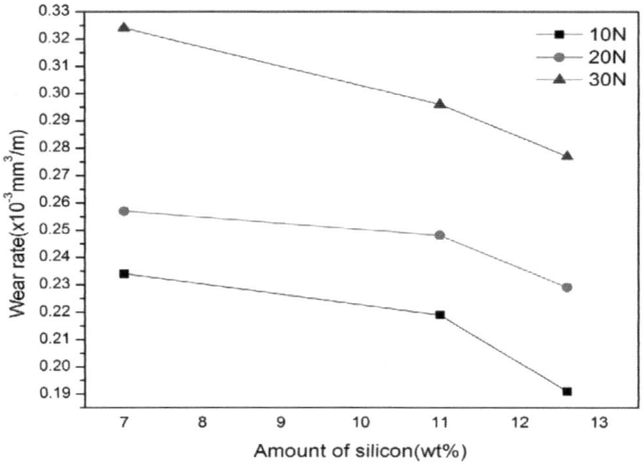

Figure.4 Wear rate of the in-situ composites as a function of amount of Si(wt%) and sliding distance

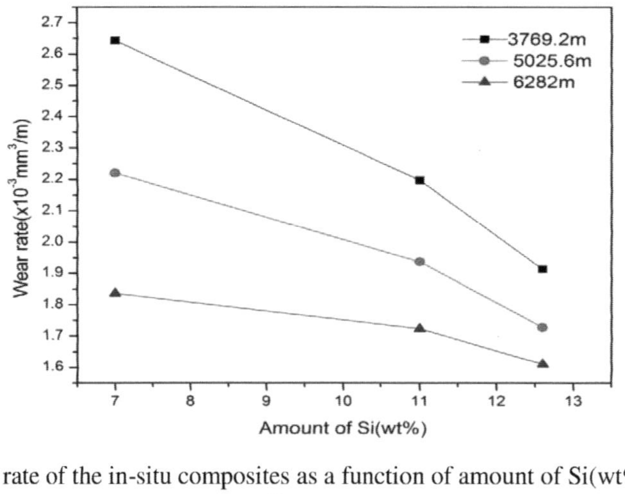

Figure.5 Wear rate of the in-situ composites as a function of amount of Si(wt%) and sliding distance

The effect of sliding speed on wear rate is shown in Figure.6 for the AMCs tested for different loads of 10, 20, 30N respectively. Wear rate was decreased with increase in sliding speed in all most the three cases ,one of the possible causes for the reduced wear rate at the start could be the surface oxidation which may increase the local hardness and strength at the interface .The drop of wear rate which causes the AMC to have increased wear resistance, i.e. the effect of oxidation increases gradually. Another thing that during wear process two opposing processes acted such as softening and hardening that took place simultaneously. The softening of surface material occurs with the increase of temperature and strength, and the hardening occurs with the formation of metal oxides .The oxidation is a continuous process at all rotational speeds .The metal oxide film thickness would possibly grow at faster rate than the rate of material removal. It means that only the strengthened and hardened metal oxide would be present at the interface and it would mask the softer matrix material underneath. But in case of 12.6 wt % Si AMC the wear rate decreases first then increases with increasing the sliding speed. This is due to severe wear for that material at higher speed. At loads of 10N, 20N, and 30N, wear rate of all samples were increases respectively. However, a large difference in wear rate was observed at 30N loads.

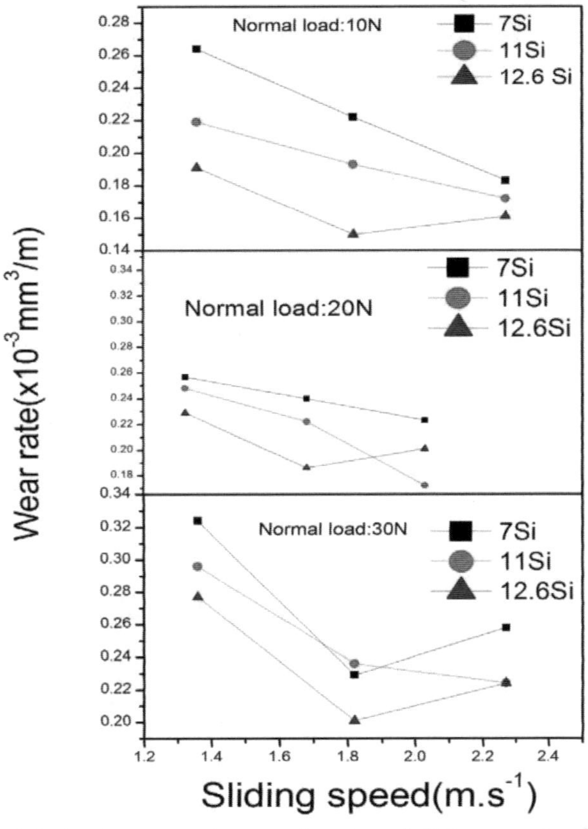

Figure.6 Wear rate of the in-situ composites as a function of sliding speed and amount of Si (wt%) in diferent load

3.4.2 Wear resistance of the composite as a function of wt % of Si and load,sliding distance

The wear resistance of the prepared composites with various parameters is shown in the Figure7-8. The composites with the higher wt % of the Si have the higher wear resistance. Similarly as the sliding distance increases the wear resistance increases.

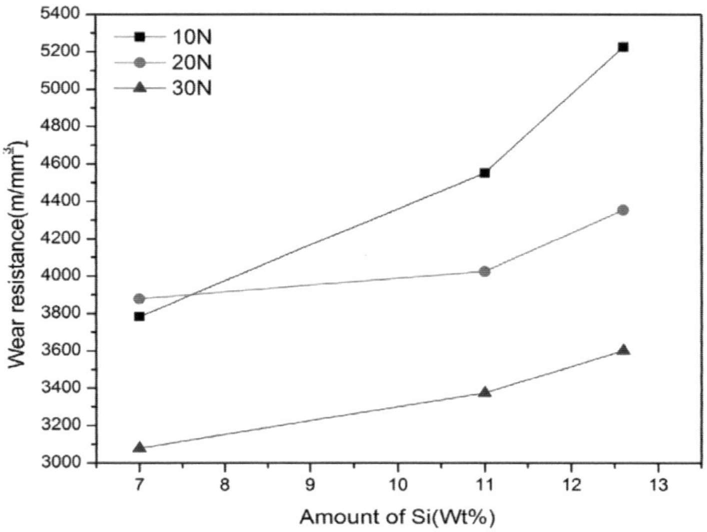

Figure.7 Wear resistance of the in-situ composites as a function of amount of Si (wt%) and load

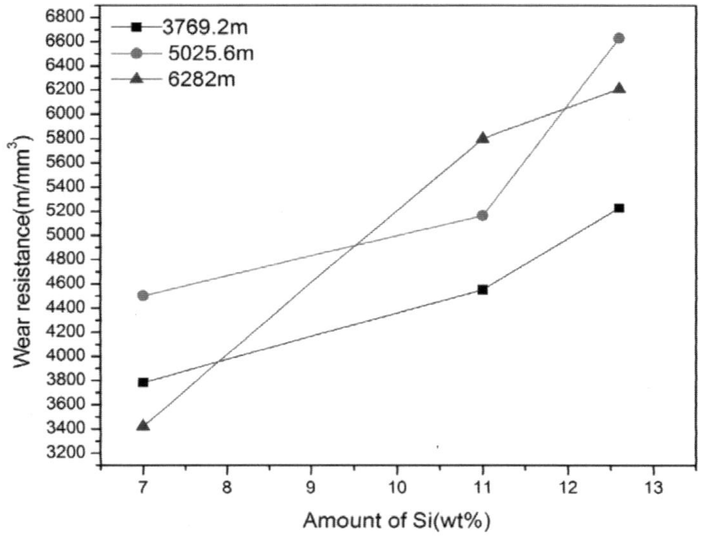

Figure.8 Wear resistance of the in-situ composites as a function of amount of Si (wt%) and sliding distance

3.4.3 Specific wear rate of the composite as a function of wt% of Si and load,sliding distance

Specific wear rate was compared with various wear testing condition such as load and sliding distance of the composite with various wt % of Si contents. Figure.9-10 proves the minimum specific wear rate in all condition. By the stir casting method the reinforcements and the Si were mixed successfully and achieved the uniform distribution to improve the wear properties. The various testing conditions were performed to observe the specific wear rate, the silicon contents were increased and the specific wear rate was decreased simultaneously.

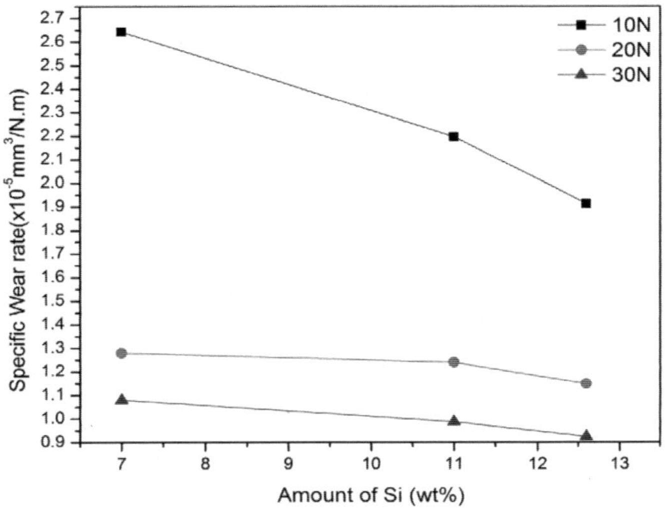

Figure.9 Specific wear rate of the in-situ composites as a function of amount of Si (wt%) and load

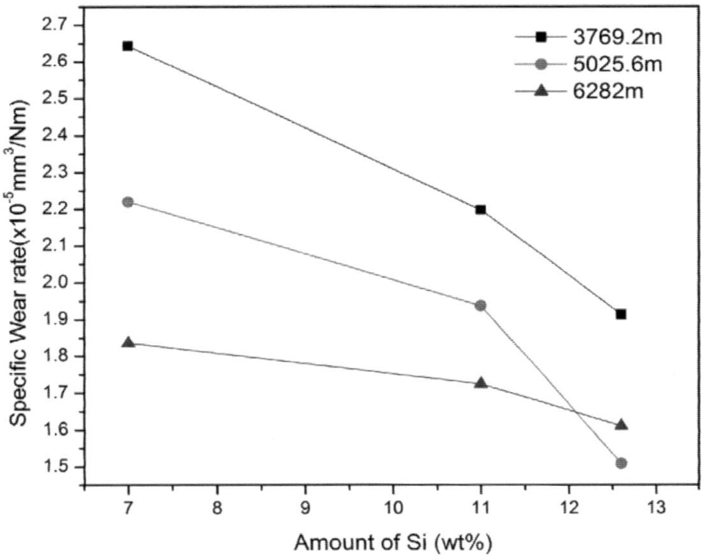

Figure.10 Specific wear rate of the in-situ composites as a function of amount of Si (wt%) and sliding distance

3.5 Worn surface analysis

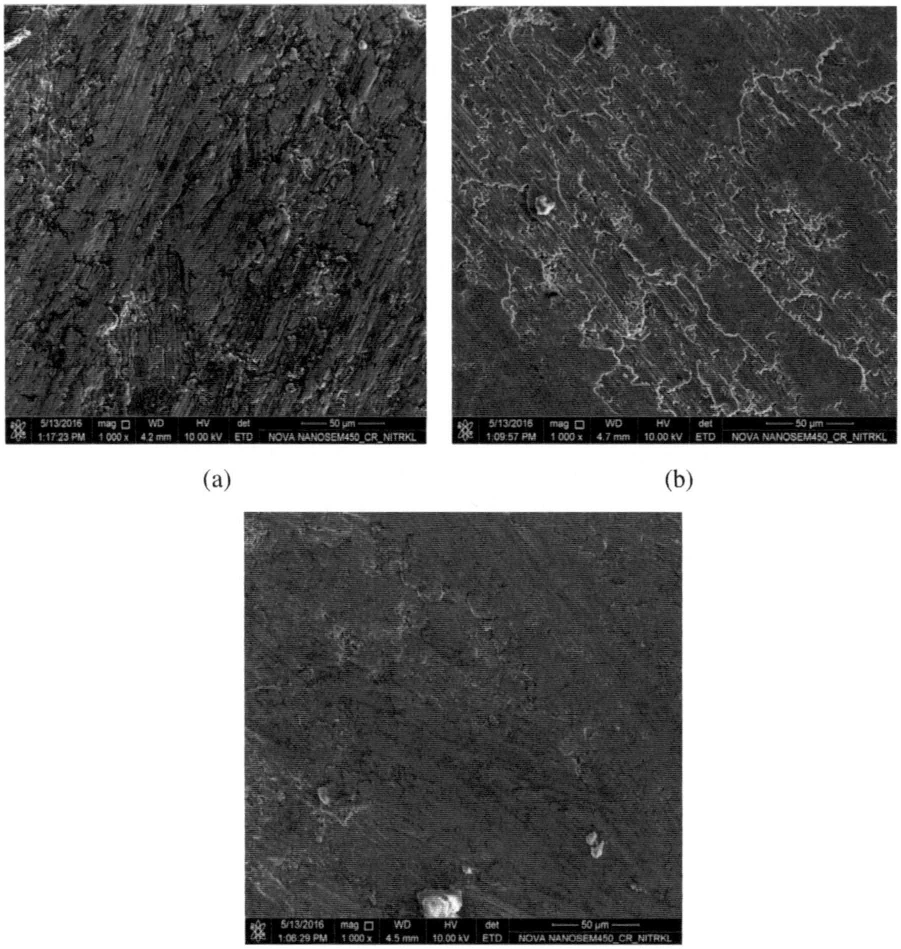

(a) (b)

(c)

Figure.11 FESEM micrographs of worn surfaces of Al-xSi-5TiB$_2$ in-situ composite (a) 7 wt% Si (b) 11% Si (c) 12.6% Si

Figure.11 shows the morphology of the worn surfaces of the AMCs with varying amounts of Si (wt %). The wear mechanism is characterized by the flow of material in wavy form and deep grooves on the wear surface. However, the material flow was less in SEM images of the worn surface of "12.6%" Si AMC. But larger grooves were clearly visible in the "7 % Si" micrographs. In all the three cases the grooves are formed on the wear surface due to ploughing action of the hard asperities of Si. This increases the temperature of the wear surface of the composite. Further, the yield strength of the composite is decreased, which results in softening of wear surface and starting of adhesive wear. It is noticed that the Si is a hard material. This may plough out from the composite during sliding of the counter face of the disc. It has also found that the hard Si improves resistance of composite against plastic deformation and material flow there by increasing the load bearing capacity. The probability of adhesive wear is less in these composite and the dominant mechanism is abrasion and de-lamination.

4. Conclusions

From the present investigations the following conclusions were obtained:

1) The Al-xSi-5TiB2 in-situ composite was synthesized successfully by stir melt casting.
2) XRD analysis confirmed that the required phases are present in the in-situ AMCs.
3) Hardness values are increases with increase in amount of Si (wt %).
4) Wear rate (mm^3/m) and specific wear rate (mm^3/Nm) decreases at all the operating condition with increase in wt % Si.
5) Wear resistance all most increases with increase in wt%Si.
6) Wear mechanism was studied from the SEM micrograph of the worn surface of the in-situ AMCs.

References

[1] Pollock. T.M, 2010, Weight Loss with Magnesium Alloys, Science, 328 (5981), 986-987.

[2] Lloyed D.J, 1994 Particle reinforced aluminium and magnesium matrix composites, Int. Mater. Rev.39, 1-23.

[3] Ibrahim.L.A, Mohamed.F.A, Lavernia. E.J, 1991, particulate reinforced metal matrix composites: a review, Matter. Sci. 26, 1137-1156.

[4] Murty. B.S, Kori.S.A, Chakraborty.M, 2002, Grain refinement of aluminium and its alloys by heterogeneous nucleation and alloying, Int. Matter. rev. 47, 3-29.

[5] Tjong.S, 2000, Microstructural and mechanical characteristics of in situ metal matrix composites, Mater. Sci .Eng. Reports. 29 , 49-113.

[6] Das.B, Roy.S,.Rai. R.N, Saha. S.C, 2016, Development of an in-situ synthesized multi-component reinforced Al-4.5%Cu-Tic metal matrix composite by FAS technique-Optimization of process parameters, Engineering Science and Technology: an international Journal 19, 1, 279-291.

[7] Emamy.M, Mahta.M , Rasizadeh.J, 2006, Formation of TiB_2 particles during dissolution of $TiAl_3$ inAl-TiB_2 metal matrix composite using an in situ technique, compos. Sci. Technol. 66, 1063-1066.

[8] Karbalaei A.M, Baharvandi.H.R, Shirvanimoghaddam.K, 2015, Tensile and fracture behavior of nano/micro TiB_2 particles reinforced casting A356 aluminium alloy composites, Mater .Des. 66, A, 150-161.

Effect of magnetostrictive material layer on the stress and deformation behaviour of laminated structure

S D Suman[1], C K Hirwani[2] A Chaturvedi[3], S K Panda[4]

Department of Mechanical Engineering, National Institute of Technology Rourkela India.

Email[1]: shardasuman92@gmail.com

Abstract. In the present investigation, static bending and strength behaviour of the laminated composite plate embedded with magnetostrictive (MS) material has been computed numerically using commercial finite element tool. For the analysis purpose, a simulation model of the laminated plate embedded with and without MS material is derived using the batch input technique popularly termed as ANSYS parametric design language (APDL). The MS bonded composite model is discretised with the help of available suitable element from ANSYS element library. The elements are chosen suitably by considering the compatibility between the mechanical and the magnetic fluxes indices. For the calculation of the necessary responses, APDL code has been prepared in ANSYS environment. Further, the stability of the model has been checked by computing the responses for various mesh sizes. Subsequently, the accuracy of the model has also been checked by conducting the comparison test. For the comparison, central deflection of the clamped Graphite/Epoxy laminated plate embedded with Terfenol-D has been evaluated and compare them with those of the available earlier published literature. The comparison study clearly indicates that, the present responses are in-line and showing good agreement with that of the reference. Further, the versatility and the applicability of the simulation model has been proven by solving the different illustrations. In the illustration, the influence of the different geometrical parameter such as thickness ratio and aspect ratio on the bending behaviour has been investigated and discussed in detail.

1. Introduction

The necessity of increasing the safety margin of the structure and the structural component of almost all the sphere of industries such as automobile, aerospace, aircraft and marine industries, increases the application as well the demand of smart composite material since last two decades. The smart materials are the material having one or more physiochemical properties which can be changed significantly due to external stimuli such as temperature, pressure, magnetic or electric field, etc. When smart materials are get stuffed or embedded with composite material or structure, the composite material or structure is known as smart composite material or smart composite structure. The smart composite structures are known for their actuating, sensing and the controlling capability of the structural responses. The embedding of the magnetostrictive material with advanced layered composite material effect on structural performance such as bending and strength characteristic significant. The study of the same has been reported by author and the few important of them are reviewed and presented in the next lines.

Noor and Burton [1] investigated the static and dynamic analysis of laminated composite plate by developing a finite element (FE) model based on first order shear deformation theory (FSDT). Anjanappa and Bi [2] established the mathematical model to analysed the vibration response of smart structure embedded with the magnetostrictive mini actuators. Pratt et al. [3] presented the active vibration control and separation phenomenon for a cantilevered beam embedded with Terfenol-D actuator. Vuksanovic [4] utilized the classical laminate plate theory (CLPT) and FSDT kinematics to examine the static and dynamic response of the laminated structure. Kant and Swaminathan [5] used MPEP Navier's technique to analyze the composite plates numerically [5]. Huang et al. [6] evaluated the stress and deflection of the composite laminated structure using hybrid stress element numerically. Kumar et al. [7] proposed the FE model to investigate the damping characteristics of the aluminum beam embedded with the layer of magnetostrictive material Terfenol-D. Lee and Reddy [8] presented the static behaviour of magnetostrictive material embedded laminated composite for various uniformly distributed load and sinusoidal load based on the third order kinematics. Park et al. [9] numerically studied the static and dynamic behaviour of both laminated composite plate and shell using four-node element. Zhen et al. [10] proposed a C^0 higher-order theory for laminated composite and sandwich plates to determine the natural frequency. The mechanical behaviour of Terfenol-D embedded laminated composite plate is investigated by Kishore et al. [11]. Sahoo and Singh [12] developed the mathematical model of the laminated composite and sandwich plates using a zigzag theory to investigate the structural behaviour. Andakhshideh and Tahani [13] utilized the minimum potential energy principle to evaluate the inter-laminar stress of the laminated plate structure analytically. The static and dynamic response of laminated structure investigated by Rango et al. [14] using a FSDT based mathematical model. Santapuri et al. [14] presented the dynamic analysis of laminated composite stuffed with the Galfenol sheets using a mathematical model based on CLPT [15].

From the survey of the literatures, it is clear that in the past sufficient amount of work have been done for the investigation of bending behaviour of laminated composite structure but for the laminated structure embedded with the magnetostrictive material are less in number. Therefore, the aim of the present work is to investigate the stress as well as deflection behaviour of the laminated structure embedded with an actuating layer of a magnetostrictive material. For the investigation, a simulation model in ANSYS environment has been prepared using APDL code. Further, the computed responses are compared with the responses of published literature. Finally, the versatility of the model has been shown by solving the new numerical example which shows the effect of adding the magnetostrictive material in pure laminate, and geometrical parameter on deflection and strength behaviour of the layered composites plate.

2. Theoretical Formulations

2.1. Displacement model

The mathematical model for the laminated composite plate embedded with magnetostrictive material as shown in figure 1 has been developed in the present investigation. The geometrical configuration of the laminate is defined as length a, width b, and the thickness as h. The laminate is composed of the N number of layer. The displacement kinematic for the same is considered according to FSDT kinematics and presented as:

$$\begin{aligned} u(x,y,z) &= u_0(x,y) + z\theta_x(x,y) \\ v(x,y,z) &= v_0(x,y) + z\theta_y(x,y) \\ w(x,y,z) &= w_0(x,y) + z\theta_z(x,y) \end{aligned} \tag{1}$$

where, u, v and w are the displacements of any point along the x, y and z coordinate axes respectively. u_0, v_0 and w_0 are corresponding displacements of a point on the mid plane and θ_x and θ_y are the rotations of normal to the mid-surface, i.e., z=0 about the y and x-axes, respectively.

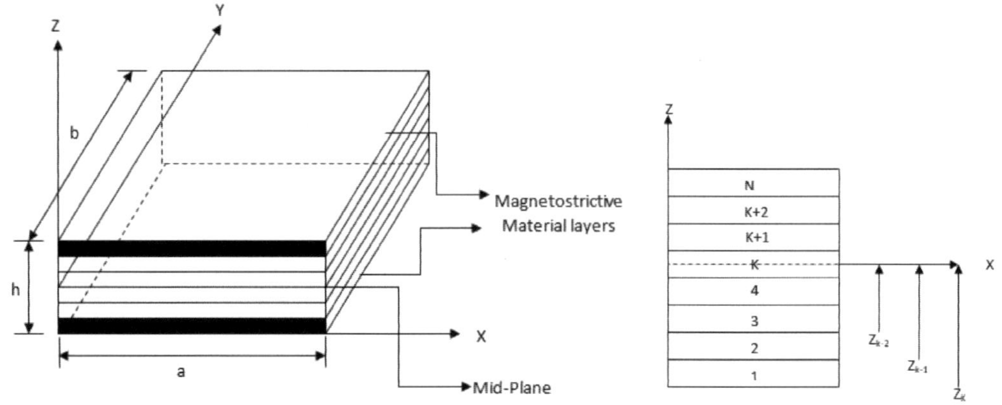

Figure 1: Geometry and stacking sequence of laminated composite plate

The magnetic field vector $\{H\}$ can be written in terms of the magnetic field potential as:

$$\{H\}=[Z_M]\{\bar{H}\} \tag{2}$$

The relation of stress and strain for any kth lamina oriented at an arbitrary angle ϕ about any arbitrary axes are presented as:

$$\{\sigma\}^k=[\bar{Q}]^k\{\varepsilon\}-[\bar{e}]^k\{H\} \tag{3}$$

$$\{B\}^k=[\bar{e}]^k\{\varepsilon\}+[\mu]^k\{H\} \tag{4}$$

where $\{\sigma\}$, $\{\varepsilon\}$, $\{H\}$ and $\overline{Q}_{ij}$ are stress vector, strain vector, magnetic field intensity and transformed reduced stiffness matrix respectively.

2.2. Finite element formulation

The displacement fields for the assumed kinematic models are expressed in terms of desired field variables and the models are discretised using suitable FEM steps. The displacement vector '*d*' at any point within the mid-plane is given by:

$$d=\sum_{i=1}^{4}N_i(x,y)d_i \tag{5}$$

$$\psi=\sum_{i=1}^{4}N_i(x,y)\psi_i \tag{6}$$

where, $\{d_i\}=\{u_{0_i}\ v_{0_i}\ w_{0_i}\ \theta_{x_i}\ \theta_{y_i}\ \theta_{zi}\}^T$ is the nodal displacement vector, $\{\psi_i\}$ is the magnetic potential and N_i is the corresponding interpolating function for the 'i^{th}' node.

The strain vector can be expressed in the matrix form after introducing the FEM steps and conceded as:

$$\{\varepsilon\}=[T]\{\bar{\varepsilon}\} \tag{7}$$

where, [*T*] is the thickness coordinate matrix and $\{\bar{\varepsilon}\}$ is the mid plane strain vector which can be further reduced as:

$$\{\bar{\varepsilon}\} = [B_L]\{d\} \tag{8}$$

$$\{\bar{H}\} = [B_M]\{\psi\} \tag{9}$$

where, $[B_L]$ is a general strain displacement relation matrix and $[B_M]$ is the magnetostrictive laminate strain displacement matrix.

The total strain energy of the laminate can be expressed as:

$$U = \frac{1}{2}\iint\left[\int_{-h/2}^{+h/2} \{\varepsilon\}^T \{\sigma\} dz\right] dxdy \tag{10}$$

The total work done due to the externally applied mechanical load (*F*) is expressed as:

$$W = \int_A \{d\}^T \{F\} dA \tag{11}$$

The stiffness matrix ([*K*]) can be further expressed as:

$$[K] = \int_A (\sum_{k=1}^{n} \int_{z_{k-1}}^{z_k} [B_L]^T [D][B_L] \tag{12}$$

2.3. Governing equations

The final form of governing equation for static analysis of flat plate is obtained using minimum potential energy principle:

$$\partial\Pi = \partial(U - W) = 0 \tag{13}$$

where, Π is the total strain energy.
Further, equation (13) is modified by substituting the equations (10) and (11) into equation (13) and expressed as:

$$[K]\{\delta\} = \{F\} \tag{14}$$

2.4. Boundary condition

To solve the above set of governing equation following boundary conditions are taken in to considerations:
Simply supported

$$v_0 = w_0 = \theta_y = \theta_z = 0 \text{ at } x\text{=0 and } a;$$

$$u_0 = w_0 = \theta_x = \theta_z = 0 \text{ at } y\text{=0 and } b; \tag{15}$$

Clamped

$$u_0 = v_0 = w_0 = \theta_x = \theta_y = \theta_z = 0 \text{ at } x\text{=0 and } a;\ y\text{=0 and } b; \tag{16}$$

Free

$$u_0 \neq v_0 \neq w_0 \neq \theta_x \neq \theta_y \neq \theta_z \neq 0 \text{ at } x\text{=0 and } a;\ y\text{=0 and } b; \tag{17}$$

3. Result and discussion

The bending behaviour of the composite plate embedded with the magnetostrictive material layer is analysed by developing a simulation model in ANSYS using APDL code. The model is discretised with the help of four-node shell element shell-181. For the convergence purpose, the central deflection responses of the Graphite/Epoxy laminated plate with Terfenol-D layers is computed for different mesh density. Further, the computed responses are compared with the response of the published literature. Finally, the effect of the magnetostrictive material on deflection and stress value are examined and discussed in detail. The material property used in the analysis has been shown in table1. The clamped condition is taken for the further investigation in this present article. The equation used for the nondimensional form of the load and deflection are as follows: $\bar{P} = q_0 a^4 / E_{22} h^4$, $\bar{w} = w / h$.

3.1 Convergence and validation study.

For the convergence study, static behaviour of clamped square symmetric ten-layer (0°/30°/-30°/30°/-30°)$_s$ laminated composite plate is analysed under uniformly distributed load for different mesh density and shown in Figure 2. The responses are evaluated using Graphite/Epoxy material property as given in table 1. From the figure, it depicts that the results are converging well for various mesh refinement. Based on that a mesh size of (8×8) is chosen for further analysis. Further the analysis is expensed for validation study. The nondimensional central deflection responses of square ten-layer (m/90°/0°/90°/0°)$_s$ Graphite/Epoxy layered plate structure embedded with Terfenol-D layers has been investigated. The material property used for the present investigation are shown in table 1. The result is evaluated for two different end condition, compared with Lee and Reddy [8] and Kishore et al. [11] and shown in table 2.

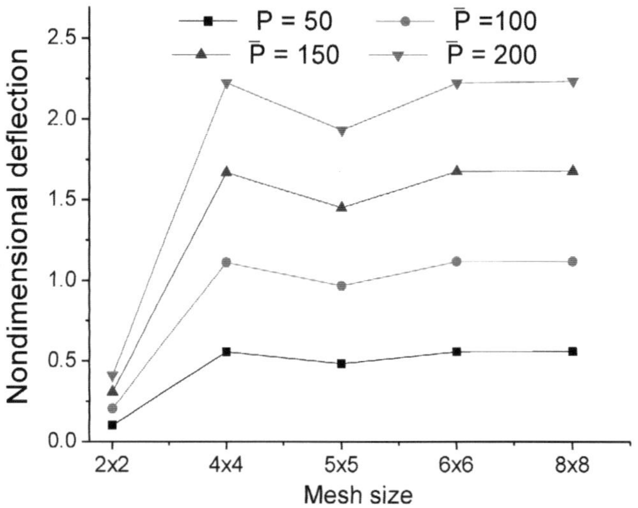

Figure 2: Nondimensional deflection at different mesh size

Table 1. Material properties of Graphite/Epoxy and Terfenol-D.

Properties	Graphite/ Epoxy	Terfenol-D
Young's modulus x direction (E_l)	138.6GPa	26.5GPa
Young's modulus y direction (E_t)	8.27GPa	26.5GPa
Young's modulus z direction (E_z)	8.27GPa	26.5GPa
Shear modulus (G_{lt})	4.96GPa	13.25GPa
Shear modulus (G_{tz})	4.12GPa	13.25GPa
Shear modulus (G_{lz})	4.96GPa	13.25GPa
Poisson's ratio (υ_{lt})	0.26	0
Poisson's ratio (υ_{lz})	0.26	0
Poisson's ratio (υ_{tz})	0.26	0
Density (ρ)	1824 Kgm^{-3}	9250 Kgm^{-3}

Table 2: Comparison study of central deflection of ten-layer Graphite/Epoxy laminated composite embedded with Terfenol-D (*a*/*h*=10).

Constraints condition	Load parameter	Present	Lee and Reddy[8]	Kishore et al.[11]
CCCC	1	0.004817	0.00482	0.0043
	5	0.024807	0.02407	0.0215
	30	0.14452	0.14292	-
	60	0.28905	0.27777	-
	100	0.48174	0.43938	0.4301
CCSS	1	0.0066438	0.00615	0.0066
	5	0.033179	0.03073	0.0331
	30	0.19103	0.18183	-
	60	0.34815	0.3507	-
	100	0.5072	0.549	0.6623

3.2 Effect of magnetostrictive layer on in-plane normal and shear stresses (σ_x, σ_y and τ_{xy})

In-plane stresses have been plotted in figure 3-5 for different lamination schemes under uniformly distributed load both for square clamped laminated plate (*a*/*h*=10) embedded with and without MS material. From the figure, it has clearly observed that the with the increase in load parameter there is an increase in the stress. In continuation to that, the effect of adding the MS material on stress behaviour is same for the normal stress (σ_x) and shear stress (τ_{xy}) i.e. stress is lower in case of adding the MS material. On the other hand the effect on normal stress (σ_y) is opposite to the previous two stresses i.e. stress is higher in case of adding the MS material.

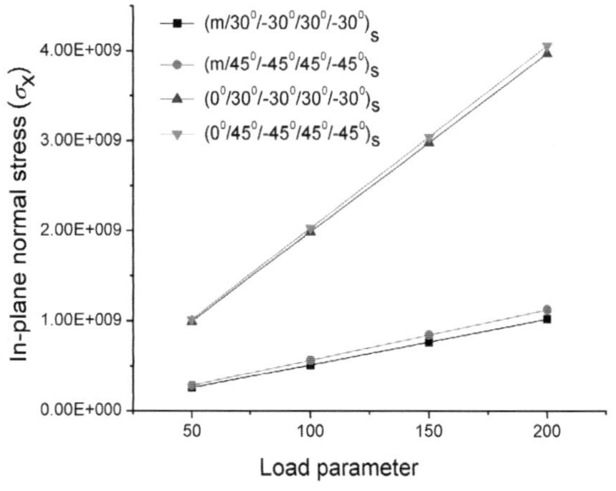

Figure 3: Effect of magnetostrictive layer on in-plane normal stress σ_x

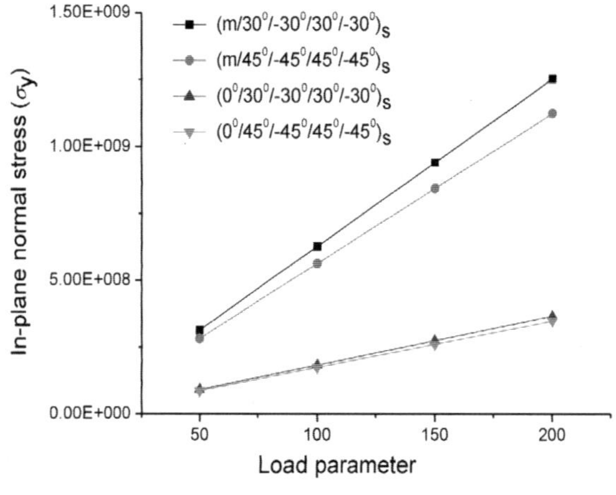

Figure 4: Effect of magnetostrictive layer on in-plane normal stress σ_y

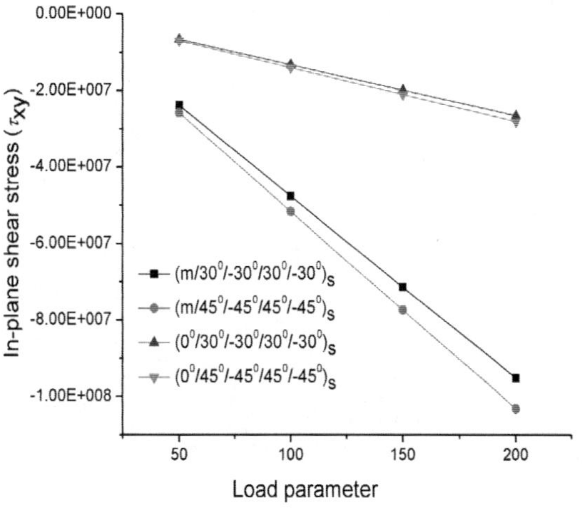

Figure 5: Effect of magnetostrictive layer on in-plane shear stress τ_{xy}

3.3 Effect of magnetostrictive layer on central deflection

The nondimensional transverse central displacement of the laminated composite plate with and without MS material is investigated in the present illustration. For the computation, square clamped ten-layer with laminated composite (a/h=10) with two lamination scheme [(m/30°/-30°/30°/-30°)$_s$ and (m/45°/-45°/45°/-45°)$_s$] has been utilised. The responses are shown in figure 6. The figure shows that deflection parameter increases for both the lamination scheme by adding the MS material. In addition, a contour plot of deflection responses with and without MS material has also been presented in figure 7 and 8.

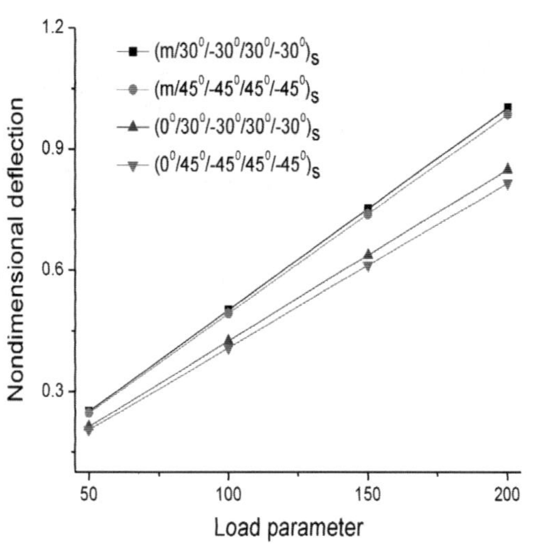

Figure 6: Effect of magnetostrictive layer on central deflection

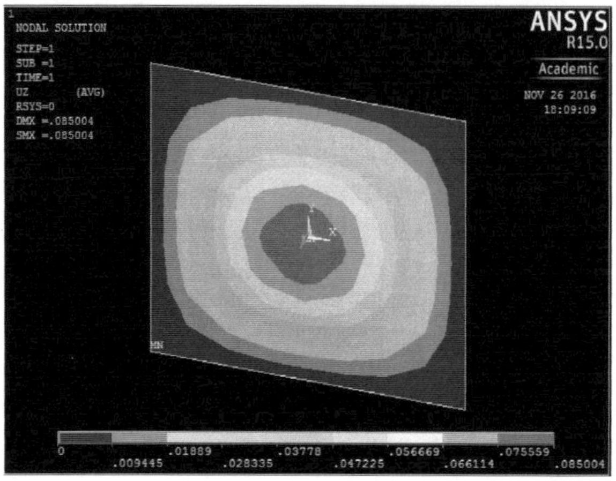

Figure 7: Contour plot of the deflection for lamination scheme $(0°/30°/-30°/30°/-30°)_s$ and load parameter of 200

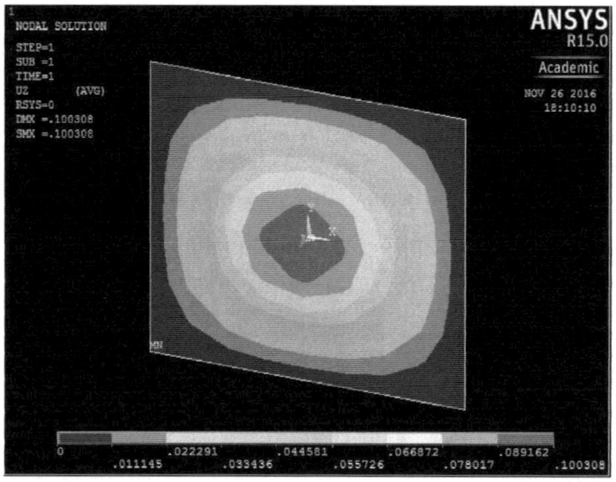

Figure 8: Contour plot of the deflection for lamination scheme $(m/30°/-30°/30°/-30°)_s$ and load parameter of 200

3.4 Effect of aspect ratio on deflection

Nondimensional central deflection of clamped square symmetric $(m/45°/-45°/45°/-45°)_s$ laminated composite plate (a/h=10) embedded with MS material is plotted for different aspect ratio and shown in figure 9. From the figure, it depicts that, with increasing the aspect ratio nondimensional deflection response increases irrespective of the loading.

IOP Conf. Series: Materials Science and Engineering **178** (2017) 012026 doi:10.1088/1757-899X/178/1/012026

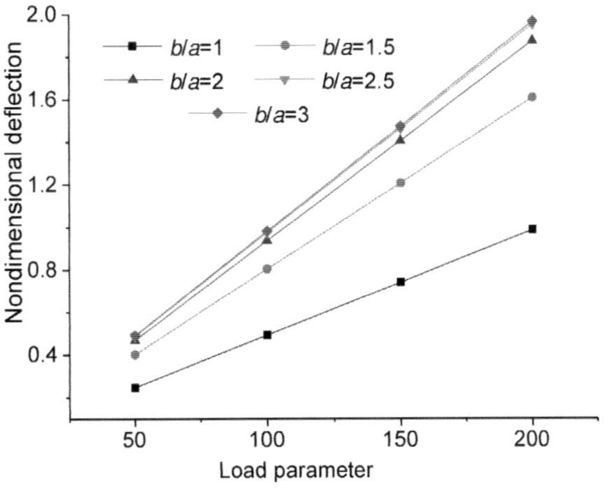

Figure 9: Variation in nondimensional deflection with change in b/a value

3.5 Effect of thickness ratio on deflection

In this illustration effect of aspect ratio of layered composite embedded with MS material have been examined. For the computation purpose, square clamped laminated composite plate embedded with MS material has been considered and analysed for four thickness ratio (*b*/*a*=10,20,50 and100). The results are evaluated using simulation model and presented in figure 10. It is observed from the figure that, with the increase in the aspect value there is a decrease in the nondimensional static deflection of the plate. It is worthy to mention that, the nondimensional form of the load is a function of the thickness ratio which decreases with increasing the thickness ratio.

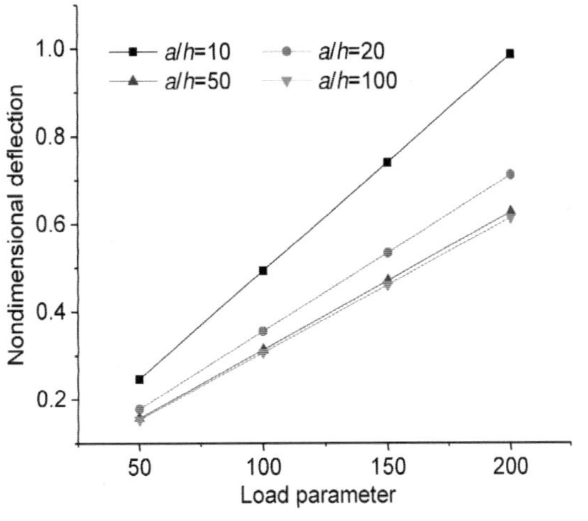

Figure 10: Variation in nondimensional deflection with change in *a*/*h* value

4. Conclusion

In the present analysis a simulation model of laminated composite plate stuffed with MS material have been developed in ANSYS environment using the APDL code to analyse the stress and deflection behaviour. The model is discretised with the help of four-noded element shell-181 from the ANSYS element library. The convergence and the comparison study have been conducted to check the stability and accuracy of the present model. Finally, the effect of adding the MS material and design parameter have been analysed by solving different illustrations and discussed in detail. On the basis of the outcome of the numerical illustrations few important conclusion has been made. The deflection parameter increases with adding the Terfenol-D layer. The normal (σ_x) and shear stress (τ_{xy}) decreases while normal stress (σ_y) increases by embedding the MS layer.

References

[1] Noor A K & Burton W S 1989 Stress and Free Vibration Analyses of Multilayered Composite Plates *Composite Structures* **11** 183-204

[2] Anjanappa M and Bi J 1994 Magnetostrictive mini actuators for smart structure applications *Smart Material Structure* **3** 383-390

[3] Pratt J, R, Shafic, S and Alu H 1999 Terfenol-D Nonlinear Vibration Absorber *Journal of Intelligent Material Systems and Structures* **10** 29

[4] Vuksanovic D 2000 Linear analysis of laminated composite plates using single layer higher-order discrete models *Composite Structures* **48** 205-211

[5] Kant T and Swaminathan K 2002 Analytical solutions for the static analysis of laminated composite and sandwich plates based on a higher order refined theory *Composite Structures* **56** 329–344

[6] Huang Y, Di S, Wu C and Sun H 2002 Bending analysis of composite laminated plates using a partially hybrid stress element with interlaminar continuity *Computers and Structures* **80** 403–410

[7] Kumar J S, Ganesan N, Swarnamani S and Padmanabhan C 2003 Active control of beam with magnetostrictive layer *Computers and Structures* **81** 1375–1382

[8] Lee S J and Reddy J N 2004 Nonlinear deflection control of laminated plates using third-order shear deformation theory *International Journal of Mechanics and Materials in Design* **1** 33-61

[9] Parka T, Kimb K and Han S 2006 Linear static and dynamic analysis of laminated composite plates and shells using a 4-node quasi-conforming shell element *Composites: Part B* **37** 237–248

[10] Zhen W, Wanji C and Xiaohui R 2010 An accurate higher-order theory and C0 finite element for free vibration analysis of laminated composite and sandwich plates *Composite Structures* **92** 1299–1307

[11] Kishore M D V H, Singh B N and Pandit M K 2011 Nonlinear static analysis of smart laminated composite plate *Aerospace Science and Technology* **15** 224–235

[12] Sahoo R and Singh B N 2013 A new shear deformation theory for the static analysis of laminated composite and sandwich plates *International Journal of Mechanical Sciences* **75** 324–336

[13] Andakhshideh A and Tahani M 2013 Interlaminar stresses in general thick rectangular laminated plates under in-plane loads *Composites: Part B* **47** 58–69

[14] Rango R F, Nallim L G and Oller S 2013 Static and dynamic analysis of thick laminated plates using enriched macroelements *Composite Structures* **101** 94–103

[15] Santapuri S, Scheidler J J and Dapino M J 2015 Two-dimensional dynamic model for composite laminates with embedded magnetostrictive materials *Composite Structures* **132** 737–745

Degradation in Mechanical and Thermal Properties of Partially Aligned CNT/Epoxy Composites due to Seawater Absorption

Sunirmal Saha[1], Smrutisikha Bal[2]
[1]Department of Physics and Astronomy, National Institute of Technology, Rourkela, India
[2]Department of Metallurgical and Materials Engineering, National Institute of Technology, Rourkela, India
[1]Email : slrsixsept@gmail.com

Abstract

This literature gives an overview of the effect of water uptake on the mechanical and thermal performances of partially aligned carbon nanotube (CNT)/ epoxy (having CNTs wt.%: 0.5, 0.75 and 1) along with pure epoxy composites. Weight change behaviour of all the composites immersed in seawater for a period of six months has been recorded. The state of CNT/matrix interface is believed to be influenced due to water absorption. Reduction in flexural modulus, strength, hardness and glass transition temperature (T_g) of the specimens have been evaluated as compared to their unexposed counterparts due to adverse effect of water absorption. Results demonstrated 35-46% reduction in flexural modulus, 43-56% decrement in flexural strength, 4-7 MPa reduction in hardness and minor deterioration in T_g values for all the composites. However, the specimen having 0.75 wt.% CNT loading, exhibited minimum reduction in properties in all aspects compared to others. Scanning electron microscopy (SEM) has been employed to study the failure mechanism of the specimens after performing the mechanical tests.

Keywords
Moisture absorption, flexural properties, hardness, glass transition temperature (T_g)

1. Introduction

Carbon nanotube (CNTs) exhibits a series of excellent properties such as high elastic modulus, strength, electrical and thermal conductivities [1-3]. CNTs are thereby being used as promising reinforcing fillers in the polymer composites over a few decades for developing high-performance multifunctional composites. However, due to the low volume fraction, poor dispersion and random orientation of the CNTs in matrices, most resultant CNT-based composites can only receive limited enhancements and exhibit properties much lower than expected [4]. To overcome these challenges, various CNT preforms such as CNT arrays, CNT yarns, functionalized CNTs have been developed to pre-arrange as well as modify the CNTs in a pre-forming structure before preparing composites [5-7]. However, as surfactants and sonication are commonly involved to achieve a good dispersion of CNTs, the CNTs dispersed would usually suffer from inevitable damages or shortenings during this process, and the CNT orientation is also uncontrollable [8]. Hence, for the purpose of getting full advantages of the anisotropic properties of individual CNTs, a number of studies focused on the developments of aligned CNT architectures and aligned CNT polymer composites. It is also well established that the well-aligned CNTs could facilitate higher CNT packing densities and yield greater property enhancements in the composites than their entangled configurations [9-11].

Evaluation of the environmental performance of polymer matrix composites on the other hand is very much essential for marine and aerospace applications [12, 13]. These composite specimens are exposed to various harsh environments which include different level of humidity, different temperatures, and immersion in several types of water etc. for a certain period of time. Depending on the type and level of exposure as well as the time exposed, the magnitude of the degrading effects on exposed composites can vary significantly. Periodic weight change of the specimens can be frequently measured to quantify the level of moisture absorption over time. Instead of extensive microscopy and measurement of weight change residual mechanical and thermal properties are often measured to determine the extent of damage occurred in the exposed specimens. Some previous researches have been reported that water absorption would cause numerous unwanted effects, such as swelling, plasticization and in certain cases degradation that may significantly affect the major properties of the composites [14, 15].

However, the trend of seawater absorption and its impact on overall properties of partially aligned CNT/epoxy composites is yet to be published that is essential for marine applications. So the main objective of this work is to explore the adverse effect of moisture absorption on the overall mechanical and thermal performances of the mentioned composites. Seawater absorption is correlated with the failure mode of the nanocomposites explaining the severe effect of water on the nanotube-matrix interfacial region with the help of microscopes. It is expected that, the adoption of CNT based epoxy composites for marine applications can be accelerated by determining the effect of seawater absorption on the overall properties of this composite for prolonged immersion in sea water.

2. Experimental details

2.1 Materials

In this investigation for fabricating the composites, epoxy resin (Ciba-Geigy, araldite LY-556 based on bisphenol A) was used as a matrix material and partially aligned CNTs as reinforcement. In particular this epoxy resin (5.3-5.4 equiv/kg) was cured with an aliphatic primary amine hardener (HY-951) in the wt. ratio 100/12. As-received CNTs were purchased from NanoAmour, USA. Here the CNTs have the diameter of 10-20 nm and their lengths varying from 5–15 μm with a carbon purity of > 95%.

2.2 Nanocomposite preparation

A schematic representation of nanocomposite presentation has been shown in Figure 1. A detailed description on the processing of the nanocomposites is given in our earlier work [16].

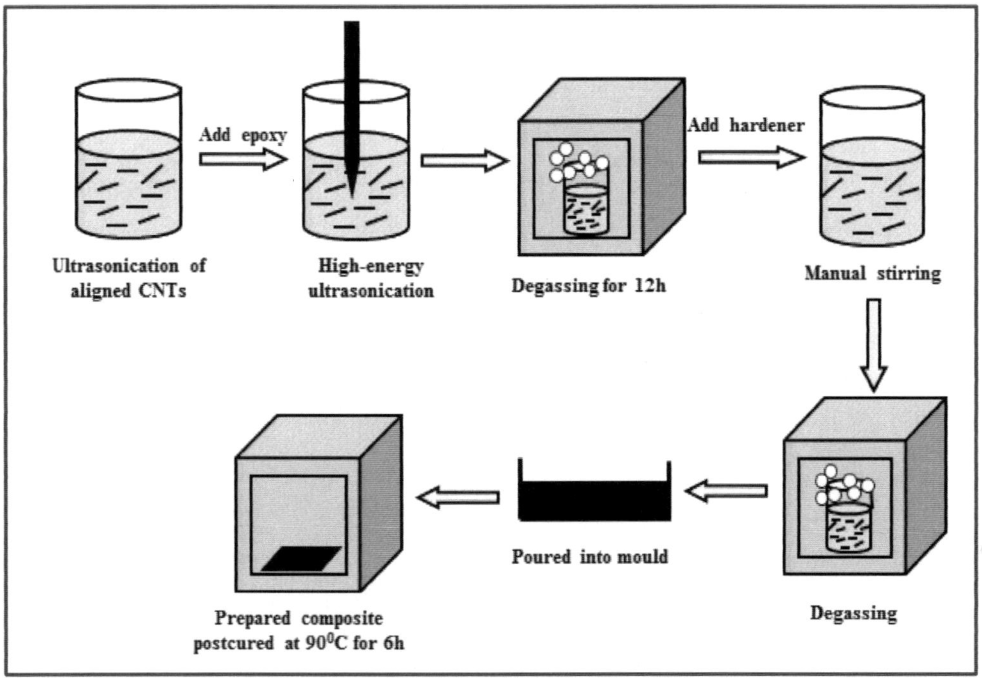

Figure 1. Schematic of nanocomposite preparation

2.3 Water absorption Tests

The moisture absorption trend of all the edge sealed specimens of dimensions 127 mm × 12.5 mm × 5 mm were studied via immersing the specimens separately into sea water up to six months. The weight gain of the samples was measured periodically using an electronic balance with an accuracy of 0.001 g after the removal of only the traveller samples from beaker. Prior to weighing, the surface of the wet samples was wiped with tissue papers. Percentage of weight gain by the samples was calculated using the formula:

$$M\% = \frac{M_t - M_0}{M_0} \times 100$$

where, M_t is the mass of the specimen after a given immersion time and M_0 is the mass of the dry specimen (before immersion). Then specimens were removed from the water bath and dried in an oven before conducting the required tests.

2.4 Characterization techniques

2.4.1 Flexural test

Flexural test was carried out according to ASTM D790–02 "Standard Test Methods for Flexural Properties of Un-reinforced and Reinforced Plastics and Electrical Insulating Materials" using universal testing machine Instron-5967 [16]. The testing mode was three point bending having a loading rate 2.0 mm/min. The dimension of the sample was 60 mm × 12.5 mm × 5 mm (span length × width × thickness) respectively.

2.4.2 Hardness test

The hardness of all the composites is studied using a micro hardness tester. The harness value is calculated using the Vickers hardness (HV) method is given by:

$$HV = 2\,F\sin(\frac{136^{\circ}}{2})/d^2\ (approximately)$$

where, F= Load in kgf and d = average length of the diagonal left by the indenter in millimetres (mm). A total of 10 points on each of the specimens were measured in order to get the average readings.

2.4.3 Thermal analysis

Differential scanning calorimetry (DSC) of all the samples had been studied to estimate the glass transition temperatures (T_g) using mettler Toledo DSC822^e differential thermal analyser; each of the sample of around 10 mg were heated from 40-150 °C with a heating rate of 5 °C/ minute in nitrogen at a flow rate of 30 ml/min.

2.4.4 Microscopic study

The surface morphology and fracture surfaces of the unexposed and seawater exposed specimens after performing the flexural test have been investigated using field emission scanning electron microscope (FESEM, Nova NANOSEM 450). The fractured surface was sputter coated with a thin gold layer in vacuum chamber for conductivity before test.

3. Results and discussion

The samples under present investigation are designated as follows: E_p- Pristine epoxy; $AC_{0.5}$– Epoxy composite with 0.5 wt.% CNTs; $AC_{0.75}$ –Epoxy composite with 0.75 wt.% CNTs; and AC_1- Epoxy composite with 1 wt.% CNTs.

3.1 Moisture absorption

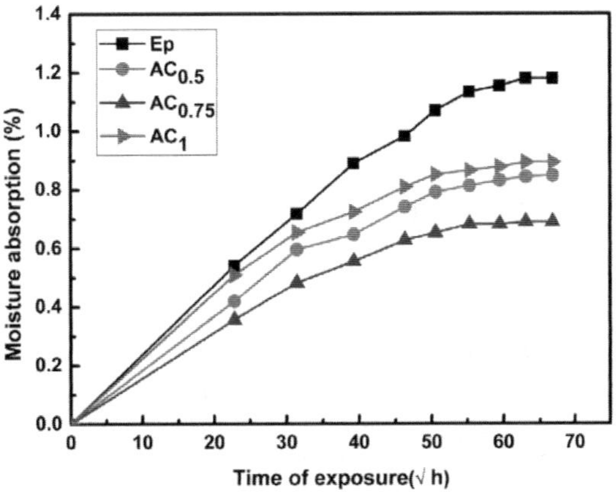

Figure 2. Moisture absorption behaviour of composites as a function of square root of time

In order to investigate the absorption behaviour of the specimens, percentage of moisture gain is plotted in Figure 2 against square root of time. The moisture absorption curves for neat epoxy and nanocomposites follow slightly different trends providing relatively lower weight gain in case of nanocomposites. Greater moisture absorption by epoxy is due to presence of hydroxyl group in it which is hydrophilic in nature and thereby very prone to water absorption. This sample gained moisture linearly and reached their saturation with equilibrium moisture content of 1.18% after 180 days of immersion in seawater. While for the nanocomposites, a mild linear increase in the moisture absorption curves was observed as compared to the pure epoxy specimens. After that, the nanocomposite specimens reached their equilibrium moisture content slowly after 110 days (approximately). The percentage of moisture absorption was 0.85%, 0.68% and 0.89% respectively for $AC_{0.5}$, $AC_{0.75}$ and AC_1 specimens.

As mentioned, the maximum water content absorbed for CNT composites is lower than that absorbed by neat epoxy resin. Present observation is well matched with some previous research reports [17-18]. The phenomenon of reduced water uptake in nanocomposites is due to the excellent barrier properties of these nano-fillers [19-20]. The presence of high aspect ratio of nano-fillers can create a tortuous pathway for water molecules to diffuse into the composites [20]. Further, when compared with other nanocomposites, $AC_{0.75}$ showed least water absorption. Indeed, naofillers occupy either additional volume or the free volume depending on dispersion quality of the nanofillers within epoxy system. When nanofillers are homogeneously distributed in the epoxy then number of nanofillers in the free volume got enhanced. These accumulated nanofillers in turn reduce the ultimate water uptake [21]. This may happen in $AC_{0.75}$ specimen and hence water penetration through the surface of this sample got seriously disturbed.

The water diffusion occurring in a material can be classified as Fickian [22], if (a) the moisture absorption graph shows an initially linear behavior with the square root of exposure time, (b) glass transition temperature (T_g) of the material is higher than the test temperature, and (c) the overall water uptake is low. Present case satisfies all the mentioned conditions; that is why the moisture absorption behaviour by the composites in this study can be categorised as Fickian nature. Similar kind of absorption phenomena also have been reported earlier by the other researchers [23, 24].

3.2 Flexural properties

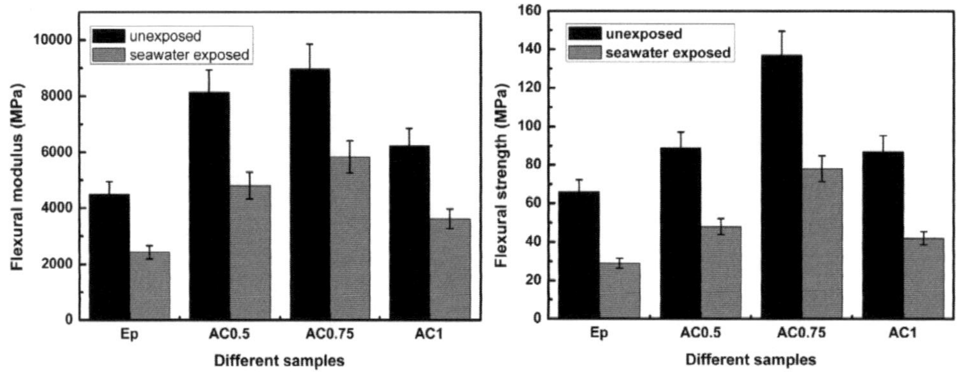

Figure 3. (a) Flexural modulus of unexposed and exposed specimens, (b) flexural strength of unexposed and exposed specimens

Among the unexposed specimens, flexural modulus & flexural strength of nanocomposites have been found to be enhanced with respect to that of neat epoxy specimen (Figure 3). The increment in modulus values were 81%, 99% and 39% respectively for $AC_{0.5}$, $AC_{0.75}$ and AC_1 specimens as compared to epoxy composite is shown in Figure 3 (a). Further, a significant amount of improvement has been noticed in the strength values also. Flexural strength of neat epoxy was 66 MPa while; the nanocomposites $AC_{0.5}$, $AC_{0.75}$ and AC_1 demonstrate the values around 89, 137 and 87 MPa respectively is shown in Figure 3 (b). This significant increase in strength & modulus of the nanocomposites compared to the neat epoxy counterparts is due to the addition of partially aligned CNTs that reinforces the performances of epoxy matrix [25]. Further, enhance effectiveness of the reinforcement and hence local stiffening occurs due to the aligned CNTs at the CNT/epoxy interface which may facilitate efficient load transfer from matrix to aligned CNTs in axial direction [26, 27]. It has been noticed that in case of higher loading of CNTs (AC_1) the increment in modulus and strength is relatively poorer as compared to others. This is probably due to the formation of CNTs' agglomeration into epoxy which causes weak bonding and thereby affecting the interfacial region [28]. However, major increment in flexural properties in case of $AC_{0.5}$ and $AC_{0.75}$ specimens are may be due to the good dispersion and occurrence of effective interfacial interaction of nanotubes into epoxy, will be further confirmed in microscopic study [29]. Obtained flexural properties were well matched with results obtained by other reports [30-31].

Flexural behaviour of all the seawater exposed specimens is shown in Figure 3. Results demonstrate that all the exposed specimens are showing reduction in flexural modulus & strength due to penetration of water molecules through the specimen. The modulus and strength values of the neat epoxy are deteriorated by 46% and 56% respectively under the action of water conditioning. It is observed that for the nanocomposites the reduction is relatively lesser as compared to neat epoxy. The decrement of moduli is 41%, 35% and 42% respectively for $AC_{0.5}$, $AC_{0.75}$ and AC_1 specimens as compared to their unexposed counterparts. Further, the reduction in strength values while comparing to non-exposed samples is 46%, 43% and 51% respectively for $AC_{0.5}$, $AC_{0.75}$ and AC_1 specimens. This reduction in flexural modulus and strength values can be attributed to the swelling and plasticization effect of water absorption in epoxy matrix [32]. Water acts as an effective plasticizer and can diffuse through the CNT-epoxy interface and weaken the bonding between them [33]. As a result, the stress transfer capability of the composites got seriously affected causing drop in flexural properties.

Apart from these, higher void content in the composites may cause degradation of composites due to higher absorption of moisture/ water and also can create poor interfacial bonding. This trapped water damaged the interface region that penetrated through the microvoids and microcracks. However, it is interesting to note that compared to neat epoxy the reduction in flexural properties is lesser in case of nanocomposites. This is due to less water absorption by the nanocomposites when compared to neat epoxy. The reason for the reduced water permeability in nanocomposites is that the nanotubes restricted the molecular dynamics of the epoxy polymer chain surrounding them by not allowing the diffusion of small molecules [34]. In addition, it is observed that reduction in properties is least in $AC_{0.75.}$ Because, at this wt.% the water absorption was also minimum among all that may be due to the achievement of strong bonding between fillers and matrix as a result of good dispersion.

3.3 Hardness

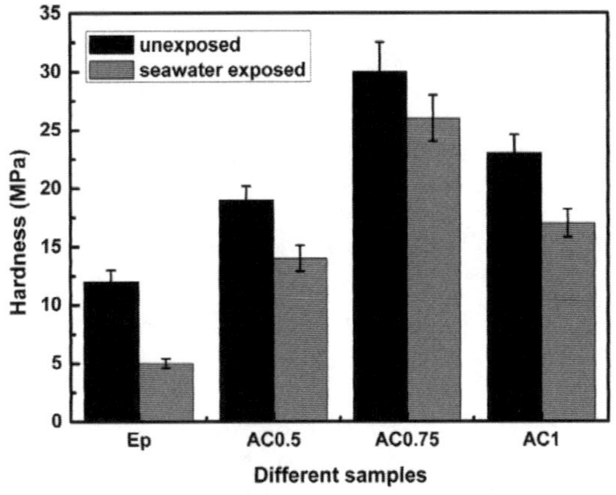

Figure 4. Hardness of unexposed and exposed specimens

In Figure 4 it has been observed that all the nanocomposite specimens demonstrated increase in hardness values as compared to the neat epoxy sample. In addition, hardness of the nanocomposites increases due to the addition of partially aligned CNTs into the epoxy matrix. An increment of 59%, 150% and 91% respectively for $AC_{0.5}$, $AC_{0.75}$ and AC_1 specimens is recorded as compared to epoxy composite. The increment in hardness value of the epoxy after the incorporation of CNTs is due to the addition of high aspect ratio, high modulus, and high strength aligned CNTs. Further, as the hardness of the composite is a localized property so that may at the point of indentation very high resistance is realized by the indenter due to the achievement strong adhesion between partially CNTs and epoxy that will be further confirmed by the micrographs.

After the water absorption test the hardness values of all the specimens got deteriorated. After the complete time of exposure to seawater (6 months) a reduction of 4-7 MPa in hardness is observed than that of unexposed one. Neat epoxy sample showed maximum decrement in hardness. The lessening of the hardness value is mainly due to the moisture absorption that causes plasticization, swelling and interfacial damages of the composite materials. Due to absorption of water molecules by the specimen, the connection between molecules of polymer may disrupt and the composite specimens become so soft that the grasp between CNT and epoxy polymer got weakened [35]. Another possible reason for the reduced hardness is the generation of pressure on CNTs. Indeed, compared to CNTs the epoxy matrix got swelled to a larger extent and thereby exerts pressure to CNTs that may sometimes breaks the nanotubes and assists in pulling out of CNTs from the matrix. These combine affects not only reduces the material properties but also generates micro cracks inside the specimen that apparently diminishes the strength and hardness of the composite material.

3.4 Glass transition temperature (T_g)

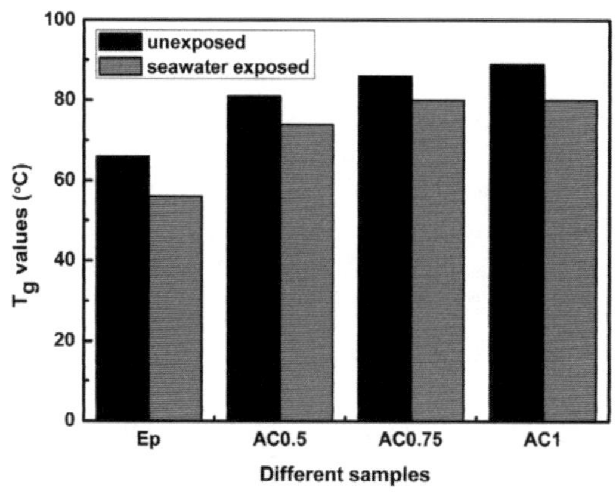

Figure 5. T_g values of unexposed and exposed specimens

Glass transition temperature (T_g) of a material is that critical temperature region in which material undergoes a transition from glassy to rubbery region, exhibiting a drastic change in the properties of the specimen. Below this temperature the material under consideration can be extensively used. Results shown in Figure 5 demonstrated an increment of around 15-23 °C in T_g value for the nanocomposites compared to neat epoxy specimen. Further T_g of the nanocomposites got slightly increased with the positive gradient of loading wt.% of CNTs. Increased values of T_g compared to neat epoxy is attributing to the cross linking reaction between epoxy and aligned CNTs [32]. Further addition of CNTs makes the epoxy matrix stiffer by restricting the mobility of chain segments of polymer and hence T_g increases [36].

T_g values of all the seawater exposed specimens are represented in Figure 5. All the specimens show reduction in T_g values (6-10 °C) after water exposure as compared to their unexposed counterparts. However, decrement of T_g value was greater in neat epoxy compared to nanocomposites. Though for the nanocomposites T_g values decreased to some extent still their values are higher than neat epoxy even after water absorption. Water exposure to these samples causes swelling (volumetric expansion) of both the matrix and fillers. Now the coefficient of volumetric expansion is higher in case of matrix rather than fillers. This difference in expansion in turn induces residual stresses at the interface. Development of residual stresses in this manner may weaken the brittle epoxy, degrade the interfacial properties, leads to generation of osmotic pressure which further creates microvoids and microcracks in the composite. This combine effects reduce the stiffness, strength and T_g of the composite [37].

3.5 Microscopical analysis

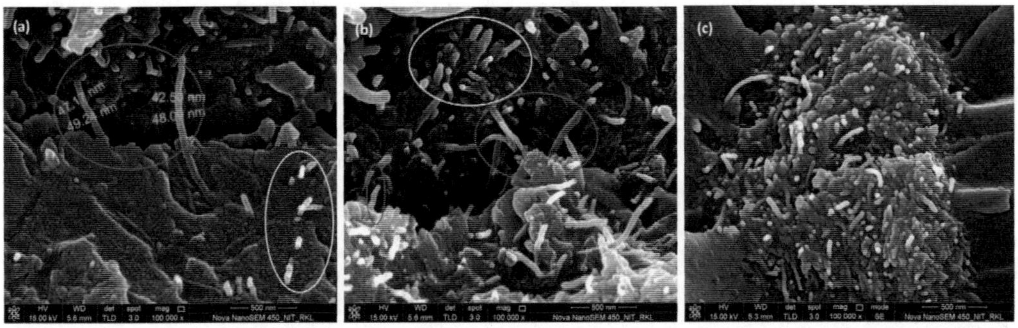

Figure 6. Fracture surface of unexposed nanocomposites showing (a) bridge formation across the crack in $AC_{0.5}$, (b) partial pull out of nanotubes in $AC_{0.75}$ and (c) agglomeration of nanotubes and weak interface in AC_1 specimen

Figure 6 (a) and (b) indicate good dispersion of CNTs in the epoxy matrix. Careful observation revealed that a major fraction of nanotubes which are in axial direction along the load got completely ruptured and partial pull out (shown by white circles) of CNTs has been noticed due to the application of load. This means that strong bonding occurs between the aligned nanotubes and host matrix. While, some of the nanotubes that are placed orthogonal to the direction of applied load try to resist the crack induced in the matrix by bridge formation (shown in red circles). The propagation of cracks are obstructed by the nanotubes is reported here which is the cause behind the increment in mechanical properties of the nanocomposites with respect to the neat epoxy resin. On the other hand, for higher wt.% (Figure 6 (c)) voids and agglomeration of nanotubes have been noticed. Due to the presence of voids moisture absorption at this wt.% was higher among the nanocomposites. Both the voids and agglomeration of CNTs are responsible for not forming good interfacial region resulting minimum values of flexural modulus and strength when compared with the other nanocomposites.

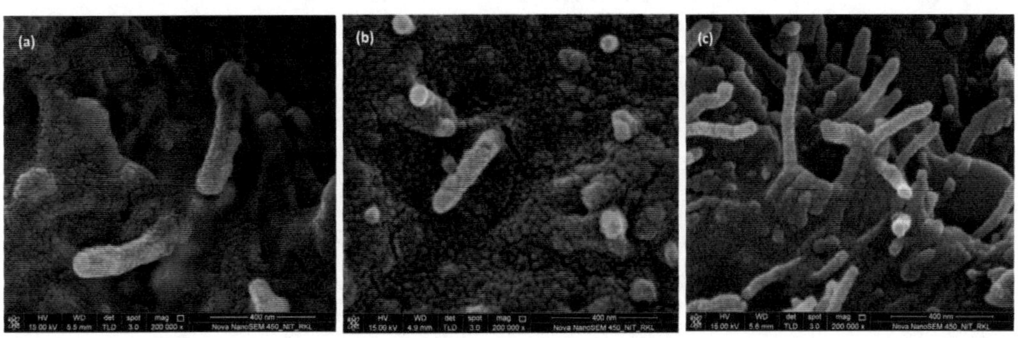

IOP Conf. Series: Materials Science and Engineering **178** (2017) 012027 doi:10.1088/1757-899X/178/1/012027

Figure 7. Fracture surface of moisture absorbed nanocomposites showing swelling of epoxy matrix and CNTs as well as partial and complete pull out of CNTs in (a) $AC_{0.5}$, (b) $AC_{0.75}$, and (c) AC_1 specimen.

Fracture surfaces of water absorbed nanocomposites are shown in Figure 7. Water absorption reduces the compatibility between the matrix and CNTs which contributes to debonding and weakening the interfacial adhesion. Matrix cracking is clearly observed in all the micrographs due to the generation of internal stresses as a result of water absorption. Further, swelling of both the matrix as well as nanotubes damages the interfacial region and hence load bearing capacity by the interface got reduced. CNTs pull out is clearly visible in the micrographs Figure 7 (a), (b) and (c) which is a clear evidence of interfacial debonding phenomena are very much responsible for the deterioration in mechanical properties. However, the degree of pull out is relatively less in $AC_{0.75}$ specimen (Figure 7 (b)) compared to others. This is why at this wt.% lesser degradation in proeprties has been noticed as compared to others.

4. Conclusions

- Flexural modulus, strength, hardness and T_g of the unexposed specimens were enhanced compared to neat epoxy specimens.
- In seawater exposed case, weight gain of neat epoxy was 1.18% while for $AC_{0.5}$, $AC_{0.75}$, and AC_1 it was 0.85%, 0.68% and 0.895 respectively i.e., inclusion of nanotubes reduced the water uptake of epoxy polymer.
- Degradation in properties of all the composites to some extent were realised due to after effects of moisture absorption. However, deterioration in properties was more pronounced in neat epoxy compared to nanocomposites. Further, among the nanocomposites $AC_{0.5}$ showed lesser degradation in all aspects.
- Enhancement in mechanical properties in nanocomposites as compared to neat epoxy was found to be dependent on the quality of dispersion of nanotubes in epoxy for the unexposed specimens. While, swelling of the matrix and fillers as well as CNT pull out were marked to be responsible for the reduction in mechanical and thermal performance of the water exposed specimens.

References

[1] Dresselhaus MS, Dresselhaus G and Saito R 1992 *Phys. Rev. B* **45(11)** 6234-6242.

[2] Li QW, Li Y, Zhang XF, Chikkannanavar SB, Zhao YH, Dangelewicz AM, Zheng LX, Doorn SK, Jia QX, Peterson DE and Arendt PN 2007 *Adv. Mater* **19(20)** 3358-63.

[3] Pop E, Mann D, Wang Q, Goodson K and Dai H 2006 *Nano. Lett* **6(1)** 96-100.

[4] Tsuda T, Ogasawara T, Moon SY, Nakamoto K, Takeda N, Shimamura Y and Inoue Y 2014 *Compos Part A Appl. Sci. Manuf* **65** 1-9.

[5] Wardle BL, Saito DS, García EJ, Hart AJ, de Villoria RG and Verploegen EA 2008 *Adv. Mater* **20(14)** 2707-14.

[6] Jiang K, Li Q and Fan S 2002 *Nature* **419(6909)** 801.

[7] Yang K, Gu M, Guo Y, Pan X and Mu G 2009 *Carbon* **47(7)** 1723-37.

[8] Liu P, Lam A, Fan Z, Tran TQ and Duong HM 2015 *Mater. Des* **87** 600-5.

[9] Sepúlveda AT, de Villoria RG, Viana JC, Pontes AJ, Wardle BL and Rocha LA 2013 *Nanoscale* **5(11)** 4847-54.

[10] Esawi AM and Farag MM 2007 *Mater. Des* **28(9)** 2394-401.

[11] Huang J and Rodrigue D 2013 *Mater. Des* **50** 936-45.
[12] Ahmad M, Singh D, Fu YQ, Miraftab M and Luo JK 2011 *Polym. Degrad. Stab* **96(8)** 1470-7.
[13] Chevali VS, Dean DR and Janowski GM 2010 *Polym. Degrad. Stab* **95(12)** 2628-40.
[14] Nogueira P, Ramirez C, Torres A, Abad MJ, Cano J, Lopez J, López-Bueno I and Barral L 2001 *J Appl. Polym. Sci* **80(1)** 71-80.
[15] Cabanelas JC, Prolongo SG, Serrano B, Bravo J and Baselga J 2003 *J. Mater. Process. Tech* **143** 311-5.
[16] Bal S and Saha S 2014 *High. Perform. Polym* **26(8)** 953-60.
[17] Mohan TP and Kanny K. 2011 *Compos. Part A: Appl. Sci. Manuf* **42(4)** 385-93.
[18] Becker O, Varley RJ and Simon GP 2004 *European Polym. J* **40** 187–95.
[19] Liu W, Hoa SV and Pugh M 2005 *Compos. Sci. Technol* **65** 2364–73.
[20] Zhao H and Li RK 2008 *Compos. Part A: Appl. Sci. Manuf* **39(4)** 602-11.
[21] Prolongo SG, Campo M, Gude MR, Chaos-Morán R and Ureña A 2009 *Compos. Sci. Technol* **69(3)** 349-57.
[22] Abot JL, Yasmin A and Daniel IM 2005 *J. Reinf. Plast. Compos* **24(2)** 195-207.

[23] Grace LR and Altan MC 2012 *Compos. Part A: Appl. Sci. Manuf* **43(8)** 1187-96.
[24] Muthirakkal S, Murthy HN, Krishna M, Rai KS and Karippal JJ 2010 *Iran Polym. J* **19(2)** 89-103.
[25] Zulfli NM, Bakar AA and Chow WS 2013 *J. Reinf. Plast. Compos* 2013, **32** 1715–1721.
[26] Thostenson ET and Chou TW 2002 *J. Phys D: Appl. Phys* 2002 **35(16)** L77.
[27]Xie XL, Mai YW and Zhou XP 2005 *Mater. Sci. Eng. R-Rep* **49(4)** 89-112.
[28] Song YS and Youn JR 2005 *Carbon* **43(7)** 1378-85.
[29] Allaoui A, Bai S, Cheng HM and Bai JB 2002 *Compos. Sci. Technol* **62(15)** 1993-8.
[30] Zhou Y, Pervin F, Lewis L and Jeelani S 2008 *Mater. Sci. Eng: A* **475(1)** 157-65.
[31] Ma PC, Mo SY, Tang BZ and Kim JK 2010 *Carbon* **48(6)** 1824-34.
[32]Lee JH, Rhee KY and Lee JH 2010 *Appl. Surf. Sci* **256(24)** 7658-67.
[33] Alamri H and Low IM 2013 *Compos. Part A: Appl. Sci. Manuf* **44** 23-31.
[34] Ladhari A, Daly HB, Belhadjsalah H, Cole KC and Denault J 2010 *Polym. Degrad. Stab* **95(4)** 429-39.
[35] Anter FH and Hasan HS 2011 *J. Univ. Anbar Pure Sci* **5** ISSN: 1991-8941.
[36] Zhu J, Wei S, Ryu J, Budhathoki M, Liang G and Guo Z 2010 *J. Mater. Chem* **20(23)** 4937-48.
[37] Schutte CL 1994 *Mater. Sci. Eng. R-Rep* **13(7)** 265-323.

IOP Conf. Series: Materials Science and Engineering **178** (2017) 012028 doi:10.1088/1757-899X/178/1/012028

Modeling of microstructure evolution in direct metal laser sintering: A phase field approach

Jyotirmoy Nandy, Hrushikesh Sarangi and Seshadev Sahoo*
Department of Mechanical Engineering, Institute of Technical Education and Research, Siksha 'O' Anusandhan University, Odisha, Bhubaneswar-751030

Email: seshadevsahoo@soauniversity.ac.in

Abstract. Direct Metal Laser Sintering (DMLS) is a new technology in the field of additive manufacturing, which builds metal parts in a layer by layer fashion directly from the powder bed. The process occurs within a very short time period with rapid solidification rate. Slight variations in the process parameters may cause enormous change in the final build parts. The physical and mechanical properties of the final build parts are dependent on the solidification rate which directly affects the microstructure of the material. Thus, the evolving of microstructure plays a vital role in the process parameters optimization. Nowadays, the increase in computational power allows for direct simulations of microstructures during materials processing for specific manufacturing conditions. In this study, modeling of microstructure evolution of Al-Si-10Mg powder in DMLS process was carried out by using a phase field approach. A MATLAB code was developed to solve the set of phase field equations, where simulation parameters include temperature gradient, laser scan speed and laser power. The effects of temperature gradient on microstructure evolution were studied and found that with increase in temperature gradient, the dendritic tip grows at a faster rate.

1. Introduction

Direct Metal Laser Sintering (DMLS) process is a type of powder-based additive manufacturing method capable of producing complex structures out of metallic powder in a layer by layer fashion. The process creates three-dimensional solid objects by bonding powdered materials using laser energy [1]. The key advantages of such a layer based fabrication technology over the conventional manufacturing are high degree of design freedom without the need for part-specific tooling, easy processing of powder material, environment friendliness due to minimization of waste and emission, less expense due to absence of support structures [2, 3]. Figure 1 shows the schematic diagram of the DMLS.

Corresponding author: S. Sahoo, Email: seshadevsahoo@soauniversity.ac.in, sahoo.seshadev@gmail.com

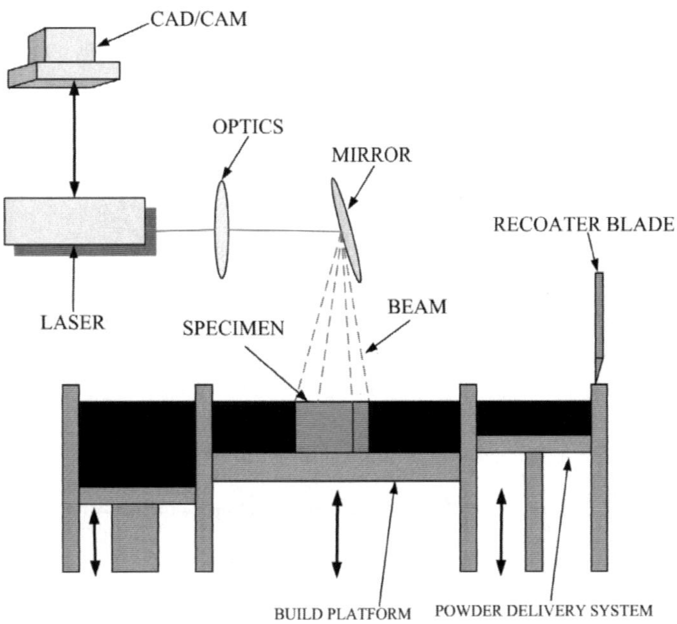

Figure 1. Schematic diagram of DMLS process

Microstructure is the central core of a material which determines the physical and mechanical properties of materials. For improving the properties of material, it is necessary to control the microstructure of the material as well as the process parameters. But in reality it is very difficult to control the microstructure of the material during experiment. Microstructure evolution takes place to reduce the total free energy that may include the bulk chemical free energy, interfacial energy, elastic strain energy, magnetic energy, electrostatic energy etc. [4]. Due to complex and nonlinear nature of microstructure evolution, numerical approaches have been employed. There are different approaches developed for microstructure modeling. During the past few years, the phase-field approach has emerged as one of the most powerful methods for modeling many types of microstructure evolution in different processes like solidification, solid-state phase transformation, grain growth and coarsening and martensitic transformation etc. [5]. The phase-field method describes the microstructure of a material by using a set of conserved and non-conserved field variables, that are continuous across the interfacial region. An important advantage of this phase-field method is that it considers diffuse interface description, there is no need to track the interfaces during microstructure evolution [6].

Over the past decades few attempts have been made to simulate the microstructure evolution in additive manufacturing processes. Fallah et al. [7] developed a quantitative phase field model for the laser deposition of Ti–Nb alloy. Their model was used to simulate the dendrite growth pattern under thermal conditions i.e., temperature gradient and solidification velocity which obtained from the previously developed thermal model. Rishi et. al. [8] developed a model for the microstructure evolution in selective laser sintering in which a discrete element approach was used, where the particle-to-particle and particle-to-wall mechanical and thermal interactions has been considered. Sahoo et al. [9, 10] developed a quantitative phase field model for the microstructure evolution of Ti-6Al-4V in electron beam additive manufacturing process. The authors used temperature gradient and the solidification velocity as the simulation parameters, which were previously extracted from a thermal model. From the simulation results, the columnar dendritic arm spacing values were estimated and compared with the experimental and analytical results. Montiel et al. [11] studied the microstructure analysis of AZ31 magnesium alloy welds by using phase field method. From the simulation they found the condition for columnar to equiaxed transition of dendrites and studied the effect of process parameters on the size and shape of dendrites.

Till date there was no such comprehensive study carried out for the microstructure evolution of laser sintering process. The objective of this study is to apply the phase field method for the microstructure evolution in Direct Metal Laser Sintering process.

2. Phase field formulation

Phase field method is employed to numerically simulate the evolution microstructure of Al-Si-10Mg in DMLS process. The governing equations, when coupled with the temperature and diffusion equations shows the behavior of the interface which is smooth but experiences brisk changes in the phase field. The following assumptions are considered for the phase field model:

i. The constant temperature term is replaced with temperature which changes with time and position, i.e. $T \rightarrow \overrightarrow{T_c}(\vec{x},t)$, where $\vec{x}$ is position vector and t is time, so as to make the system non isothermal.
ii. Latent heat of fusion has been neglected and temperature field has been introduced in the form of temperature gradient.
iii. Thermo-physical properties of the system are taken as constants.
iv. Solid/liquid interface is taken to be in local equilibrium.

By using the conserved and non-conserved phase field variables, the governing equations are given by

$$\tau A^2(\varphi)\frac{\partial\varphi}{\partial t} = W_\varphi^2\nabla.(A^2(\varphi)\nabla\varphi) - W_\varphi^2\partial_x\left[A(\varphi)A'(\varphi)\partial_y\varphi\right] + W_\varphi^2\partial_y[A(\varphi)A'(\varphi)\partial_x\varphi] - \frac{d_g(\varphi)}{d\varphi} - \frac{L(T_c - T_m)}{HT_m}\frac{dp(\varphi)}{d\varphi} + \eta'(x,T_c) \qquad (1)$$

$$\frac{\partial T}{\partial t} = \nabla.(\alpha(\varphi)\nabla T) + \frac{Lh'(\varphi)}{C_p}\frac{\partial\varphi}{\partial t} \qquad (2)$$

where,
$A'(\varphi)$ = derivative of $A(\varphi)$ with respect to the order parameter
$\eta'(x,T_c)$= stochastic noise function emulating thermal fluctuations

By rescaling space and time as $(\bar{x} \rightarrow x/W_\emptyset)$ and $(\bar{t} \rightarrow t/\tau)$, we get

$$\varphi^{n+1} = \varphi^n + \frac{\Delta\bar{t}}{A^2(\theta)}\left\{\nabla.\left(A^2(\theta)\nabla\varphi - \partial_{\bar{x}}\left[A(\theta)A'(\theta)\partial_{\bar{y}}\varphi\right] + \partial_{\bar{y}}[A(\theta)A'(\theta)\partial_{\bar{x}}\varphi]\right) - \frac{dg(\varphi)}{d\varphi} - \bar{\lambda}H\frac{dP(\varphi)}{d\varphi} + \eta'(x,T_c\right\} \qquad (3)$$

$$H^{n+1} = H^n + \frac{\bar{D}\Delta\bar{t}}{\nabla^2\bar{x}}\Delta^2H^n + \Delta\bar{t}h'(\varphi^n)\left(\frac{\varphi^{n+1} - \varphi^n}{\Delta\bar{t}}\right) \qquad (4)$$

Where$\bar{D} = a_2\bar{\lambda}$, a_2 is a constant, $\bar{\lambda}$ is a rescaled λ

In the current phase field model, the anisotropy effects are included as given in the following equations,

$$A(\theta) = 1 - 3\varepsilon_4 + 4\varepsilon_4 \frac{\left(\partial\varphi/_{\partial x}\right)^4 + \left(\partial\varphi/_{\partial y}\right)^4}{|\nabla\varphi|^4} \ldots\ldots\ldots\ldots\ldots\ldots\ldots (5)$$

As mentioned in the assumptions, the temperature field was introduced in the form of temperature gradient. The temperature of the molten pool was derived from Echebarria et al. [12]

$$T(x,t) = T_0 + G(x - V_p t) \ldots\ldots\ldots\ldots\ldots\ldots\ldots (6)$$

where,

T_o= reference temperature,

V_p= pulling speed in the x-direction

G= temperature gradient.

3. Numerical procedures for the phase field simulation

For microstructure simulation the temperature gradient was extracted from the thermal model and integrated in the phase field model. Finite difference method with the central difference scheme was used to solve the governing equations with zero flux boundary condition. To solve the phase field equations, a MATLAB code was developed. The phase field model was carried out by considering the top surface of the powder layer in the build geometry which is in X and Y directions. The heat flux affects the top surface of the powder layer and the microstructural changes occur in this surface. The computational domain was taken with 200 μm × 200 μm in both X and Y direction. The domain was divided into 1000 × 1000 number of grids with grid spacing of dx=dy=0.2 μm. The simulations were initialized with a circular seed of radius 5μm at the center of the computational domain.

The iterations were based on the following procedure:

i. Compute the initial parameters and the phase field variable in the defined domain.
ii. Prepare the order parameters field and start the modeling by enabling the adaptive mesh refinement
iii. Build the anisotropy functions of the order parameters, differentiate them anisotropic order and fix the boundary conditions.
iv. Build the temperature field with order parameters and differentiate them.
v. Solve the phase field equations using the central difference scheme of the finite difference method.
vi. Provide boundary conditions to the temperature field and add the stochastic noise.
vii. Update the fields using new values and then return to steps (iii) and (iv) for next iteration until the field reaches the boundary.

4. Results and Discussion

There are a number of process parameters which determine the microstructure evolution of the build sample in DMLS process. It is very important to control all these parameters namely solidification growth rate, temperature gradient, chemical composition and the undercooling. Among all these parameters, temperature gradient is one of the most important parameter on which the microstructure evolution depends. The thermo-physical properties of Al-Si-10Mg and the process parameters used for simulation are given in Table 1 and Table 2. The temperature gradient is calculated from the thermal simulation profile shown in Figure 2.

Table-1: Thermo-physical properties of AlSi10Mg alloy [13]

Properties	**Values**
Thermal conductivity (K)	113 W/mK
Specific heat capacity (C_p)	940J/kgK
Convective heat transfer coefficient (h)	80 W/m^{2}K
Density (ρ)	2.67 g/m^3
Emissivity (ε)	0.19
Solidus temperature (T_s)	830 K
Liquidus temperature (T_l)	869 K

Table-2: Process parameters of DMLS process

Parameters	**Values**
Laser Power	130 Watts
Scanning speed	500 mm/s
Laser spot size	0.2 mm
Thickness of layer	1mm
Laser absorptivity	0.95

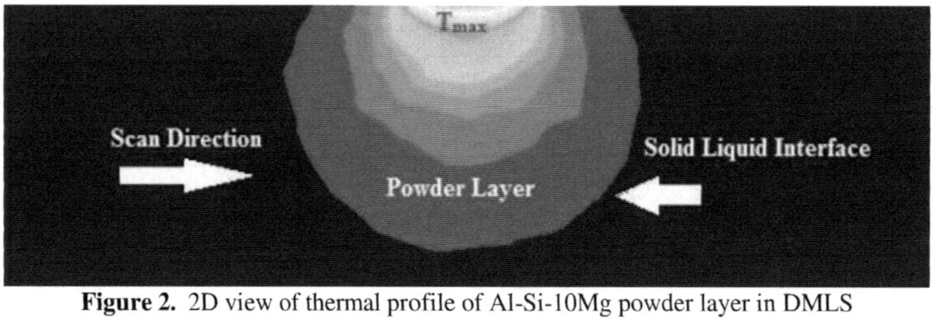

Figure 2. 2D view of thermal profile of Al-Si-10Mg powder layer in DMLS

The temperature gradient was calculated based on the correlation [8]

$$G = \frac{T_{max} - T_l}{r} \quad \ldots\ldots (7)$$

The main factors for the calculation of the temperature gradient are
T_{max} = the maximum temperature in the melt pool
T_l = liquidus temperature of the alloy
r = distance between the liquidus line and the spot of the maximum temperature
Based on the above calculation the temperature gradient for a scan speed of $500\ mms^{-1}$ was found to be $528.27 Kmm^{-1}$. Taking the temperature gradient and the scan speed at $500\ mms^{-1}$, the phase field simulation is carried out. Figure 3 shows the phase profile of Al-Si-10Mg at different time. At time t=0 sec, the computational domain is initialized with a small solid nucleus having a size of 5 μm placed in the center of the domain. The colour bar of the simulation domain represents constant values, so it has no

unit. In the colour bar, '0' represents completely liquid state and 100 represents completely solid state. At time t = 50 ms, the small solid nucleus grows and forms the dendritic arms. With increase in solidification time the dendritic arms grow and finally it grows like columnar structure which is clearly seen in the simulation results. The laser sintering process involves partial melting and solidification phenomena. When the laser beam scans the powder bed, a large temperature gradient created on the powder bed due to the high energy laser beam. This temperature gradient helps to melt the powder particles in the powder bed, then solidify and sinter the powder particles within few fraction of seconds. Due to the melting and solidification, a large temperature gradient exists on the powder bed which helps for the growth of the dendritic arm.

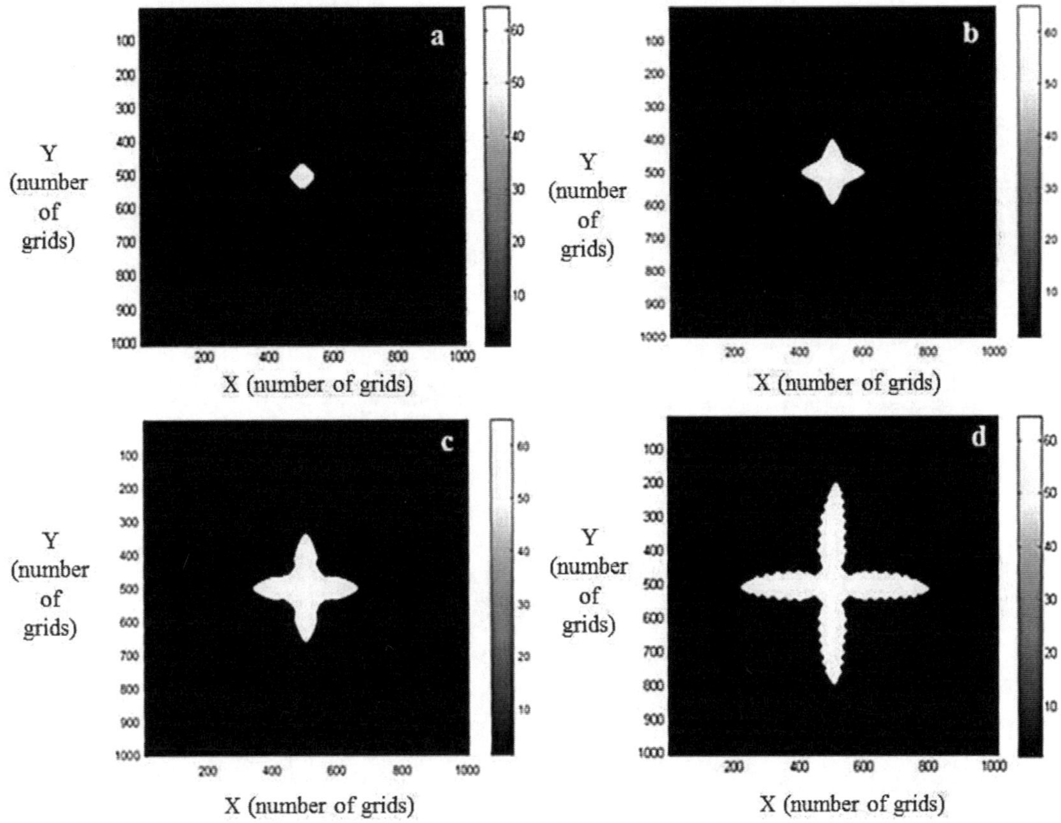

Figure 3. Phase profile of Al-Si-10Mg at different solidification time (a) t= 50 ms (b) t=300ms (c) t=1500 ms (d) t=13000 ms

4.1 Effect of temperature gradient on microstructure

The phase field simulations of sintered Al-Si-10Mg are carried out using different temperature gradients. All of these temperature gradients are found using the thermal simulations and the melt pools inhibited by their results. As the temperature gradients have been used prominently in the simulations, one important factor plays a vital role in the final microstructure formation of the materials i.e., cooling rate. The cooling rate which can be described as,

$$T^* = GR \quad \text{......} \quad (8)$$

Where G= temperature gradient (Kmm^{-1})

R = solidification growth rate (mms^{-1})

The cooling rate, being directly proportional to the temperature gradient and the solidification growth rate has a great impact on the microstructure. Any change amongst both of the parameters leads to a change in the final dendritic structure. Increase in the cooling rate tends to extract more heat from the liquid which results in the increase in growth of the dendrites as shown in Figure 4.

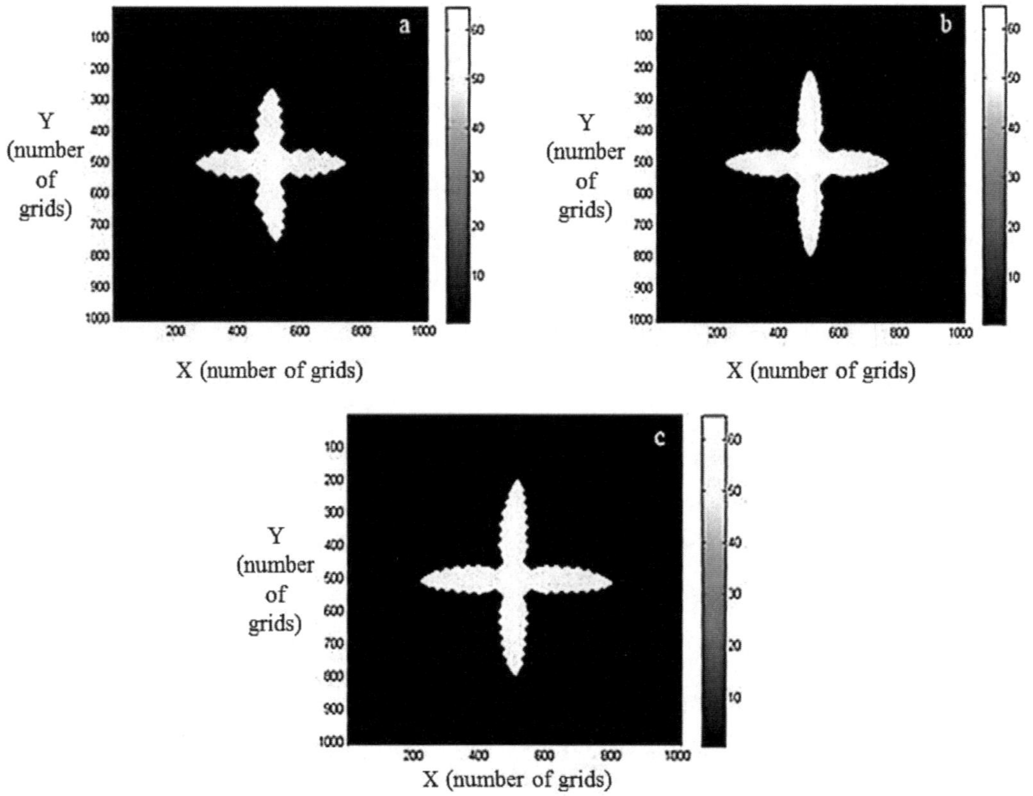

Figure 4. Phase profiles of Al-Si-10Mg at different temperature gradients (a) 263.33 K/mm (b) 318.18 K/mm(c) 362.66 K/mm

The phase profile shows at low temperature gradient; the growth of the columnar dendrites is slower as compared to high temperature gradient. With increase in temperature gradient the growth becomes faster and more columnar structure is observed. This is due to the increase in cooling rate with increase in temperature gradient as the scan speed remains constant. As the cooling rate increases, the extraction of heat from the liquid increases which results faster growth.

5. Conclusion

Phase field method was successfully applied for the simulation of microstructure of Al-Si-10Mg alloy in DMLS processes. MATLAB code was used to solve the phase field equations, which incorporates the temperature gradient and laser scan speed as the simulation parameters. As the input to the phase field model i.e., the temperature gradient was extracted from the thermal model and incorporated in the phase field model. From the simulation results it was observed that the temperature gradient plays an important role for the evolution of microstructure. With increase in temperature gradient, the tip of the dendrites grows faster.

References

[1] Simchi A. 2006, Materials Science and Engineering A.**428**148.
[2] Kruth JP, Wang X, Laoui T,Froyen L.2003, Assembly Automation.**23**357.
[3] Qian B, Shen Z. 2013, J. Asian Ceramic Societies. **1(4)**315.
[4] Grong Ø, Shercliff HR. 2002,Progress in Materials Science.**47** 163.
[5] Chen L. 2002, Annual Review of Material Research.**32** 113.
[6] Steinbach I. 2009, Modeling and Simulation in Materials Science and Engineering.**17** 1.
[7] Fallah V, Amoorezaei M, Provatas N, Corbin SF, Khajepour A. 2012, ActaMaterialia. **60**1633.
[8] Ganeriwala R, Zohdi TI. 2014Procedia CIRP **14** 299.
[9] Sahoo S. 2014,Int. J. Nano and Biomaterials. **5(4)**228.
[10] Sahoo S,Chou K. 2016, Additive Manufacturing **9** 14.
[11] Montiel D, Liu L, Xiao L, Zhou Y, Provatas N. 2012, Acta Mater. **60** 5925.
[12] Echabarria B, Folch R, Karma A, Plapp M. 2005, Physical Review E.**72** 061604.
[13] Hu H, Ding X, Wang L. 2016, Optik **127** 8883.

Development of Ceramic Coating on Metal Substrate using Industrial Waste and Ore Minerals

S K Bhuyan[1], T K Thiyagarajan[2] and S C Mishra[1]
[1]Metallurgical & Materials Engineering Department, N.I.T. Rourkela.
[2]Laser and Plasma Technology Division, BARC, Mumbai.

Email: subratbhuyan2014@gmail.com

Abstract
The technological advancement in modern era has a boon for enlightening human life; but also is a bane to produce a huge amount of (industrial) wastes, which is of great concern for utilization and not to create environmental threats viz. polution etc. In the present piece of research work, attempts have been made to utilize fly ash (wastes of thermal power plants) and along with alumina bearing ore i.e. bauxite, for developing plasma spray ceramic coatings on metals. Fly ash and with 10 and 20% bauxite addition is used to deposit plasma spray coatings on a metal substrate. The surface morphology of the coatings deposited at different power levels of plasma spraying investigated through SEM and EDS analysis. The coating thickness is measured. The porosity levels of the coatings are evaluated. The coating hardness isalso measured. This piece of research work will be beneficial for future development and use of industrial waste and ore minerals for high-valued applications.

Keywords: Plasma Spraying; Industrial Wastes; Ceramic Coating

1. Introduction

Coal-fired thermal power stations generate vast quantities of fly ash as a by-product [1].In present years, fly ash is utilized for manufacture of cement, bricks, concrete, and tiles along with road embankment, mine filling and so forth [2]. Since last two decades, industrial wastes have been investigated for making high valued utilization [3, 4]. In this context fly ash; the waste of captive power plant and red mud, the waste of alumina companies are a subject for creating environmental threat and pollution. Some studies have already been made to use these wastes to develop plasma spray coatings [5,6]. It has been observed that, these coatings can be useful for wear resistant purpose of machine tool components and for thermal barrier applications [7-11].The present piece of research work is aimed at, not only to use an industrial waste but also addition of an ore mineral with the industrial waste and to evaluate its functionality and usefulness. Fly ash with 10% and 20% bauxite additions are plasma sprayed on metals to develop overlay coatings.

The use of Plasma spray coating is limited due to high cost of plasma spray grade powders. The objective of the present work is to develop a plasma spray ceramic coating on metal substrate using an industrial waste.

2. Materials and Methods

The raw material used are, fly ash from captive power plant of Rourkela Steel Plant (CPP-2) and bauxite ore obtained from National Aluminium Company (NALCO). Particle size analyses of both the materials

are done by sieve analysis. In case of fly ash, the particle size was ranging from 20 to 110 micron and most of the particle are in the range of 40-80 micron, where as in case of bauxite powder, although the particle are in the range of 20-90 micron but majority of particles are below 60 micron.
Apart from raw fly ash, the other two compositions are made by adding 10 and 20 wt. % of bauxite ore to the fly ash. Prior to plasma spraying the raw materials are mixed thoroughly. Compositional analysis of fly ash and bauxite ore fines were done by wet chemical analysis and given in the table 2.1.Loss on ignition values were calculated by heating the raw materials at 1000^{0}C for 1 hour.

Table 2.1Weight Chemical analysis of raw materials.

Raw materials	Constituents (Oxides) in Wt. %					LOI
	SiO_2	Al_2O_3	Fe_2O_3	TiO_2	CaO/K_2O etc.	
Fly ash	61.2	31.5	2.6	1.25	2.75	0.7
Bauxite	4.0	43.0	23.0	3.0	5.0	22.0

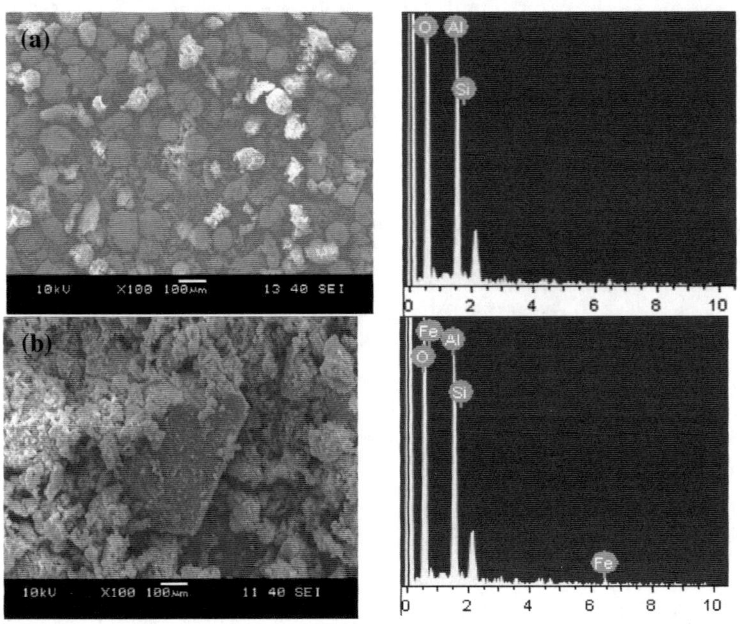

Element	Wt. %	At. %
O K	56.45	69.16
Al K	15.48	11.24
Si K	28.07	19.59

Element	Wt. %	At. %
O K	52.84	68.95
Al K	32.30	24.99
Si K	1.37	1.02
Fe L	13.50	5.05

Figure. 2.1 Scanning electron micrographs with EDS spectra of a) Fly ash and b) bauxite.

The substrate chosen for the spraying is copper plates having 3mm thickness and a cross sectional area of 25.4mm x 25.4 mm. The substrates were cleaned properly and were sand blasted to get a surface roughness of ~4.0 Ra. A non-transferred arc plasma torch was used for plasma spraying. The power levels maintained are 10, 12, 16 and 20 kW DC. Argon is used as the plasma forming gas, with a flow rate of 30 L/min and with nitrogen at a flow rate of 2 L/min. Argon was also used as carrier gas with a flow rate of 6 L/min. The powder feed rate was maintained at 10gm/min. Torch to substrate distance 70mm is maintained. The surface morphology was studied using JEOL (JSM-6480LV) scanning electron microscope. The coating thickness was measured on the polished cross section of the specimen with an optical microscope. The porosity of the coatings was evaluated using AXIO vision software. The micro hardness of the coatings was measured with Vickers hardness tester with a load of 50 gf and dwell time of 20 sec. (as per machine specifications).

3. Results and discussion

The fly ash coatings without bauxite addition made at different operating power levels of plasma gun i.e. 10 kW to 20 kW are shown in figure 3.1.

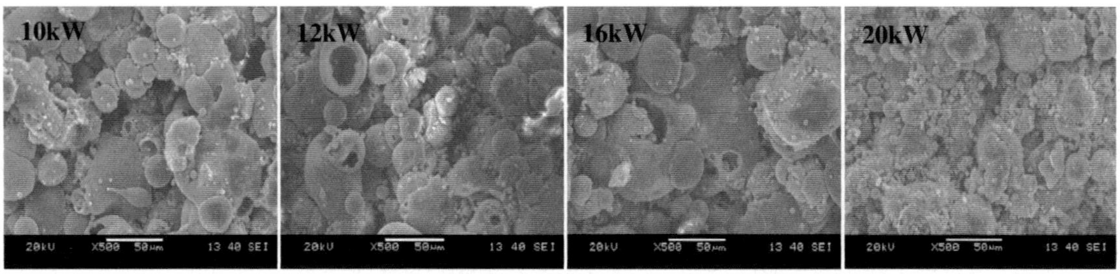

Figure 3.1 Scanning electron micrographs of fly ash coated samples
10, 12, 16 & 20 kW power level.

From the coatings it can be seen that, at lower power level there is a lot of porosity, though some semi-molten globular particles are observed. With increase in power level, the flatness on coating (surface) is observed i.e. the semi-molten particle adhere to each other. That's why showing flatten regions. With further increase in power level, although the globular particles are seen, the porosity level of the coating is reduced; but there are some small fragmentations of particles, observed on the spheroids bigger particles/masses. Whereas, further increase in power level i.e. at 20kW, although very small amount of fragmented particles are seen but there is an increase in porosity (as evidenced from figure 3.3). From the EDS analysis it is observed that, there is a variation in silicon, aluminum and oxygen content with varying the power levels.

Scanning electron micrographs of fly ash+10%Buxite coated samples 10, 12, 16 & 20 kW power levels are shown in figure.3.2.

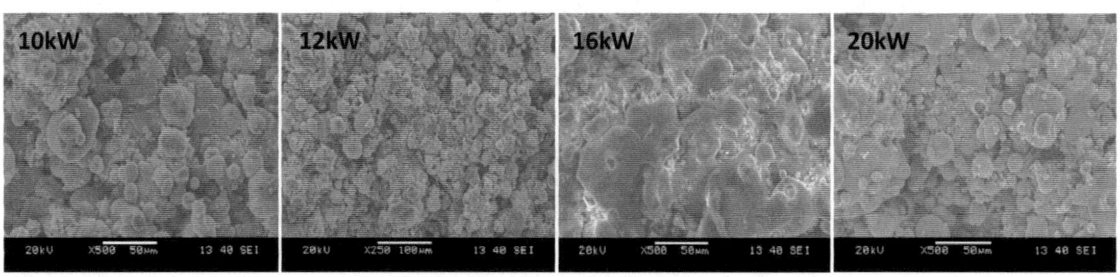

Figure 3.2 Scanning electron micrographs of fly ash coated samples
10, 12, 16 & 20 kW power level.

The surface morphology of coating is developed with 10% bauxite addition with fly ash as sheen in figure 3.2; similar type of observations is also made. Although the coating becomes denser with increase in power level, but flattening zone is more at 16kW than that of the coating deposited at 20kW. But at 20kW power level some small spherical particles are observed. It may be due to at high power level, fragmentations of particles have occurred and the small size particles are semi-molten and directly deposited. As because it a ceramic body, the temp zone at axial region of the plasma is in terms of few thousands of degree, whereas the flame end temp is of few hundred degrees. So quenching rate is very fast that's why small fragmented particles also got deposited at some regions. Also there is an increase in porosity level at 20kW power level. Similar types of observations are also noticed in case of coatings deposited with 20% Bauxite addition.

The porosity of the coatings was measured/evaluated using AXIO vision software is shown in Figure. 3.3.

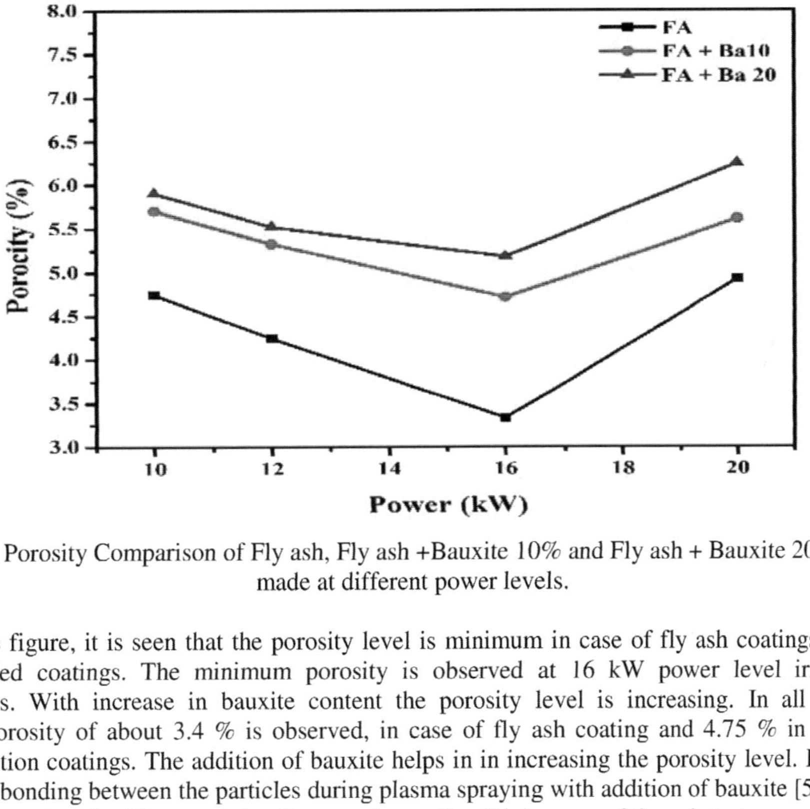

Figure.3.3 Porosity Comparison of Fly ash, Fly ash +Bauxite 10% and Fly ash + Bauxite 20% coatings made at different power levels.

From the figure, it is seen that the porosity level is minimum in case of fly ash coatings than that of bauxite added coatings. The minimum porosity is observed at 16 kW power level irrespective of compositions. With increase in bauxite content the porosity level is increasing. In all the coatings minimum porosity of about 3.4 % is observed, in case of fly ash coating and 4.75 % in case of 10% bauxite addition coatings. The addition of bauxite helps in in increasing the porosity level. It may be due to improper bonding between the particles during plasma spraying with addition of bauxite [5].

Coating is deposited for 30 seconds in all cases; the coating thicknesses of the substrates are measured and are shown in figure 3.4.

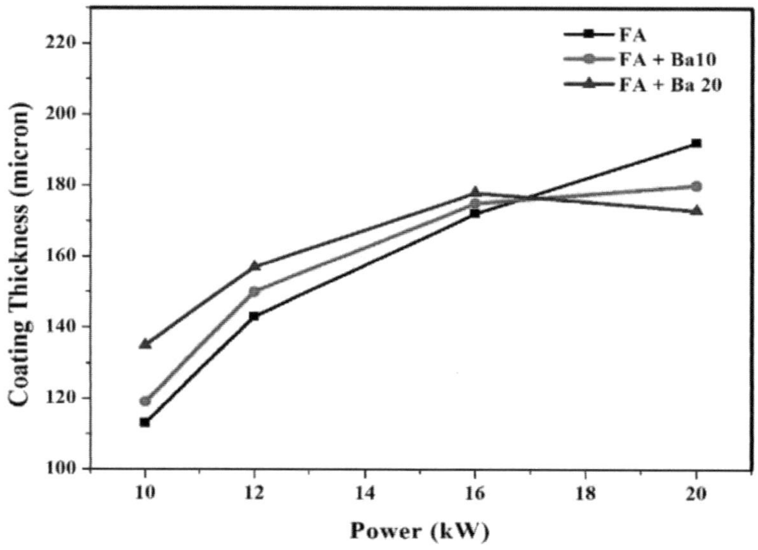

Figure. 3.4 Coating Thickness Comparison of Fly ash, Fly ash + Bauxite 10% and Fly ash + Bauxite 20% coatings made at different power levels.

Here, also similar things are observed that, addition of bauxite gives the higher/improved coating thickness. With 10% bauxite addition, the thickness is increased slightly and after 16 kW the rate of increase is less. When the bauxite addition is 20%, the thickness is further increased. It may be due to with higher amount of bauxite addition more alumino-silicate phase is forming which is helping in increasing the thickness where as in lower power level the particles are not able to form any complex phase. But after 16 kW power level the coating thickness is decreased. Whereas with only fly ash coating the coating thickness is increasing continuously with increase of power level and provides a maximum thickness of 192micron at 20 kW power level. It may be due to the fact that, at higher power level the high enthalpy plasma flame is favoring the melting of SiO_2/other oxides present, formed other/mixed compound phases which might give rise to a higher thickness [10].

Micro hardness of Fly ash, Fly ash + Bauxite 10% and Fly ash + Bauxite 20% at different power levels is shown in figure 3.5.

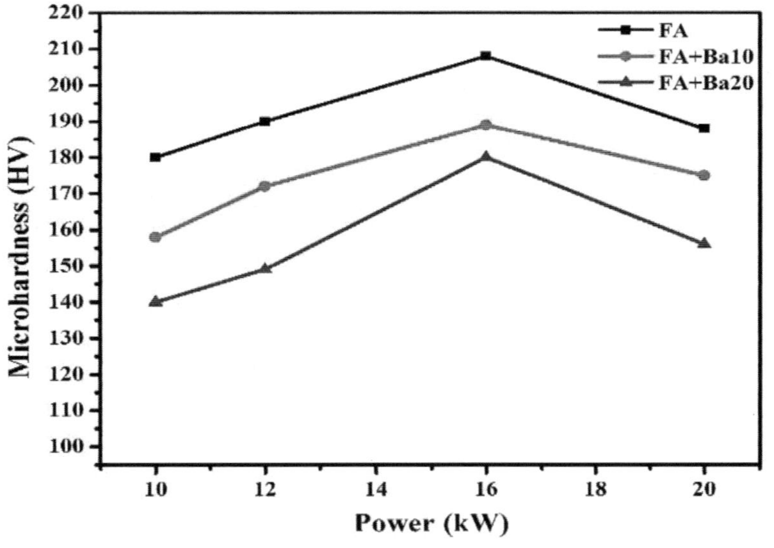

Figure. 3.5 Micro hardness of Fly ash, Fly ash + Bauxite 10% and Fly ash + Bauxite 20% coatings made at different power levels.

From the above figure it can be well visualized that, with only fly ash coating the micro hardness is highest among all other compositions, with a value of 208 HV at 16 kW power level. With increase in power level the micro hardness increases up to 16 kW and then decreases in all the cases. This may be due to, at 20kW power level the particle velocity is too high and it finds less time to melt/solidify and adhere to each other. The addition of bauxite is decreasing the micro hardness value. This may be because of, with increase in the porosity level with bauxite addition.

Conclusions

1. From the scanning electron micrograph we observed that the particles are melted efficiently at 16 kW operating power level and further increase in power level restricts the proper bonding/melting and deposition etc.
2. The porosity is always high at 20 kW power level in all cases.
3. Increase in power level favors the increase in coating thickness in all cases except fly ash with 20% bauxite addition at 20 kW. This may be due to reason that at 20 kW operating power and with this composition, the particles are not melted properly and some of the particles may be fragmented or/and fly off happened/favored for smaller particles.
4. The fly ash coating without bauxite addition gives the best micro hardness value at 16 kW power level. It is found that 16 kW operating power level is the optimum condition for deposition of best coatings using these raw materials.

References

[1] Openshaw SC 1992 *Utilization of coal fly ashPh.D. Thesis University of Florida.*
[2] Haque Md 2013*Waste Resources* **3** 22-25.
[3] Mishra S C, Mishra P C, Rout K Cand Anantapadmanabahn P V 1999*Fly ash utilization for value added products NML. Jamshedpur* 131-135.

[4] Mishra S C andPraharaj S2011 *Orissa Journal of Physics* **18** 199-206.
[5] Sathpathy A 2005 *Thermal Spray Coating of Red Mud on Metals PhD Thesis, NIT Rourkela, India.*
[6] Sathpathy A and Mishra S C 2003*National Conference on Materials and Related Technology, TIET.*
[7] Satapathy A, Mishra SC, Mohanty U, Mishra TK, Ananthapadmanabhan P V and Sreekumar K P 2004 *Proceedings of the International Conference on Industrial Tribology, Tribological Society of India, Mumbai.*
[8] Mishra S C, Praharaj S andSatapathy A 2009*Journal of Manufacturing Engineering* **4** 241-246.
[9] Pati P R andSatapathy A 2015 *Journal of Material Cycles and Waste Management* **17** 135-143.
[10] Behera, A and Mishra S C 2012 *Journal of Materials & Metallurgical Engineering* **2** 23-30.
[11] Prasad N, Sutar H, Mishra S C, Sahoo S K &Acharya S. K 2013 *International Research Journal of pure and applied chemistry* **3** 59.
[12] Mishra S C 2012 *Thermal Plasma Application in Metallurgy LAMBART Academic Publication* ISBN: 9783659235221 Germany.

IOP Conf. Series: Materials Science and Engineering **178** (2017) 012030 doi:10.1088/1757-899X/178/1/012030

Study of process parameter on mist lubrication of Titanium (Grade 5) alloy

Kalipada Maity and Swastik Pradhan
Department of Mechanical Engineering,
National Institute of Technology, Rourkela, Odisha, India – 769008

Corresponding author Email: kpmaity@gmail.com

ABSTRACT: This paper deals with the machinability of Ti-6Al-4V alloy with mist cooling lubrication using carbide inserts. The influence of process parameter on the cutting forces, evolution of tool wear, surface finish of the workpiece, material removal rate and chip reduction coefficient have been investigated. Weighted principal component analysis coupled with grey relational analysis optimization is applied to identify the optimum setting of the process parameter. Optimal condition of the process parameter was cutting speed at 160 m/min, feed at 0.16 mm/rev and depth of cut at 1.6 mm. Effects of cutting speed and depth of cut on the type of chips formation were observed. Most of the chips forms were long tubular and long helical type. Image analyses of the segmented chip were examined to study the shape and size of the saw tooth profile of serrated chips. It was found that by increasing cutting speed from 95 m/min to 160 m/min, the free surface lamella of the chips increased and the visibility of the saw tooth segment became clearer.

Keywords: weighted principal component analysis, grey relational analysis, serrated chip formation, cutting forces, tool wear

1. Introduction

Titanium alloy is known to have the most amazing properties towards the strength to weight ratio. Now a days titanium alloy is used in most of the field such as aerospace and jet engine component, turbine blades for marine application. Artificial hips and biological implants are also made from titanium alloy material. Apart from this, titanium is also known as difficult to machine material due to its low thermal conductivity and high chemical reactivity with cutting tool material. This causes rapid tool wear rate and adhesion of the material between cutting tool and titanium alloy [1-4]. A number of studies had been done on the machinability of titanium alloys. According to the existing literature survey that most for turning titanium alloy at low cutting speed is preferable [5-7]. Cutting tool material of tungsten carbide was used to machine titanium alloy, Cemented carbide (WC-Co) with 6 wt.% of cobalt (Co) and 0.8 and 1.4 μm grain size of tungsten carbide (WC) prevail good cutting condition for machining titanium alloy [8-10]. Various type of coating layers are created on the carbide cutting tool using chemical vapour deposition (CVD) coating technique such as TiN, TiC and Al_2O_3 so that it will act as a protective layer. This layer provides protection against abrasion, corrosion and oxidation on the tool surface [11-14]. Shaji et al. and Routara et al.[15,16] used taguchi method to study the effect of workpiece speed, longitudinal feed, radial infeed and modes of dressing on the surface grinding with graphite as lubricant. Dhavlikar et al.[17] determined the roundness error of the workpiece while performing centre-less grinding operation by using taguchi and dual response method. Saglam et al.[18] studied the surface roughness and roundness error in cylindrical grinding using orthogonal array taguchi method. Now a day's multi response optimization has become an essential concern in the modern manufacturing practice where more than one correlated responses must be judged simultaneously by avoiding quality loss to other quality attributes. Most of the optimization responses are uncorrelated, so it does not consider the possible correlation among the responses and contradicts with the basic assumption of the taguchi optimization to overcome such type of problem during the optimization. Principal component analysis (PCA) coupled with Grey relational taguchi method (GRA) has to be applied [19,20]. In PCA method, the weight of the responses are evaluated and the correlation between the responses are made through the correlation matrix, so that their relative significance should be described properly and objectively, further the grey relational analysis method was applied to determine the optimal selection of machining parameters [21,22]. In the present paper the mist lubrication turning operation of the titanium alloy (Grade 5) has been carried out by using KC5010 cutting tool to study the effects of several responses such as cutting forces, material removal rate, tool wear, cutting temperature, surface roughness and chip reduction coefficient by varying the cutting speed, feed and depth of cut. The multivariate statistical weighted principal component analysis coupled with grey relational analysis has been done to find out the optimum parameter setting for the L27 orthogonal taguchi layout. The formation of different types of saw-tooth chips, tool wear and the effects of all the responses on the cutting speed, feed and depth of cut were studied.

2. Experimental details

The experimental work is carried out on a high precision NH 26 lathe with 11 *kW* spindle power with a maximum spindle of 1020 *rpm*. The work piece material used for the present investigation is titanium alloy of grade 5 (Ti-6Al-4V). The size of the work piece material was of 50 *mm* diameter and 600 *mm* in length. The properties of the titanium alloy are shown in Table 1. KC5010 of Kennametal make (SNMG120408) cutting tool inserts were selected for machining of the titanium alloy. The machining operation was carried out in air oil mixed lubrication environment. During machining, the cutting forces i.e. thrust force (N), radial force (N), tangential force (N) and cutting temperature (°C) were measured and after machining, surface roughness (μm), tool wear (mm), material removal rate (mm^3/s) and chip reduction coefficient were measured. The cutting forces were measured using piezoelectric tool dynamometer. Non-contact infrared thermometer was used to measure temperature. The surface finish was measured by Taylor Hobson Surtronic 3+ surface roughness tester. The tool wear was measured using Tool maker's microscope and optical microscope was used to analyse the chip produced during machining.

3. Design of experiment

The design of experiment is done to effectively utilize the cutting variables by reducing the number of experimental run. The cutting variables used for the experiment are cutting speed, feed and depth of cut. In this research, three cutting variable with three levels are selected for the experiment as shown in Table 2. L27 orthogonal array layout was designed by using Minitab 16 software with cutting variable and their levels. The total numbers of experiment were conducted according to the cutting variables with their levels.

Table 1. Chemical Composition and Properties of Titanium Alloy (Ti–6Al–4V)[23]

Chemical composition *Wt.%*	*C* 0.1 (max)	*V* 4	*Al* 6	*Fe* 0.037	*Ti* Remaining
Properties	Density (*g/cm3*)	Tensile strength (*MPa*)	Thermal conductivity (*W/m°C*)	Hardness (*HRC*)	
	4.42	950	6.7	36	

Table 2. Process Parameter and Levels

Cutting parameter	Units	Levels of the cutting parameter		
		I	II	III
Cutting speed (*v*)	m/min	95	124	160
Feed rate (*f*)	mm/rev	0.04	0.08	0.16
Depth of cut (*d*)	mm	0.4	0.8	1.6

4. WPCA Coupled with grey-based taguchi analysis method

Optimization done using various hybrid taguchi based method such as grey taguchi, desirability function based method, utility concept based taguchi method are irrelevant to generate a correlation among the responses [16]. Weighted principal component analysis (WPCA) method is carried out to overcome the correlation error between the responses. In this study, WPCA (Weighted principal component analysis) combined with grey taguchi method was used. In WPCA response correlation is eliminated and converted to uncorrelated quality indices called as principal component and further these principal components are aggregated to calculate the multi response performance index (MPI). After that a combined quality loss (CQL) is calculated i.e. absolute deviation of the calculated MPI. The signal to noise ratio (S/N ratio) of the CQL was calculated by taguchi method to find the overall effect of the cutting parameter on the output responses. Detailed stepwise discussion of the hybrid WPCA combined GRA method has been used. To avoid the problem of dissimilarity in scales and units, the data are normalized corresponding to the smaller the better criteria and higher the better criteria as shown in Equation (1) and Equation (2). Material removal rate and chip reduction coefficient are normalized with higher the better criteria and the others responses are normalized using smaller the better criteria. The normalized data are shown in Table 3.

$$x_i(k) = \frac{y_i(k) - \min y_i(k)}{\max y_i(k) - \min y_i(k)} \tag{1}$$

$$x_i(k) = \frac{\max y_i(k) - y_i(k)}{\max y_i(k) - \min y_i(k)} \tag{2}$$

where, $y_i(k)$ is the original sequence; $x_i(k)$ is the sequence after normalization; $\min y_i(k)$ is the smallest value of $y_i(k)$ for the k_{th} variables; and $\max y_i(k)$ is the largest value of $y_i(k)$ for the k_{th} response.

Table 3. Normalized Value of the Responses

Run	*Thrust force (Fx)*	*Radial force (Fy)*	*Tangential force (Fz)*	*Surface roughness (Ra)*	*Temper-ature*	*Tool wear*	*Chip reduction coefficient (ξ)*	*Material removal rate (MRR)*
Ideal sequence	1	1	1	1	1	1	1	1
1	0.343	0.441	0.514	0.293	1.000	1.000	0.364	0.027
2	0.680	0.084	0.153	0.572	0.477	0.426	0.636	0.041
3	0.108	0.164	1.000	0.664	0.854	0.470	0.727	0.072
4	0.395	0.569	0.528	0.472	0.621	0.512	0.455	0.058
5	0.607	0.102	0.142	0.601	0.363	0.479	0.682	0.096
6	0.106	0.211	0.288	0.638	0.325	0.643	0.682	0.192
7	0.238	1.000	0.259	0.428	0.661	0.532	0.455	0.153
8	0.486	0.189	0.133	0.546	0.500	0.724	0.568	0.229
9	0.091	0.307	0.128	0.342	0.304	0.508	0.659	0.499
10	0.337	0.507	0.594	0.550	0.804	0.600	0.455	0.046
11	0.739	0.090	0.170	0.472	0.398	0.488	0.455	0.058
12	0.095	0.175	0.150	0.669	0.554	0.438	1.000	0.097
13	0.515	0.732	0.514	0.697	0.547	0.563	0.455	0.085
14	1.000	0.120	0.149	0.561	0.353	0.448	0.682	0.132
15	0.123	0.241	0.358	0.301	0.279	0.479	0.591	0.269
16	0.062	0.137	0.184	0.352	0.641	0.198	0.500	0.259
17	0.102	0.248	0.107	0.430	0.345	0.272	0.614	0.312
18	0.024	0.225	0.055	0.419	0.261	0.174	0.841	0.692
19	0.486	0.614	0.623	0.790	0.707	0.506	0.455	0.050
20	0.548	0.096	0.168	0.847	0.342	0.453	0.909	0.086
21	0.132	0.196	0.352	1.000	0.323	0.472	0.909	0.148
22	0.262	0.632	0.333	0.731	0.526	0.458	0.682	0.118
23	0.354	0.152	0.167	0.692	0.350	0.578	0.682	0.184
24	0.060	0.228	0.152	0.241	0.248	0.419	0.636	0.358
25	0.047	0.141	0.317	0.125	0.436	0.085	0.295	0.135
26	0.024	0.103	0.073	0.205	0.281	0.089	0.636	0.448
27	0.028	0.073	0.049	0.127	0.205	0.033	0.886	1.000

After the normalization of the experimental data, a check has been done to know whether the responses are correlated or not. The Pearson's correlation coefficients between the input parameters are presented in Table 4.

Table 4. Pearson Correlation Coefficient Data to Check Correlation Between Responses

Responses	*Fz*	*Fy*	*Fx*	Tempe -rature	Surface roughness	Tool wear	chip reduction coefficient
Fy	0.422						
Fx	0.014	0.02					
Temperature	0.739	0.499	0.128				
Surface roughness	0.297	0.142	0.386	0.152			
Tool wear	0.395	0.368	0.402	0.523	0.399		
chip reduction coefficient	-0.343	-0.449	-0.184	-0.484	0.362	-0.252	
Material removal rate	-0.51	-0.273	-0.523	-0.579	-0.538	-0.594	0.383

Since all the correlation coefficients between the responses are non-zero value that means the features are correlated to each other. In order to eliminate the correlation of the responses PCA has been applied. By using Minitab 16 software principal component analysis had been done. The PCA results i.e. Eigen value, Eigen vector, accountability and cumulative accountability proportion were presented in Table 5. Now the correlated cutting parameters are converted to uncorrelated quality indices. These uncorrelated quality indices are known as principal components. All the individual principal components are shown in Table 6.

Table 5. Eigen Analysis of the Correlation Matrix

	$\Psi 1$	$\Psi 2$	$\Psi 3$	$\Psi 4$	$\Psi 5$	$\Psi 6$	$\Psi 7$	$\Psi 8$
Eigen Value	3.5644	1.6609	1.0393	0.6515	0.4752	0.297	0.2269	0.0848
Eigen vector	0.395	-0.170	-0.415	0.388	0.164	-0.279	0.589	0.201
	0.319	-0.298	-0.164	-0.803	0.271	-0.049	-0.068	0.240
	0.236	0.419	0.608	-0.045	0.223	-0.576	0.047	0.116
	0.437	-0.237	-0.160	0.260	-0.137	-0.337	-0.691	-0.224
	0.241	0.587	-0.361	-0.148	0.268	0.146	0.047	-0.591
	0.401	0.163	0.067	-0.222	-0.836	0.064	0.235	-0.005
	-0.280	0.483	-0.500	-0.049	-0.162	-0.266	-0.253	0.524
	-0.445	-0.214	-0.142	-0.249	-0.183	-0.615	0.218	-0.463

AP	0.446	0.208	0.130	0.081	0.059	0.037	0.028	0.011
CAP	0.446	0.653	0.783	0.865	0.924	0.961	0.989	1.000

N.B. : AP Accountability proportion, CAP Cumulative accountability proportion

Table 6. Individual Principal Component

Sl.No	PC1	PC2	PC3	PC4	PC5	PC6	PC7	PC8
Ideal	-0.7250	0.2690	-0.6420	-0.2980	-0.3450	-0.8410	0.1830	0.0610
1	1.2192	0.1911	-0.4612	-0.2003	-0.6780	-0.6919	0.1904	0.0244
2	0.5678	0.8208	-0.2421	-0.1354	-0.1791	-0.6575	0.1887	-0.0033
3	0.9592	0.4258	-1.0947	0.2174	-0.2304	-0.7186	0.3864	0.0148
4	0.9207	0.3243	-0.5427	-0.3284	-0.1423	-0.6489	0.3250	0.0794
5	0.4941	0.8500	-0.3043	-0.2125	-0.2306	-0.6113	0.2591	-0.0027
6	0.4838	0.6225	-0.6975	-0.2964	-0.4284	-0.3987	0.3799	-0.0640
7	0.8873	0.1242	-0.6003	-0.7831	-0.1556	-0.5811	0.1045	0.0834
8	0.6069	0.6670	-0.3363	-0.3185	-0.4801	-0.6382	0.2311	-0.1183
9	0.1821	0.3477	-0.5858	-0.4417	-0.4484	-0.5788	0.2534	-0.0481
10	1.0522	0.3267	-0.6455	-0.2309	-0.2368	-0.6677	0.2634	-0.0111
11	0.6007	0.7186	-0.0729	-0.1511	-0.2036	-0.6504	0.3119	-0.0176
12	0.3928	0.7567	-0.8479	-0.2122	-0.3490	-0.4520	0.0151	0.0401
13	1.0253	0.4802	-0.5609	-0.5412	-0.0500	-0.6774	0.3851	-0.0001
14	0.5532	0.9715	-0.0621	-0.2406	-0.1275	-0.8671	0.2776	0.0583
15	0.3487	0.3343	-0.5684	-0.2342	-0.3516	-0.5010	0.4210	0.0866
16	0.3201	0.2271	-0.5642	-0.0600	-0.2065	-0.5173	0.0317	-0.1336
17	0.1981	0.3949	-0.5664	-0.3040	-0.2085	-0.4675	0.1780	-0.0628
18	-0.1594	0.4043	-0.7449	-0.4068	-0.2562	-0.6714	0.1213	-0.1190
19	1.1095	0.4994	-0.6629	-0.3538	-0.0138	-0.7076	0.3366	-0.0840
20	0.4689	1.0800	-0.5489	-0.2398	-0.1858	-0.5883	0.2345	-0.0222
21	0.4837	0.9311	-0.9546	-0.2771	-0.2048	-0.4125	0.3409	-0.1246
22	0.7413	0.5464	-0.7581	-0.5262	-0.1068	-0.5424	0.2080	-0.0005
23	0.4766	0.7798	-0.5128	-0.2920	-0.3426	-0.5054	0.2942	-0.1072
24	0.1438	0.3127	-0.5318	-0.3110	-0.3880	-0.4740	0.2899	0.0594
25	0.2932	0.0212	-0.4017	0.0356	-0.0693	-0.3957	0.2649	0.0235
26	-0.1025	0.2467	-0.5270	-0.1754	-0.1974	-0.5176	0.1483	-0.0162
27	-0.5106	0.2268	-0.6766	-0.3058	-0.3141	-0.8974	0.1091	-0.0888

The accountability proportions of the individual principal components are also considered as individual priority weights of the corresponding responses. Finally MPI value of each experimental run is calculated by aggregating the individual principal components of the output responses multiplied with their respective individual priority weights respectively. The deviation of the MPI from its ideal value was determined. It is also termed as CQL. Now the taguchi method has been performed to find out the S/N ratio of the CQL value using the higher the better criteria as tabulated in Table 7. Optimal setting plot has been shown in the Figure 1. From the main effects plots, the optimal setting was found to be cutting speed 160 *m/min*, feed 0.16 *mm/min* and depth of cut 1.6 *mm*. According to the taguchi prediction present in the Minitab software, the predicted value of the S/N ratio for the CQL with respect to the main effects plot optimal setting was 19.6407. The S/N ratio for the same configuration setting performed in the experiment was 22.1916. The quality has improved by 12.9878 % by using the optimal setting. The response for S/N ratio was shown in Table 8. Referring to the response, it can be concluded that the overall outputs are effected by mostly feed then followed by depth of cut.

Table 7. Calculated MPI, CQL and S/N ratio of CQL

Sl.No	MPI	CQL	S/N ratio
Ideal	-0.4207	0.0000	-
1	0.4473	0.8680	1.2297
2	0.3519	0.7726	2.2413
3	0.3625	0.7831	2.1233
4	0.3585	0.7792	2.1674
5	0.3114	0.7321	2.7088
6	0.2005	0.6211	4.1364
7	0.2533	0.6739	3.4277
8	0.2931	0.7138	2.9286
9	0.0003	0.4210	7.5152
10	0.4032	0.8239	1.6828
11	0.3681	0.7888	2.0609
12	0.1687	0.5894	4.5917
13	0.4232	0.8439	1.4747
14	0.3900	0.8107	1.8229
15	0.1056	0.5263	5.5751
16	0.0799	0.5006	6.0106
17	0.0469	0.4676	6.6028
18	-0.1547	0.2660	11.5024
19	0.4654	0.8860	1.0509
20	0.3166	0.7373	2.6477
21	0.2437	0.6644	3.5518

22	0.2825	0.7032	3.0583
23	0.2526	0.6732	3.4367
24	0.0032	0.4239	7.4555
25	0.0748	0.4954	6.1003
26	-0.1039	0.3167	9.9870
27	-0.3430	0.0777	22.1916

Table 8. Response Table for Signal to Noise Ratios

Level	Cutting speed	Feed	Depth of cut
1	3.164	2.353	2.911
2	4.592	3.537	3.826
3	6.609	8.474	7.627
Delta	3.445	6.121	4.716
Rank	3	1	2

The Analysis of variances (ANOVA) table is shown in Table 9. From the ANOVA analysis, the factor having higher influence and its percentage of contribution over the responses can be analyzed. From the ANOVA table, it can be analyzed that feed, depth of cut, cutting speed and combination of both cutting speed and feed are the most significant parameters as per the 95% of acceptance label i.e. the P-value will be less than 0.05. The interaction plot of the S/N ratio according to smaller to better criteria has been shown in the Figure 2.

Table 9. Analysis of Variance for SN ratios

Source	DF	Seq SS	Adj SS	Adj MS	F	P
Cutting speed	2	53.920	53.920	26.958	8.320	0.011
Feed	2	189.710	189.710	94.853	29.280	0.000
Depth of cut	2	112.560	112.560	56.279	17.370	0.001
Cutting speed * Feed	4	53.070	53.070	13.268	4.100	0.043
Cutting speed * Depth of cut	4	23.380	23.380	5.846	1.800	0.221
Feed * Depth of cut	4	43.080	43.080	10.770	3.320	0.070
Residual Error	8	25.920	25.920	3.240		
Total	26	501.630				

The interaction plot shows that with increase in cutting speed, feed and depth of cut, the S/N ratio also increases. With the cutting speed value 160 *m/min*, the CQL is maximized when the feed and depth of cut are 0.16 *mm/min* and 1.6 *mm* respectively.

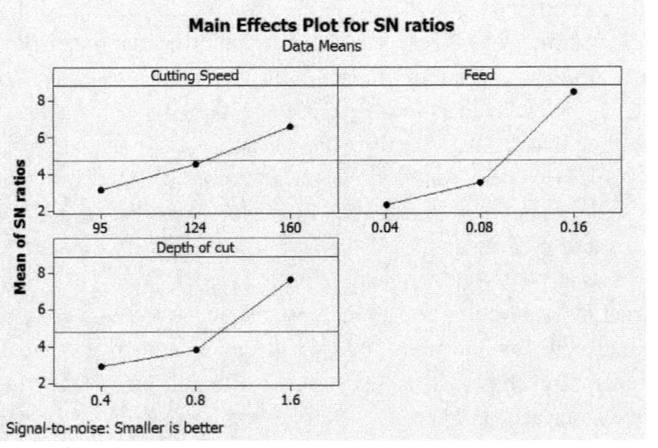

Figure 1. Main effects plot for S/N ratio

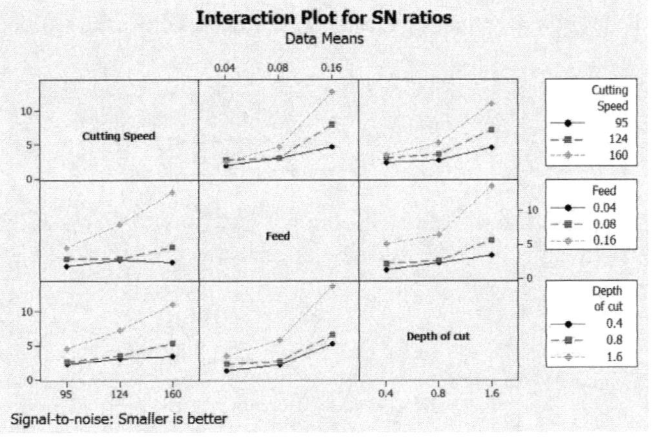

Figure 2. Interaction plot of the S/N ratio

5. Results and discussion

5.1 Effects of Cutting Speed, Feed and Depth of cut

The effect of cutting speed, feed and depth of cut with respect to axial force, radial force, tangential force, surface roughness, temperature, tool wear, chip coefficient ratio and material removal rate have been studied. The normalized value of all the responses were plotted with respect to cutting speed, feed and depth of cut, so that all the responses can be plotted in a single graph as shown in Figure 3 (a), (b) and (c). It can be observed from the Figure 3(a) that with increase in cutting speed, cutting

IOP Conf. Series: Materials Science and Engineering **178** (2017) 012030 doi:10.1088/1757-899X/178/1/012030

forces and surface roughness decrease, whereas temperature, tool wear and material removal rate are increasing. But chip reduction coefficient increases rapidly from low to medium cutting speed but from medium to high cutting speed, it increases slowly. While increasing the cutting speed, the amount of heat generated during machining will be more and the titanium alloy have the ability to withstand at high elevated temperature during machining. With this, the ease of machining enhances, with decrease in cutting forces. A maximum percentage of the heat is taken away by the chips produced during machining operation. Therefore, the surface damage of the workpiece decreased and consequent smoothing of surface peaks and feed marks enhanced the quality of the surface finish [24]. The chips carrying a large amount of heat comes in contact with the tool and due to the abrasion mechanism of the chips and cutting tool, the wear rate of the cutting tool increases. Referring to Figure 3(b), it can be observed that tangential cutting force, axial force, temperature, surface roughness, tool wear and material removal rate increase with increase in feed, but radial cutting force and chip reduction coefficient first increase and than decrease with respect to feed. The tangential force, radial force, temperature, chip coefficient ratio and material removal rate increase with increase in depth of cut, whereas, surface roughness and tool wear decrease with increase in depth of cut within the lower range. But, in the higher range surface roughness and tool wear increase. The axial cutting force increases first and than decreases with respect to depth of cut as shown in Figure 3(c). With increase in depth of cut, the contact surface area between the workpiece and cutting tool increases, thereby, increasing the friction that results in more heat formation. This causes wears in the tool surface. Increase in depth of cut decreases the stability and accuracy of the machining by creating undesired chatter and vibration, thereby, decreasing the surface quality of the machined surface.

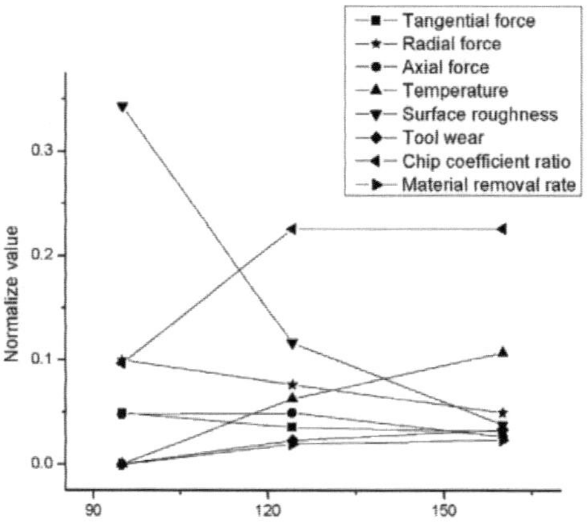

(a) Effects of cutting speed on process responses

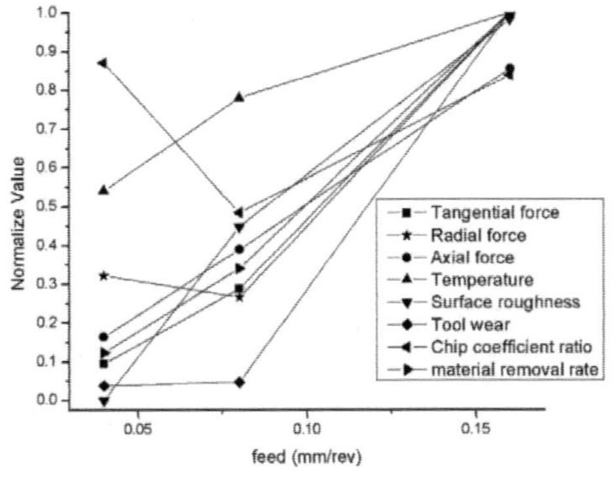

(b) Effects of feed on process responses

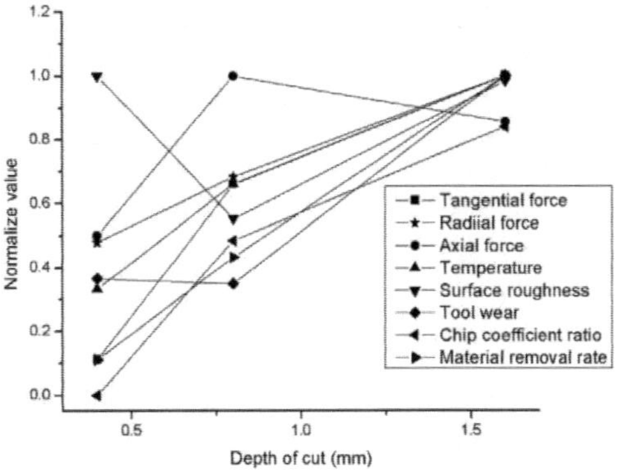

(c) Effects of depth of cut on process responses

Figure 3. Effects of process parameter on the process responses

5.2 Tool Wear Analysis

Tool wear of the cutting tool depends on various types of cutting parameter, shape of the cutting tool, machining environment and material of the workpiece. Tool wear can be measured by crater wear or flank wear. Crater wear of the tool can be measured on the top face of the cutting tool, while flank wear can be measured on the side face of the cutting tool. The crater wear increases due to the temperature, load and pressure exerted on the rake face of the cutting tool and mostly caused by diffusion wear. Flank wear is mainly caused due to the abrasion. The flank wear of the cutting insert KC5010 was measured by using tool maker's microscope. The images of the cutting inserts shown in the Figure 4 were taken by using the optical microscope. Referring to Figure 4, it can be observed that, flank wear increases rapidly with increase in cutting speed from 95 m/min to 160 m/min. The crater wear of the inserts was also very high due to the high speed machining of cutting tool with respect to the workpiece.

When the feed rate is increasing from 0.04 mm/rev to 0.16 mm/rev with the constant cutting speed and depth of cut, the flank wear of the cutting inserts increases gradually as shown in the Figure 5. While Figure6 shows that increase in the depth of cut from 0.4 mm to 1.6 mm at constant cutting speed and feed rate, the flank wear slowly decreases initially and then increases with increase in the depth of cut. Since increasing the depth of cut from 0.4 mm to 1.6 mm, the contact between the workpiece and tip of the cutting tool increases, so the tool wear starts initially due to attrition and micro chipping formation at the sharp edge of the cutting tool. Afterwards the failure of the cutting tool or catastrophic mechanical breakage and plastic deformation of the cutting tool occurred due to the intensive pressure, temperature and dynamic loading at the tip of the cutting tool [25].

(a) Optical tool wear image of cutting insert at *V*- 95 *m/min f*- 0.16 *mm/rev d* - 0.4 *mm*

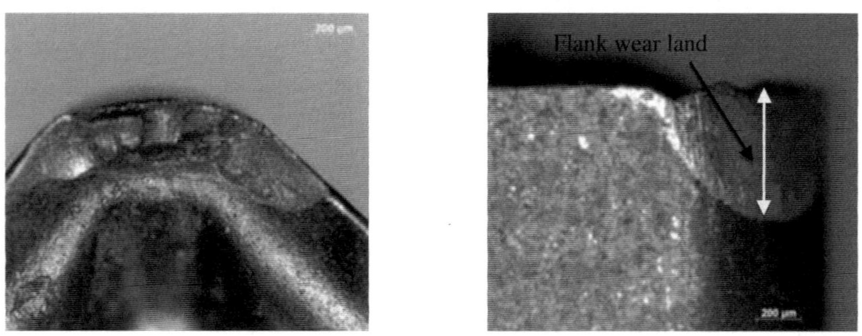

(b) Optical tool wear image of cutting insert at *V* - 124 *m/min f* - 0.16 *mm/rev d* - 0.4 *mm*

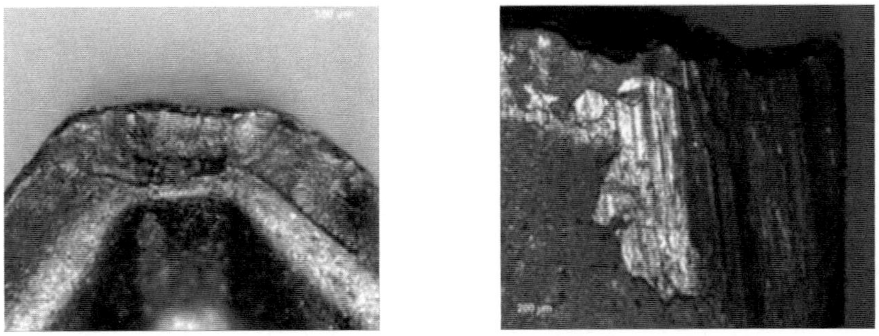

(c) Optical tool wear image of cutting insert at *V*- 160 *m/min f* - 0.16 *mm/rev d* - 0.4 *mm*

Figure 4. Optical tool wear image of the cutting inserts at cutting speed 95-160 *m/min*, feed - 0.16 *mm/rev* and depth of cut - 0.4 *mm*

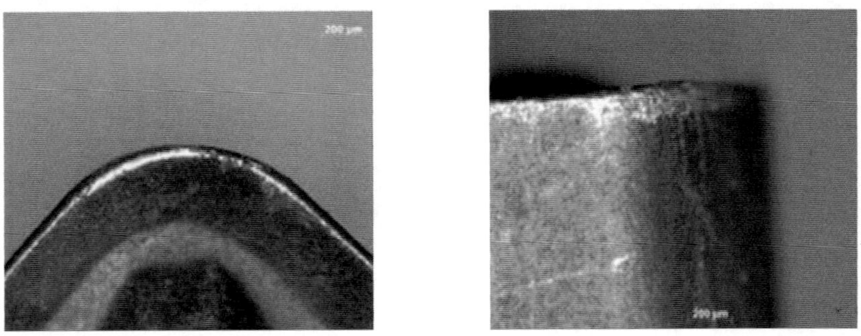

(a) Optical tool wear image of cutting insert at *V* - 124 *m/min f* - 0.04 *mm/rev d* - 0.4 *mm*

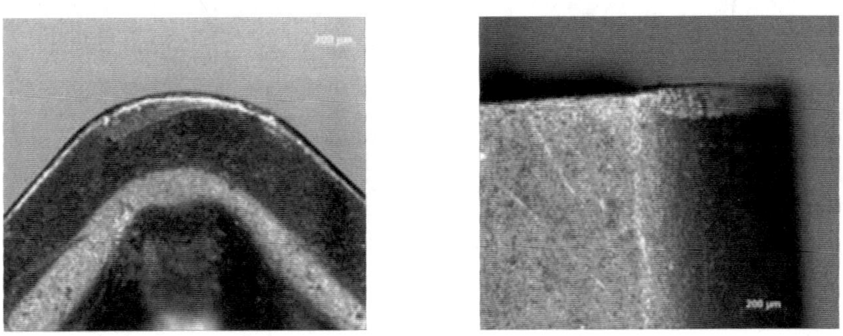

(b) Optical tool wear image of cutting insert at *V* - 124 *m/min f* - 0.08 *mm/rev d* - 0.4 *mm*

(c) Optical tool wear image of cutting insert at V - 124 *m/min* f - 0.16 *mm/rev* d - 0.4 *mm*

Figure. 5. Optical tool wear image of cutting inserts at cutting speed 124 *m/min*, feed - 0.04-0.16 *mm/rev* and depth of cut - 0.4 *mm*

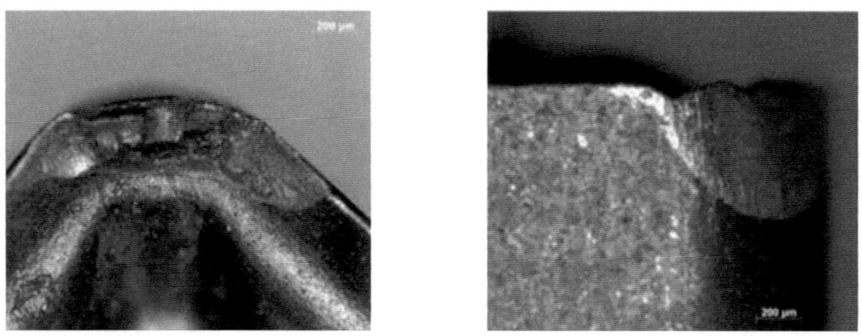

(a) Optical tool wear image of cutting insert at V - 124 *m/min* f - 0.16 *mm/rev* d - 0.4 *mm*

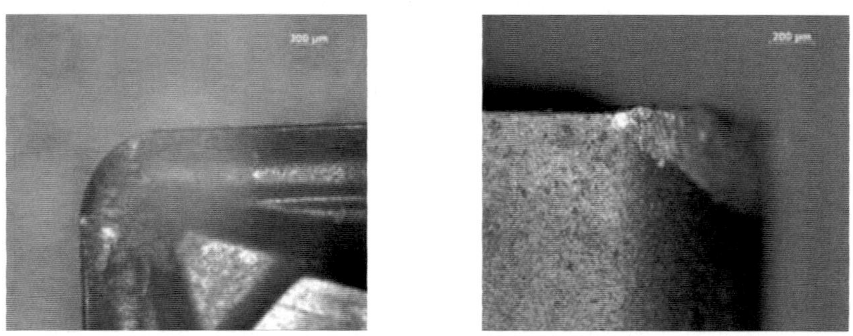

(b) Optical tool wear image of cutting insert at V - 124 *m/min* f - 0.16 *mm/rev* d - 0.8 *mm*

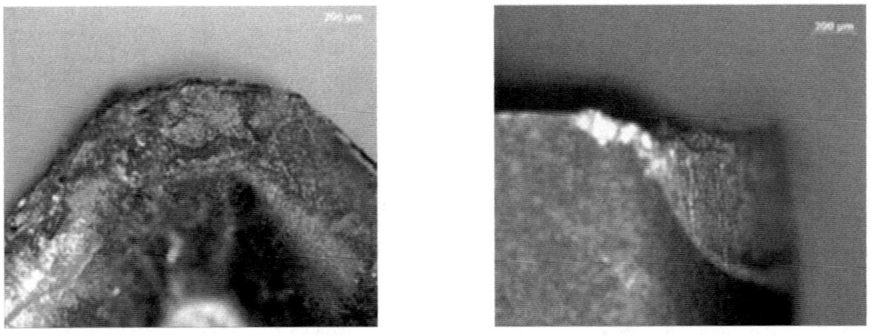

(c) Optical tool wear image of cutting insert at *V* - 124 *m/min f* - 0.16 *mm/rev d* -1.6 *mm*

Figure 6. Optical tool wear image of cutting insert at cutting speed 124 *m/min*, feed - 0.16 *mm/rev* and depth of cut - 0.4-1.6 *mm*

5.3 Analysis of Chips

5.3.1 Analysis of Chips in Macroscale

After the turning of the titanium alloy at different cutting parameter the chips are collected and the images of the chips are taken by digital camera without any magnification and considered as macroscale image. It can be seen from the Figure 7 that during the machining of the titanium alloy various type of chips are formed. Mostly the development of the different types of chip depends upon the various factors such as orthogonal rake angle, cutting variables, inherent material properties, friction between area of contact of chip and rake face [26]. Normally most of the chips forms are of helical shape and tubular type of chips. By increasing cutting speed from 95 to 160 *m/min* and feed from 0.04 *mm/rev* to 0.16 *mm/rev*, long helical type of chips changes to long tubular type of chips. By increasing depth of cut at low cutting speed and constant feed long helical type of chips are formed. At high cutting speed and constant feed long tubular type of chips are formed.

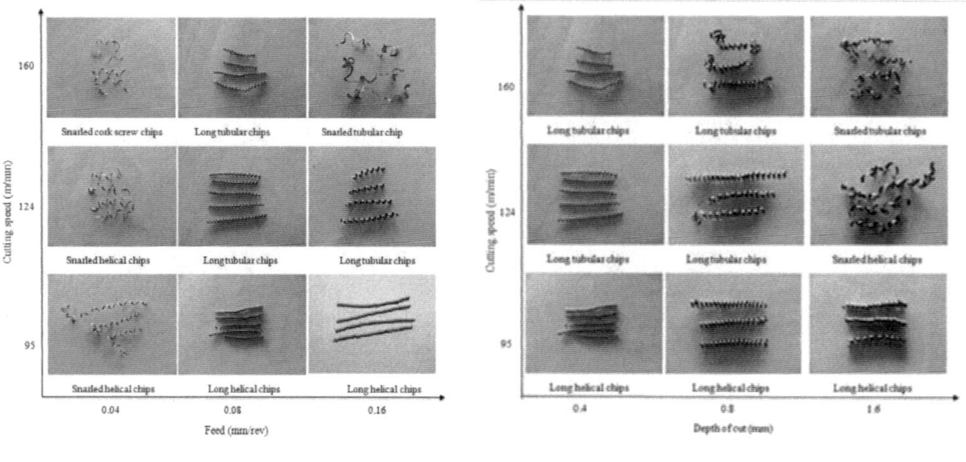

(a) (b)

Figure 7. Macroscale images of the chips obtained from machining with variation of process parameter

5.3.2 Comparison of the Chips in Microscale

Segmented serrated chips formed are during the machining of titanium alloy. The image are obtained and measured by using optical microscope. When the cutting parameter feed and depth of cut will be low, then the formation of serrated chips during the machining are less. In other case, the most of the chips formed during machining of titanium alloy are serrated type. But due to the variation in the cutting parameter, the serrated tooth of the chip formed is neither in proper shape nor arranged in an even manner. The specification of serrated chips where, H is maximum thickness of saw tooth chip, h_l is the chip thickness at local shear deformation, P_C is the distance between sawtooth segmentation, φ_l is saw tooth chip angle, θ is slipping angle are shown in Figure 8. Single sawtooth chip was shown by dashed quadrilateral boundary. The serrated chips formed while varying cutting speed from 95 *m/min* to 160 *m/min* with constant feed and depth of cut of 0.16 *mm/rev* and 1.6 *mm* respectively are shown in Figure 9.

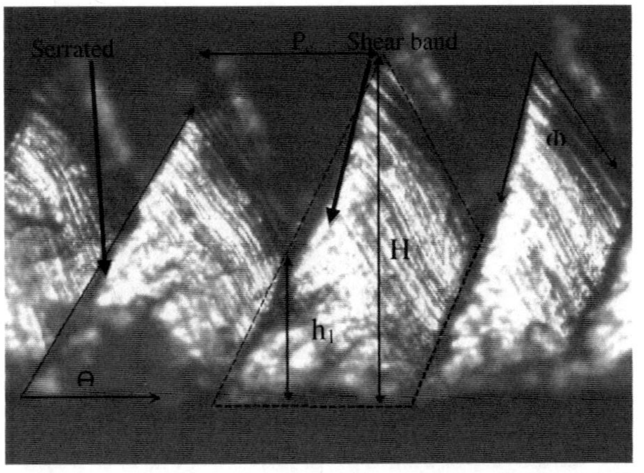

Figure 8. Sawtooth chip morphology image of Ti-6Al-4V

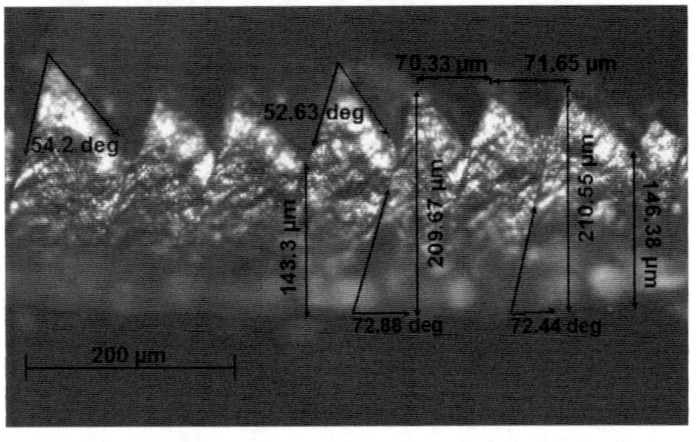

(a) V - 95 *m/min*, f - 0.16 *mm/rev*, d - 1.6 *mm*

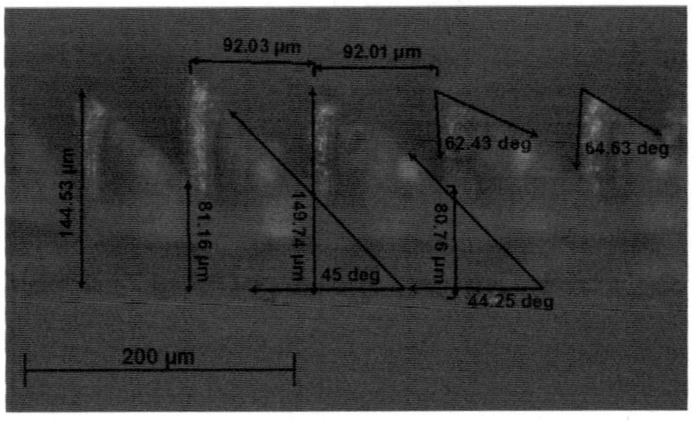

(b) V - 124 *m/min*, f - 0.16 *mm/rev*, d - 1.6 *mm*

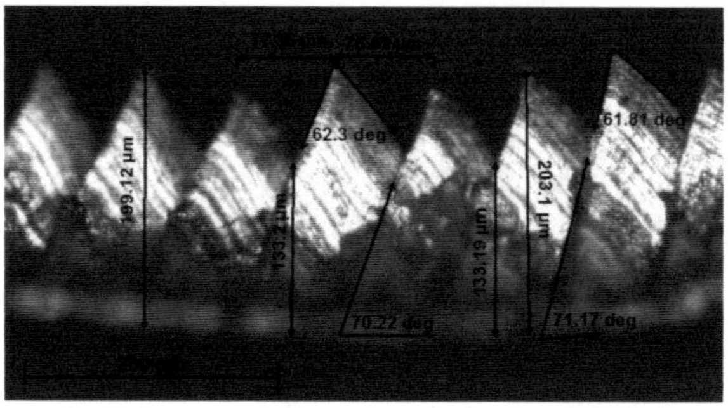

(c) *V* - 160 *m/min*, *f* - 0.16 *mm/rev*, *d* - 1.6 *mm*

Figure 9. Measurement of chip parameter of the sawtooth chip

From Figure 10 it can be analyzed that the maximum thickness of the saw tooth chip, chip thickness at local shear deformation and slipping angle are decreasing whereas the distance between saw tooth chips and saw tooth chip angle are increasing by increasing the cutting speed. The chip morphology on the free surface of the chip is shown in Figure 11. Larger magnification of the chips at cutting speed 95 and 124 m/min was made to clearly analyze the chips free surface. The free surfaces of the chips are recognized as lamella structures. The lamella structures consist of two sections: one is major section and another is the corner section. Both the sections are formed due to shearing of side cutting edge and vertical influence of the tool nose radius respectively [27,28]. Change in the cutting speed during machining produced different types of chip morphology, so the lamella structures formed are also varied. With the increase in the cutting speed, the numbers of lamella present in the same measuring area decreases. When the visibility of the lamella becomes clearer, the saw-tooth generated in the chips will be prominently visible.

IOP Conf. Series: Materials Science and Engineering **178** (2017) 012030 doi:10.1088/1757-899X/178/1/012030

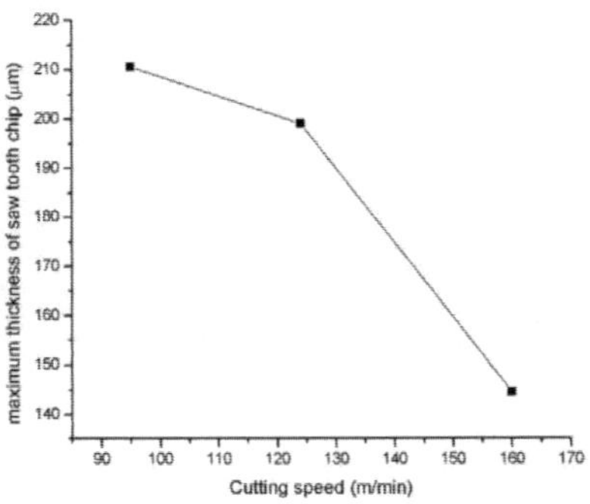

(a) Maximum thickness of saw tooth chip *vs.* cutting speed

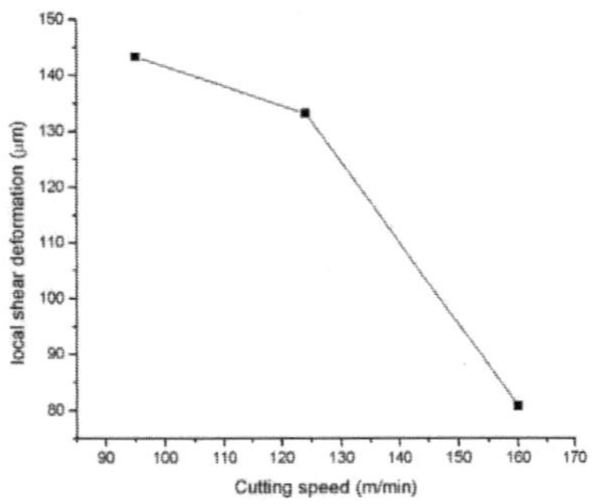

(b) Local shear deformation *vs.* cutting speed

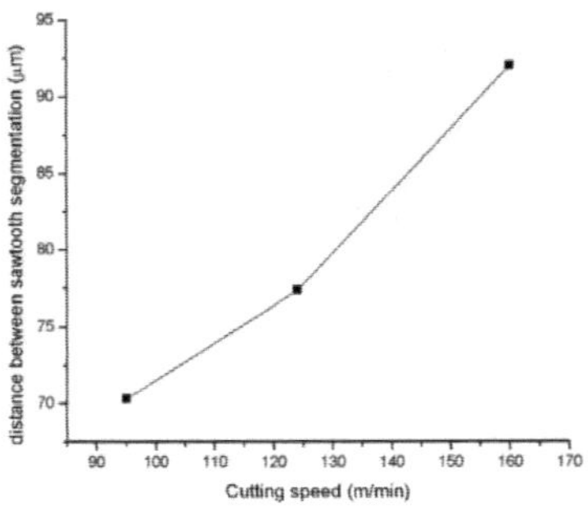

(c) Distance between saw tooth segmentation *vs.* cutting speed

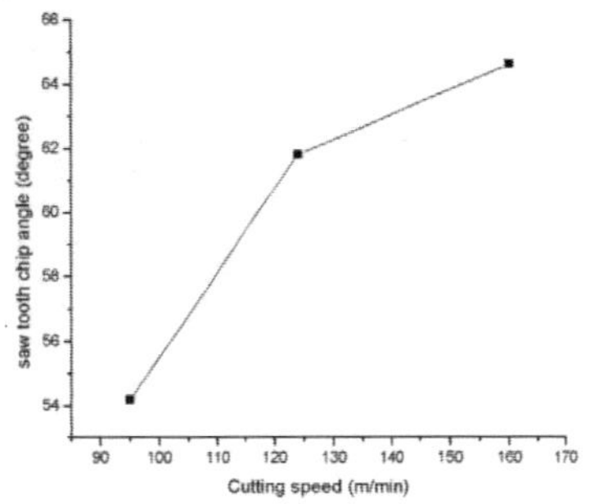

(d) Saw tooth chip angle *vs.* cutting speed

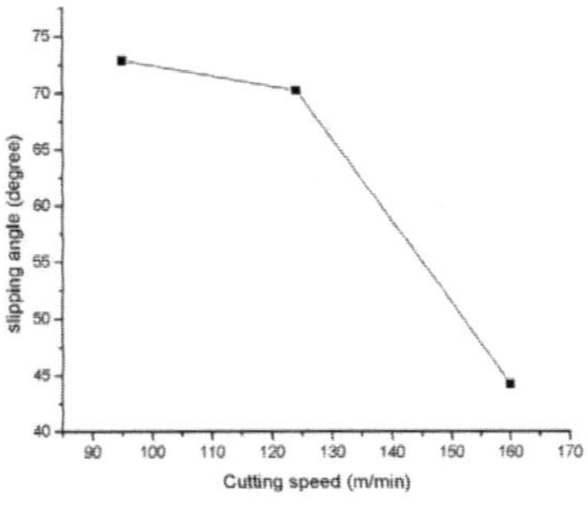

(e) Slipping angle *vs.* cutting speed

Figure 10. Serrated chip geometry variation with respect to Cutting speed

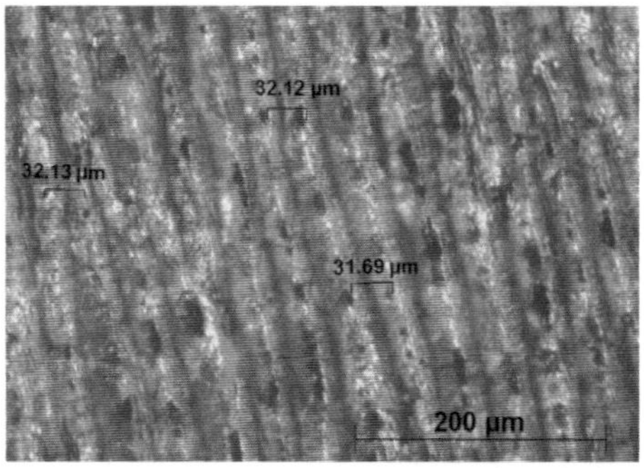

(a) *V* - 95 *m/min*, *f* - 0.16 *mm/rev*, *d* - 1.6 *mm*.

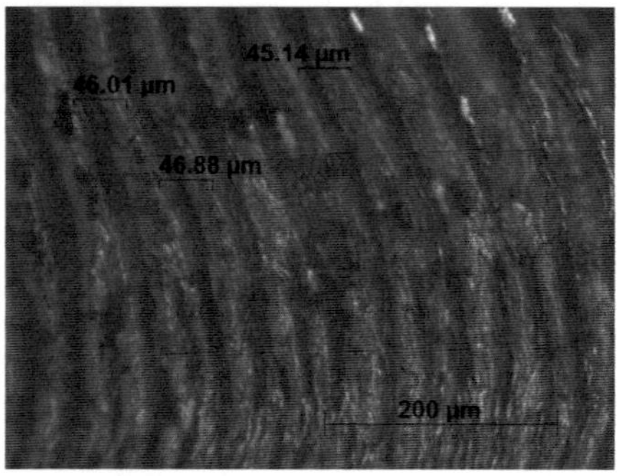

(b) *V* - 124 *m/min*, *f* - 0.16 *mm/rev*, *d* - 1.6 *mm*.

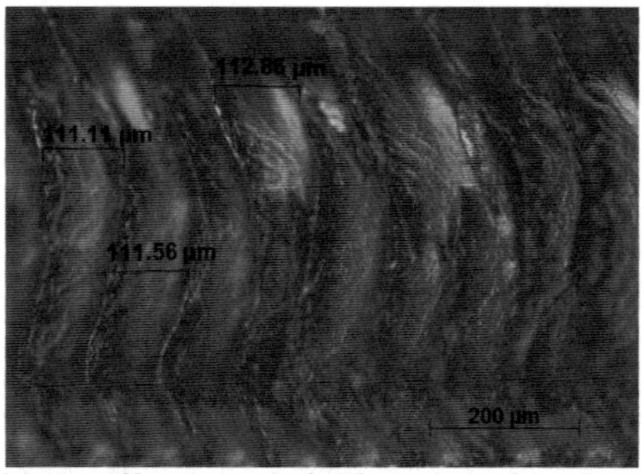

(c) *V* - 160 *m/min*, *f* - 0.16 *mm/rev*, *d* - 1.6 *mm*.

Figure 11. Visible free surface lamella of serrated chip

6. Conclusion

This study deals with the mist turning machining of the titanium alloy with KC5010 inserts using L27 orthogonal array design to find out the effect of output responses by varying the input factors. The weighted principal component analysis coupled with grey relational analysis was done to find out the optimal setting of the factors. The effect of tool wear, type of chips formed and the effect of the factors were being discussed. The succeeding conclusions may be drawn from the experimental results and discussion.

- The response correlation has been eliminated in the optimization so that, the principal component can be considered as independent response variables.
- With increase in the interaction between the cutting speed, feed and depth of cut, the S/N ratio increases.
- The optimal settings of the cutting parameter are cutting speed 160 *m/min*, feed 0.16 *mm/rev* and depth of cut 1.6 *mm*.
- Tool wear, surface roughness, temperature, chip reduction coefficient and material removal rate increase with increase in cutting variables.
- Typically chips forms during machining were long tubular and long helical type.

- With increase in the cutting speed, free surface lamella of the chips increases. As a result, the shape of the saw tooth becomes more clear and prominent.

References

[1]. Ezugwu E, Wang Z 1997 Titanium alloys and their machinability—a review. *Journal of Materials Processing Technology* **68** (3) 262-274

[2]. Arrazola P-J, Garay A, Iriarte L-M, Armendia M, Marya S, Le Maitre F 2009 Machinability of titanium alloys (Ti6Al4V and Ti555. 3). *Journal of Materials Processing Technology* **209** (5) 2223-2230

[3]. Cui C, Hu B, Zhao L, Liu S 2011 Titanium alloy production technology, market prospects and industry development. *Materials & Design* **32** (3) 1684-1691

[4]. Revankar GD, Shetty R, Rao SS, Gaitonde VN 2014 Analysis of surface roughness and hardness in titanium alloy machining with polycrystalline diamond tool under different lubricating modes. *Materials Research* **4** (17) 1010-1022

[5]. Komanduri R, Von Turkovich B 1981 New observations on the mechanism of chip formation when machining titanium alloys. *Wear* **69** 179-188

[6]. Bonifacio M, Diniz A 1994 Correlating tool wear, tool life, surface roughness and tool vibration in finish turning with coated carbide tools. *Wear* **173** (1) 137-144

[7]. Ramesh S, Karunamoorthy L, Palanikumar K 2008 Surface Roughness Analysis in Machining of Titanium Alloy. *Materials and Manufacturing Processes* **23** (2) 174-181.

[8]. Hartung PD, Kramer B, Von Turkovich B 1982 Tool wear in titanium machining. *CIRP Annals-Manufacturing Technology* **31** (1) 75-80

[9]. Narutaki N, Murakoshi A, Motonishi S, Takeyama H 1983 Study on machining of titanium alloys. *CIRP Annals-Manufacturing Technology* **32** (1) 65-69

[10]. Ramesh S, Karunamoorthy L, Palanikumar K Surface quality investigation in machining of titanium alloy with round CVD coated inserts by neural network approach. In: *Proceedings of the 9th International Symposium on Measurement and Quality Control (ISMQC 2007), IIT Madras*, 2007. pp 55-60

[11]. Siekmann HJ 1955 How to machine titanium. *Tool Engineering* **34** 78-82

[12]. Schintlmeister W, Wallgram W, Kanz J, Gigl K 1984 Cutting tool materials coated by chemical vapour deposition. *Wear* **100** (1) 153-169

[13]. Sarin V 1995 Systematic development of customized CVD coatings. *Surface and Coatings Technology* **73** (1) 23-33

[14]. Fitzsimmons, Sarin VKM 2001 Development of CVD WCCo coatings. *Surface and Coatings Technology* **137** 158-163

[15]. Shaji S, Radhakrishnan V 2003 Analysis of process parameters in surface grinding with graphite as lubricant based on the Taguchi method. *Journal of Materials Processing Technology* **141** (1) 51-59

[16]. Routara BC, Mohanty SD, Datta S, Bandyopadhyay A, Mahapatra SS 2010 Combined quality loss (CQL) concept in WPCA-based Taguchi philosophy for optimization of multiple surface quality characteristics of UNS C34000 brass in cylindrical grinding. *The International Journal of Advanced Manufacturing Technology* **51** (1-4) 135-143

[17]. Dhavlikar M, Kulkarni M, Mariappan V 2003 Combined Taguchi and dual response method for optimization of a centerless grinding operation. *Journal of Materials Processing Technology* **132** (1) 90-94

[18]. Saglam H, Unsacar F, Yaldiz S 2005 An experimental investigation as to the effect of cutting parameters on roundness error and surface roughness in cylindrical grinding. *International Journal of Production Research* **43** (11) 2309-2322

[19]. sibalija TV, Majstorovic VD 2010 Novel approach to multi-response optimisation for correlated responses. *FME Transactions* **38** (1) 39-48

[20]. Ananthakumar.P, Dr.Ramesh.M, Parameshwari 2013 Optimization of Turning Process Parameters Using Multivariate Statistical Method-PCA Coupled with Taguchi Method. *International Journal of Scientific Engineering and Technology* **2** (4) 263-267

[21]. Huang J-T, Liao Y-S 2003 Optimization of machining parameters of wire-EDM based on grey relational and statistical analyses. *International Journal of Production Research* **41** (8) 1707-1720

[22]. Gupta M, Kumar S 2013 Multi-objective optimization of cutting parameters in turning using grey relational analysis. *International Journal of Industrial Engineering Computations* **4** (4) 547-558

[23]. Satyanarayana K, Gopal AV, Babu PB 2014 Analysis for optimal decisions on turning Ti–6Al–4V with Taguchi–grey method. *Proceedings of the Institution of Mechanical Engineers, Part C: Journal of Mechanical Engineering Science* **228** (1) 152-157

[24]. Thakur A, Gangopadhyay S 2016 Dry machining of nickel-based super alloy as a sustainable alternative using TiN/TiAlN coated tool. *Journal of Cleaner Production* **129** 256-268

[25]. Khan M, Mithu M, Dhar N 2009 Effects of minimum quantity lubrication on turning AISI 9310 alloy steel using vegetable oil-based cutting fluid. *Journal of Materials Processing Technology* **209** (15) 5573-5583

[26]. Jawahir I, Oxley P 1988 The tool restricted contact effect as a major influencing factor in chip breaking: an experimental analysis. *CIRP Annals-Manufacturing Technology* **37** (1) 121-126

[27]. Li A, Zhao J, Zhou Y, Chen X, Wang D 2012 Experimental investigation on chip morphologies in high-speed dry milling of titanium alloy Ti-6Al-4V. *The International Journal of Advanced Manufacturing Technology* **62** (9-12) 933-942

[28]. Sun J, Guo Y 2008 A new multi-view approach to characterize 3D chip morphology and properties in end milling titanium Ti–6Al–4V. *International Journal of Machine Tools and Manufacture* **48** (12) 1486-1494

Flexural Strength of Functionally Graded Nanotube Reinforced Sandwich Spherical Panel

Trupti R Mahapatra[1,*] Kulmani Mehar[2], Subrata K Panda[3], S Dewangan[4] and Sushmita Dash[5]

[1,4,5]KIIT University-Bhubaneswar, Odisha, India, 751024

[2,3]National Institute of Technology-Rourkela, Odisha, India, 769008

*E-mail: trmahapatrafme@kiit.ac.in

Abstract: The flexural behaviour of the functionally graded sandwich spherical panel under uniform thermal environment has been investigated in the present work. The face sheets of the sandwich structure are made by the functionally graded carbon nanotube reinforced material and the core face is made by the isotropic and homogeneous material. The material properties of both the fiber and matrix are assumed to be temperature dependent. The sandwich panel model is developed in the framework of the first order shear deformation theory and the governing equation of motion is derived using the variational principle. For the discretization purpose a suitable shell element has been employed from the ANSYS library and the responses are computed using a parametric design language (APDL) coding. The performance and accuracy of the developed model has been established through the convergence and validation by comparing the obtained results with previously published results. Finally, the influence of different geometrical parameters and material properties on the flexural behaviour of the sandwich spherical panel in thermal environment has been investigated through various numerical illustrations and discussed in details.

1. Introduction

Conventional laminated composites consist of various layers of lamina that are homogenous and are held together to achieve enhanced mechanical properties. Thus, they are light in weight and at the same time possess higher strength/stiffness to weight ratio. However, since there is abrupt changes in material properties in between different layers of lamina, high inter-laminar stresses are induced those cause failure of structure due to delamination. This failure is more common when these structures serve under severe thermal load. To overcome this problem, Functionally Graded Materials (FGMs) are used in which the properties of materials are varied smoothly and continuously along thickness direction. In general, the FGMs are constructed by ceramic and metal, in such a way that properties are varied by changing volume fraction of constituent materials along thickness of plate. Owing to excellent thermal resistance property FGMs are widely used in various high performance engineering structures/structural components as in aerospace, automotive, marine and nuclear industries.

Recently, in the family of advanced materials, carbon nanotubes (CNTs) are very much appreciated because of its distinct mechanical, thermal and electrical properties in comparison to other existing advanced materials. The **CNT Reinforced Functionally Graded materials (FGCNTs)** are preferably used as plates, beams or shells, in advanced structural components. The sandwich type construction has also become much more attractive due to their outstanding properties. Therefore, it is highly essential to understand the mechanical behavior of the sandwich structures made up of FGCNT for their better design in real life application. Many researchers have attempted to investigate the linear and nonlinear behaviour of CNT reinforced FG sandwich plates and shells numerically/analytically by using various existing and refined theories. In this regard, the first order shear deformation theory (FSDT) and higher order shear deformation theory (HSDT) are more popular in comparison to refined/layer wise theories due to simplicity in formulation [1]. Zenkour [2] presented the bending analysis of simply supported FG sandwich plates under sinusoidal loading using sinusoidal, third-order, first-order and classical shear deformation theories. Shen [3] studied the nonlinear bending behavior of FGCNT reinforced composite (FGCNTRC) plates in thermal environments using the HSDT and von-Karman nonlinear kinematics. Ke et al. investigated the nonlinear free vibration [4] and dynamic stability [5] behaviour of FGCNTRC beams in the framework of the Timoshenko beam theory and von-Karman type strains. Analytical solutions for the nonlinear bending and free vibration responses of simply supported CNTRC [6] and sandwich plates [7] in thermal environment are presented by Wang and Shen using the HSDT and general von-Karman type equations and considering temperature dependent material properties. Zhu et al. [8] presented the static and free vibration analysis of CNTRC plates using the FSDT and finite element method (FEM). Lei et al. [9] studied the nonlinear deflection behaviour of FGCNTRC plates using element-free kp-Ritz method. The formulation is based on the FSDT mid-plane kinematics and von-Karman nonlinearity. Zhang et al. [10] studied the flexural strength of FGCNTRC cylindrical panels in the frame work of the FSDT. The static and dynamic responses of sandwich plates with CNT reinforced face sheets have been investigated by Natarajan et al. [11] using the HSDT mid-plane kinematics. Mehar et al. [12] examined the nonlinear free vibration of CNTRC flat panel under uniform thermal environment using the HSDT and Green-Lagrange geometrical nonlinearity.

It is evident from the brief review that, many efforts have already been made in past on the theoretical developments for the numerical or analytical solutions of linear/nonlinear flexural behaviour of CNT reinforced FGM sandwich structures. To the best of the authors' knowledge, no work has been reported in the open literature on flexural behaviour of FGCNT reinforced sandwich spherical panel. The aim of the present investigation is to study the bending behavior of sandwich spherical panels with FGCNT face sheets. In order to do so, a simulation model for the FG sandwich panel has been developed using APDL code and discretized using Shell 281 element from ANSYS elemental library. Numerical illustrations are presented to show the effect of various parameters on the flexural response of FGCNT reinforced sandwich spherical panel and discussed in detail.

2. Methodology

In the present study, two types of CNT reinforced sandwich panels have been considered with length "*a*", width "*b*", thickness "*h*" and principal radius of curvature "*R*", as shown in Figure 1 (a) and (b). The notations UD and FG stands for uniform distribution and functionally graded distribution of the carbon nanotubes along the thickness direction of the sandwich panel. Due to convenience, the Mori-Tanaka scheme [8] is employed to estimate the material properties of two-phase nano-composites (mixture of CNTs and isotropic polymer) by using CNT efficiency parameter. Thus, the material properties can be written as

$$E_{11} = \eta_1 V_{CNT} E_{11}^{CNT} + V_m E^m \tag{1}$$

$$\frac{\eta_2}{E_{22}} = \frac{V_{CNT}}{E_{22}^{CNT}} + \frac{V_m}{E^m} \tag{2}$$

$$\frac{\eta_3}{G_{12}} = \frac{V_{CNT}}{G_{12}^{CNT}} + \frac{V_m}{G^m} \tag{3}$$

where, E_{11}^{CNT}, E_{22}^{CNT} and G_{12}^{CNT} represents the Young's moduli and shear modulus of single walled CNTs, respectively and E^m and G^m indicate the corresponding properties of the isotropic matrix.

The UD and FG carbon nanotubes reinforcement composite are related as follows:

$$V_{CNT} = V_{CNT}^* \qquad \text{(UD)} \tag{4}$$

$$V_{CNT}(z) = 2\left(\frac{2|z|}{h}\right)V_{CNT}^* \qquad \text{(FG)} \tag{5}$$

where, $$V_{CNT}^{*} = \frac{W_{CNT}}{W_{CNT} + (\rho^{CNT} / \rho^{m}) - (\rho^{CNT} / \rho^{m}) W_{CNT}} \tag{6}$$

W_{CNT} = mass fraction of the carbon nanotube in the composite panel
ρ^{m} = density of matrix
ρ^{CNT} = density of carbon nanotube

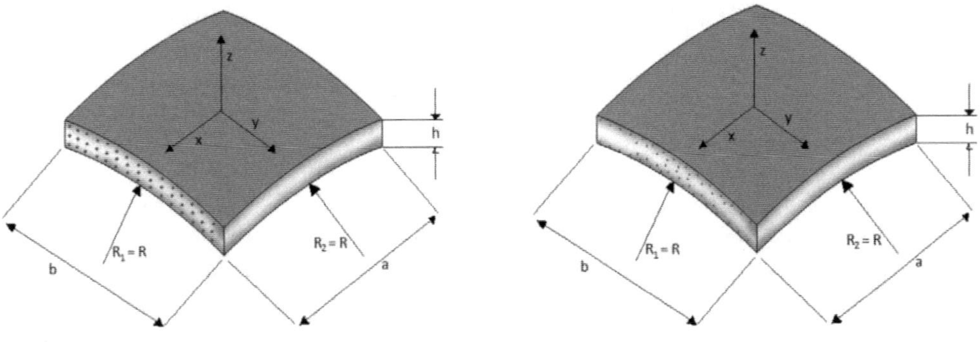

Figure 1. Configuration and gradation of FGCNT sandwich spherical panel **(a)** UD **(b)** FG

In order to compute the desired responses a simulation model for the sandwich spherical panel has been developed using APDL code in ANSYS 15.0 environment. In ANSYS mechanical APDL, various element types are available for modeling of layered structures. For the present analysis, Shell 281 element has been taken. Shell 281 is known for its robustness and suitability for the analysis of thin to slightly thick shell structures. The considered element has total eight nodes having six degrees of freedom per each node say, translation along x, y, and z axes, and rotation about x, y, and z axes. This element gives satisfactory results for linear, large rotation, and/or large strain nonlinear solutions. It also accounts for load stiffness effects of distributed pressure. It frameworks and carry the solutions based on the FSDT and logarithmic strain and true stress measures. Thus, the displacements at any point in the panel u, v and w along x, y and z directions, respectively can be represented as:

$$\begin{aligned} u(x, y, z) &= u^{0}(x, y) + z\theta_{x}(x, y) \\ v(x, y, z) &= v^{0}(x, y) + z\theta_{y}(x, y) \\ w(x, y, z) &= w^{0}(x, y) + z\theta_{z}(x, y) \end{aligned} \tag{7}$$

where, u^0, v^0 and w^0 are the displacements of any point in the mid-plane along x, y and z directions, respectively and θ_x, θ_y and θ_z are the shear rotations. Figure 2 represents the geometrical parameters such as node location and element coordinate system for this element. The element is defined by shell section information and by eight nodes (I, J, K, L, M, N, O and P).

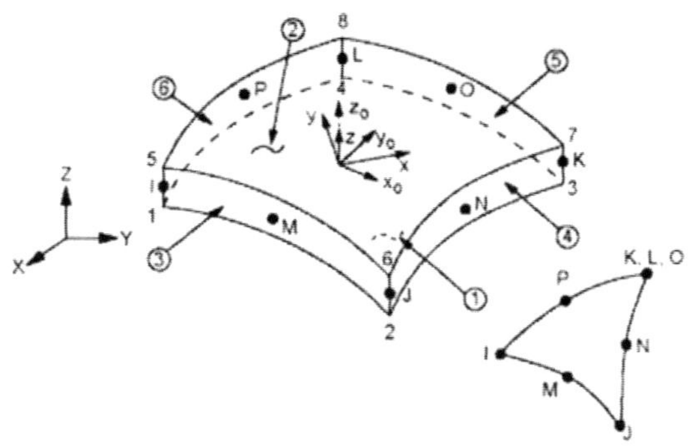

Figure 2. SHELL281 geometry (ANSYS15.0)

3. Results and Discussions

In this section, the flexural responses of FGCNT sandwich spherical panel under thermal environment have been investigated using the present simulation model. The material properties used for the computation purpose have been provided in Table 1. For the computation purpose PMMA and the single walled carbon nanotube (SWCNT) of armchair (10, 10) configuration are considered as the matrix material and the reinforcement phase, respectively. The effective properties and the effective parameters for the PMMA and SWCNT can be seen in [12]. The ability and efficacy of the present model has been established through the convergence and validation studies. Subsequently, various parametric studies have been provided to bring out their significance. If not stated otherwise, uniformly distributed mechanical load q_0=1MPa and temperature of 300K is assumed throughout for the present study. To restrict the rigid body motion and to reduce the number of unknowns for finding the solution, the various boundary conditions used in the present analysis are given as:

1. All sides simply supported (SSSS)

$v^0 = w^0 = \theta_y = \theta_z = 0$ at $x = 0$ and a; $u^0 = w^0 = \theta_x = \theta_z = 0$ at $y = 0$ and b.

2. All sides clamped (CCCC)

$u^0 = v^0 = w^0 = \theta_x = \theta_y = \theta_z = 0$ at both $x = 0$ and a; $y = 0$ and b.

3. All sides free (FFFF)

$u^0 \neq v^0 \neq w^0 \neq \theta_x \neq \theta_y \neq \theta_z \neq 0$ at both $x = 0$ and a; $y = 0$ and b.

The transverse central deflection of the FGCNT sandwich spherical panel in nondimensional form is expressed using the formula as: $w_{\text{nondimensional}} = w_{\text{central}}/h$.

3.1 Convergence Study

The convergence behavior of the present simulation model has been tested in this example. In order to do so, FGCNT reinforced square (*a*/*b*=1) sandwich spherical panels (*h*=2mm *a*/*h* =50 and *R*/*a*=2, V_{CNT}=0.11) subjected to uniformly distributed load (q_0=1MPa) with both UD and FG grading condition and different core to face thickness ratios (0.5, 1 and 2) are considered. The nondimensional central deflections values computed over various mesh refinements under SSSS and CCCC support conditions are plotted in Figure 3 (a) and (b), respectively. It is clearly observed that the responses computed using the present simulation model are converging well with mesh refinement. Based on the convergence study a 10×10 mesh has been chosen for determining the responses further throughout the present study.

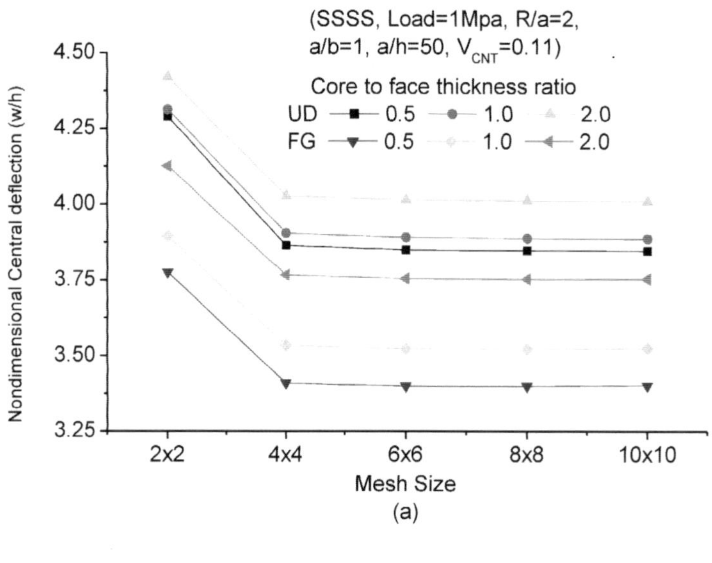

(a)

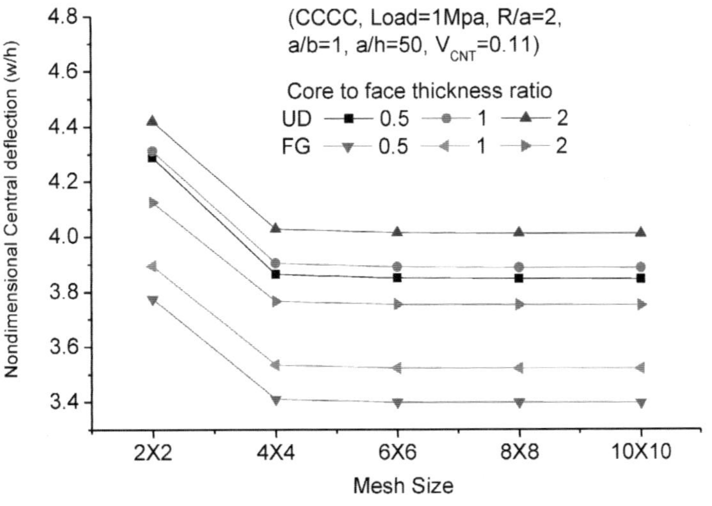

Figure 3. Convergence behaviour of nondimensional central deflection (*w*/*h*) of FGCNT reinforced sandwich spherical panel subjected to uniformly distributed load under **(a)** SSSS and **(b)** CCCC boundary condition.

3.2 Comparison Study

In order to validate the present simulation model, FGCNT reinforced sandwich flat panels as in [8] have been considered. This is because only flat panel examples are available in the concerned domain and the flat panels are considered to be the simplest form of the shell panels. The responses are computed for uniformly distributed load (q_0=0.1MPa), diverse volume fraction of CNT (V_{CNT} = 0.11, 0.14 and 0.17), SSSS and CCCC boundary conditions and considering both UD and FG gradient. For the sake of computation, the geometry, material properties, support conditions and nondimensional formula for the maximum deflection used are same as considered reference [8]. The comparison of results presented in Figure 4 depicts that the responses computed using the present model are as good as the reference values.

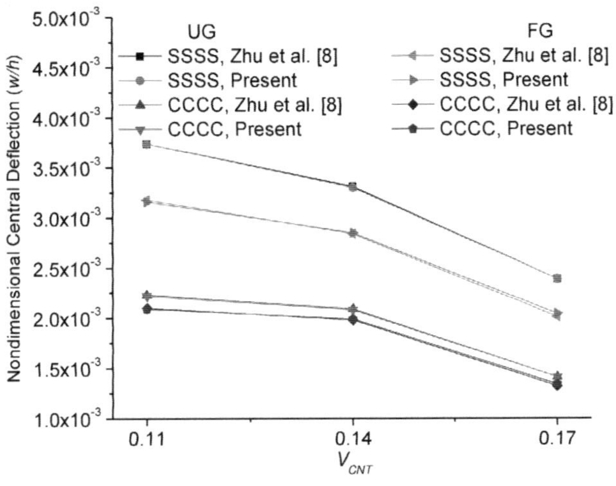

Figure 4. Comparison of nondimensional central deflection of FGCNT reinforced composite flat panel under uniform distributed load.

3.3 Numerical Illustrations

The stability and accuracy of the present developed model has been confirmed in the convergence and the comparison study. In this section, some more numerical investigations have been carried out to demonstrate the applicability of the proposed model and to bring out the effect of parameters (curvature ratio, thickness ratio, boundary conditions and core thickness to face thickness ratio) on the flexural behavior of FGCNT reinforced sandwich spherical panel under thermal environment.

3.3.1 Effect of curvature ratio (R/a) and temperature

It is a fact that membrane strength of the panel structure greatly depends on its curvature. The effect of curvature ratio on the bending behavior of FGCNT reinforced sandwich spherical panel in thermal environment is investigated in this example. Responses are computed for different curvature ratios (R/a = 5, 20, 50 and 100) with both UD and FG gradient composition separately for different volume fraction of CNT (V_{CNT} = 0.11, 0.14 and 0.17) and temperature (T=300K, 500K and 700K). The geometrical parameters and support condition considered as: a/b= 1, a/h= 10, core thickness to face thickness ratio = 0.5, SSSS boundary condition and the results are presented in Table 1. It is observed that the central deflection values increase with curvature ratio whereas decrease with increasing volume fraction of CNT. It is also worthy to note that as the temperature increases the maximum deflection value also increases.

Table 1. Effect of curvature ratio and temperature on nondimensional central deflection of FGCNT reinforced sandwich spherical panel.

Temperature		300K		500K		700K	
V_{CNT}	R/a	UD	FG	UD	FG	UD	FG
0.11	5	36.6750 x 10^{-3}	31.6410 x 10^{-3}	36.994 x 10^{-3}	31.882 x 10^{-3}	37.154 x 10^{-3}	32.003 x 10^{-3}
	20	37.8450 x 10^{-3}	32.5160 x 10^{-3}	38.184 x 10^{-3}	32.771 x 10^{-3}	38.353 x 10^{-3}	32.898 x 10^{-3}
	50	37.9085 x 10^{-3}	32.5635 x 10^{-3}	38.248 x 10^{-3}	32.819 x 10^{-3}	38.418 x 10^{-3}	32.947 x 10^{-3}
	100	37.9175 x 10^{-3}	32.5705 x 10^{-3}	38.257 x 10^{-3}	32.826 x 10^{-3}	38.428 x 10^{-3}	32.954 x 10^{-3}
0.14	5	32.7145 x 10^{-3}	28.6445 x 10^{-3}	32.972 x 10^{-3}	28.837 x 10^{-3}	33.101 x 10^{-3}	28.933 x 10^{-3}
	20	33.6790 x 10^{-3}	29.3825 x 10^{-3}	33.952 x 10^{-3}	29.586 x 10^{-3}	34.088 x 10^{-3}	29.687 x 10^{-3}
	50	33.7315 x 10^{-3}	29.4230 x 10^{-3}	34.005 x 10^{-3}	29.626 x 10^{-3}	34.142 x 10^{-3}	29.728 x 10^{-3}
	100	33.7390 x 10^{-3}	29.4285 x 10^{-3}	34.013 x 10^{-3}	29.632 x 10^{-3}	34.150 x 10^{-3}	29.734 x 10^{-3}
0.17	5	25.9980 x 10^{-3}	22.3620 x 10^{-3}	26.204 x 10^{-3}	22.516 x 10^{-3}	26.307 x 10^{-3}	22.594 x 10^{-3}
	20	26.8470 x 10^{-3}	22.9990 x 10^{-3}	27.066 x 10^{-3}	23.162 x 10^{-3}	27.175 x 10^{-3}	23.244 x 10^{-3}
	50	26.8925 x 10^{-3}	23.0330 x 10^{-3}	27.112 x 10^{-3}	23.197 x 10^{-3}	27.222 x 10^{-3}	23.279 x 10^{-3}
	100	26.8985 x 10^{-3}	23.0380 x 10^{-3}	27.118 x 10^{-3}	23.202 x 10^{-3}	27.228 x 10^{-3}	23.284 x 10^{-3}

3.3.2 Effect of thickness ratio (a/h)

It is well known that the thickness ratio of the panel structures contributes significantly their strength and stiffness. In order to study the effect of thickness ratio on the flexural behavior of FGCNT reinforced sandwich spherical panel under uniform distributed load, all sides clamped square sandwich spherical panels (R/a = 10, core thickness to face thickness ratio = 1) with different volume fraction (V_{CNT} = 0.11, 0.14 and 0.17) are considered. The responses are computed for various thickness ratio (a/h = 10, 20, 40 and 80) and presented in Figure 5. It is observed that the nondimensional central deflection increases with increase in thickness ratio. It is generally expected as the stiffness of the structure decrease as the panel tends to become thin. It is also noted that the maximum deflection decreases with increasing volume fraction of the CNT.

3.3.3 Effect of support conditions

Strength and stiffness behavior of any structure/structural component largely depends on the support conditions which in turn affects their flexural responses. In this example, the effects of various support conditions (SSSS, CCCC, CSCS and CFCF) on the nondimensional maximum deflection of FGCNT reinforced sandwich spherical panel have been investigated. For the sake of computation, the geometrical parameters are taken as: R/a = 10, core thickness to face thickness ratio = 1.5, a/b = 1 and *a*/*h*= 10. Results are computed for both UD and FG gradient with three different CNT volume fractions (V_{CNT} = 0.11, 0.15 and 0.17) and shown in Figure 6. It is noticed that the transverse central deflection values increase as the number of constraints at the support decrease.

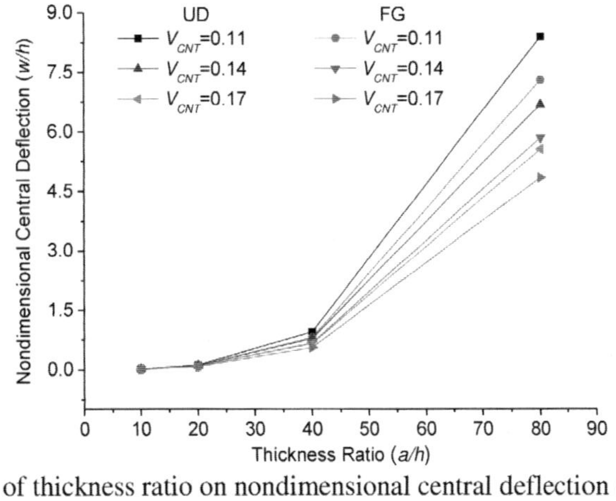

Figure 5. Effect of thickness ratio on nondimensional central deflection of functionally graded nanotube reinforced sandwich spherical panel

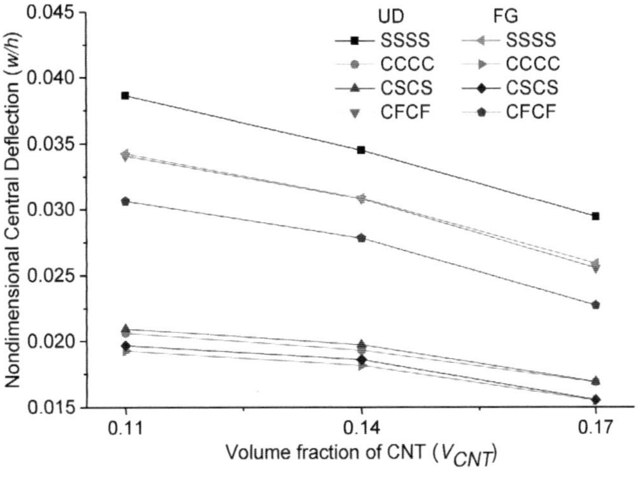

Figure 6. Nondimensional maximum deflection FGCNT sandwich spherical panel subjected to uniformly distributed load for various support conditions.

3.3.4 Effect of core to face thickness ratio

In this example the flexural behavior of FGCNT reinforced sandwich spherical panel (*R*/*a*= 40, *a*/*h*= 10, *a*/*b*= 1 and under CSCS support condition) due to variation of the core thickness to face thickness ratio has been investigated. The nondimensional central deflection values are computed for four different core to face thickness ratio (0, 0.5, 1, 1.5 and 2), three different CNT volume fractions (V_{CNT} = 0.11, 0.14 and 0.17) and for UD and FG gradient separately. The results plotted in Figure 7 depict that in both UD and FG gradient the transverse central deflection values increase with increase in core to face thickness ratio, which is within the expected line.

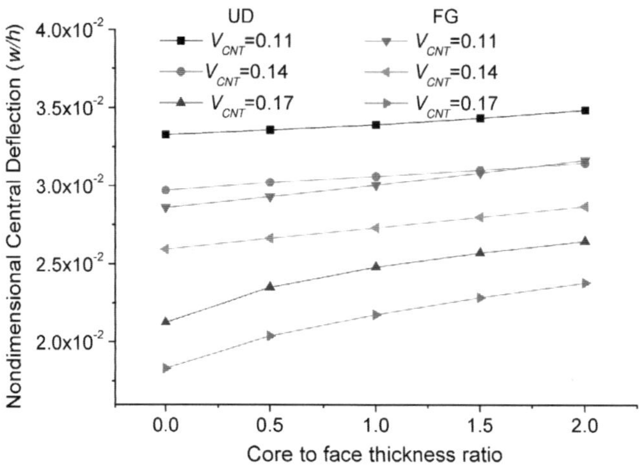

Figure 7. Effect of core thickness to face thickness ratio on nondimensional central deflection of functionally graded nanotube reinforced sandwich spherical panel.

4. Conclusion

In this article, the flexural behavior of FGCNT reinforced sandwich spherical panel of two different grading (UD and FG) under uniformly distributed load and thermal environment has been investigated. The sandwich panel is made of FGCNT reinforced face sheets and isotropic and homogeneous material core. The properties of the panel structure are considered to vary in the thickness direction based on the volume fraction of the CNT and are governed by Mori-Tanaka scheme. In addition to this the properties of the CNT is also assumed to be temperature dependent. For the computation of the desired responses a simulation model for the FGCNT sandwich panel has been developed in ANSYS 15.0 using APDL code. The convergence and validation study shows the efficacy of the present model in terms of performance and accuracy to obtain the desired responses. Few new examples are solved to investigate the effect of various parameters on the flexural behavior of the FGCNT reinforced sandwich spherical panel. Based on the parametric studies, it is observed that, the central deflection values decrease with increase in volume fraction of CNT. It is worthy to note that the stiffness of the spherical sandwich panel decreases with increase in temperature. For a particular value of CNT volume fraction, the FG grade show lesser deflection values as compared to UG grade. It is also seen that the transverse central deflection values decrease as the number of constraints at the support increases.

References

[1] Jha D K, Kant T, and Singh R K 2013 *Composite Structures* **96** 833–849
[2] Zenkour A M 2005 *International Journal of Solids and Structures* **42(18-19)** 5224–5242
[3] Shen H S 2009 *Composite Structures* **91(1)** 9–19
[4] Ke L L, Yang J and Kitipornchai S 2010 *Composite Structures* **92(3)** 676–683
[5] Ke L L, Yang J and Kitipornchai S 2013 *Mechanics of Advanced Materials and Structures* **20(1)** 28–37
[6] Wang Z X and Shen H S 2011 *Computational Materials Science* **50(8)** 2319–2330
[7] Wang Z X and Shen H S 2012 *Composites: Part B* **43(2)** 411–421
[8] Zhu P, Lei Z X and Liew K M 2012 *Composite Structures* **94(4)** 1450–1460
[9] Lei Z X, Liew K M and Yu J L 2013 *Computer Methods in Applied Mechanics and Engineering* **256** 189–199
[10] Zhang L W, Lei Z X, Liew K M and Yu J L 2014 *Composite Structures* **111** 205–212
[11] Natarajan S, Haboussi M and Ganapathi M 2014 *Composite Structures* **113** 197–207
[12] Mehar K and Panda S K 2016 *Composite Structures* **143** 336–346

IOP Conf. Series: Materials Science and Engineering **178** (2017) 012032 doi:10.1088/1757-899X/178/1/012032

Dielectric, Ferroelectric and Piezoelectric study of BNT-BT solid solutions around the MPB region

T. Badapanda[1*], S. Sahoo[2], P. Nayak[3]

[1]Department of Physics, C.V. Raman College of Engineering, Bhubaneswar, Odisha, India-752054
[2]Department of Physics, Centurion University of Technology & Management, Bhubaneswar, Odisha, India-751020
[3]Department of Physics and Astronomy, National Institute of Technology, Rourkela, Odisha-769008, India
Corresponding author email: badapanda.tanmaya@gmail.com

Abstract:

The (1-x)$(Bi_{0.5}Na_{0.5})TiO_3$- x$BaTiO_3$ (x=0.01-0.1) has been prepared by solid state reaction route. The structural characterizations of the solid solutions were done by X-ray diffraction and Raman spectroscopy, which shows that the morphotropic phase boundary exist between x=0.05-0.1. The dielectric, ferroelectric and piezoelectric properties were studied around the MPB region. It is observed from the temperature and frequency dependent dielectric study that the dielectric constant increase with x values up to x=0.08 and then after decreases with x values. The variation of polarization with electric field for all compositions have been done and it is observed that polarization value is high at x=0.08. The electric displacement of the compositions having MPB with various electric field has been carried out and highest piezoelectric coefficient has been observed at x=0.08. All the characterizations concluded that the composition x=0.08 will be very suitable for actuator application.

Keywords: BNT-BT solid solution, MPB, X-Ray diffraction, Dielectric, ferroelectric, Piezoelectric

1. Introduction:

In development of electronic industry, piezoelectric materials play an important role in the applications of actuators, sensors, resonators and ultrasonic transducers [1]. Apart from the piezoelectric properties, ideal ceramic capacitor is associated with some basic requirements such as high dielectric constant, low dielectric loss, and high hysteresis free strain near room temperature over the wide frequency conditions and the performance. These requirements were fulfilled by the lead-based piezoelectric materials with superior piezoelectric, ferroelectric, dielectric properties and are most widely used in practical applications [2]. High piezoelectric and ferroelectric properties are found in compositions close to the morphotropicphase boundary (MPB) [2]. As lead is considered as toxic element and cause a serious risk to humanhealth and environmental pollution, there isahunt to replace lead based compositions with lead-free and environment friendly ceramic materials. Recently, extensive research has been done on the development of the lead-free ceramics, which possess high dielectric constant, large field-induced strain, polarization, piezoelectric properties. Among the lead-free piezoelectric materials, sodium bismuth titanate($Bi_{0.5}Na_{0.5}TiO_3$, abbreviated as NBT) and Barium titanate ($BaTiO_3$,abbreviated as BT) are accepted as an important material for the fabrication of lead-free piezoelectric materials. The BNT shows strong ferroelectric property along with high Curie temperature, high remnant polarization and a coercive field at room temperature [3–6]. However, the large coercive field and large conductivity makes pure NBT as hard to pole, which hinders its industrial application.

In order to obtain useful lead-free piezoelectric ceramics a lot of work has been done on the NBT-based solid solutions to modify and improve the electrical properties. Recent studies have been focused on improving piezoelectric properties of NBT-based ceramics to use the solid-solution technique to substitute the individuals and seek for the merits of the selected materials. Among these NBT based solid solutions, the (1-x) $(Bi_{0.5}Na_{0.5})TiO_3$-x$BaTiO_3$ is the most attractive one due to their excellent piezoelectric properties at morphotropic phase boundary (MPB) composition as studied by several researchers [7-14]. Morphotropic phaseboundary (MPB) is the boundary between rhombohedralphase of NBT and tetragonal phase of BT. $BaTiO_3$ is chosen because of its superior dielectric, ferroelectric, piezoelectric, [15–18] that make it an extremely important system for fundamental electro-ceramics research. It was reported that NBT–BT 94/6 ceramics shows high piezoelectric constant d_{33}=125 pC/N and an electromechanical coupling coefficient of k_{33}=0.55 near the rhombohedral/tetragonal MPB region. These properties are suitable for high frequency ultrasonic uses or piezoelectric actuator applications. As for the NBT–BT crystals, the highest value of d_{33} is 160 pC/N obtained in the NBT–BT 94/6 single crystals grown by Bridgman method [18]. The piezoelectric properties of ⟨001⟩ poled NBT–BT 95/5 show good piezoelectric properties with d_{33}=280 pC/N. However, the d_{33} values of the ⟨111⟩ oriented NBT–BT 95/5 crystals are 90 pC/N [19]. Ching et al. have reported that NBT–BT 94.5/5.5 single crystal on the ⟨001⟩ exhibit d_{33} value up to 450 pC/ N, which is close to some PZT ceramics [20].The enhanced piezoelectric and ferroelectric properties are observed around the MPB region in BNT-BT composition. However, there is still a ambiguity in the exact MPB composition and lack of fundamental understanding of the structure-property relations and mechanisms for high piezoelectric properties near the MPB composition. .In this paper, we report our work on the preparation and characterization of (1-x) BNT-xBT ceramics with special emphasis on searching the morphotrophic phase boundary (MPB) of this system via structural, dielectric and piezoelectric properties.

2. Experimental

Solid-state synthesis method was adopted to powder samples (1-x) BNT-xBT with $0 \leq x \leq 0.1$. The raw materials used were Barium Carbonate - $BaCO_3$(reagent grade Meark, India), Titanium Oxide -TiO_2 (reagent grade, Merck), Bismuth Oxide - Bi_2O_3(reagent grade Meark, India) and Sodium Carbonate - Na_2CO_3 (reagent grade, Merck).

All the powders are weighed according to the stochiometry ratio and powders were ball milled with Zirconium media in ethyl alcohol for 12 hrs. The mixed dry powders were calcined respectively at 950^{0}C for 3hrs in a programmable furnace. The phase purity of the samples was investigated using X-ray diffraction (XRD), (Xport MPD, Philips, UK). The granules of all the compositions were made by adding 5% polyvinyl alcohol as a binder and pressed to obtain discs. The discs were sintered at 1150^{0}C for 3 hrs. Silver paste was applied on both faces followed by heat treatment at 500^{0}C (20mins) of the samples for electrical measurements. The relative permittivity against the temperature was measured at various frequencies using an LCR meter. The polarization versus electric field (P–E) loops of the specimens, were studied using modified (Radiant Technologies) radiant P–E loop tracer circuit by applying AC field.

3. Result and Discussion:

3.1 X-ray diffraction Study

Fig 1 (a) shows the XRD patterns of (1-x) $(Bi_{0.5}Na_{0.5})TiO_3$ - x $BaTiO_3$ ceramics with $0 \leq x \leq 0.1$ sintered at 1150°C for 3 hrs. All the compositions exhibit a pure perovskite structure and no second phases are observed, which implies that $BaTiO_3$ (BT) ceramic has diffused into the $Bi_{0.5}Na_{0.5}TiO_3$ lattices to form a solid solution. All the peaks in the XRD pattern of (1-x) BNT-xBT are correspond to the BNT phase with rhombohedral structure as reported by different researchers [21]. With the increasing of BT content the peak positions shift towards lower angle and the peaks are also get broaden, it implies that crystallite size was decreasing with addition of barium titanate. Fig. 1(b) shows the XRD patterns of the ceramics in the 2θ range of 46–48^0. The rhombohedral symmetry of BNT ceramic at room temperature can be characterized by a (0 0 3)/(0 2 1) peak splitting between 39 and 41^0 and a single peak of (2 0 2) between 42 and 48^0. It shows that the (2 0 2) peak became broad and asymmetric up to x = 0.04 and a distinct (0 0 2)/(2 0 0) peak splitting appears after x =0.05, referring to a tetragonal symmetry.This characterizes a coexistence of rhombohedral (R) and tetragonal (T) phases, which is consistent with a MPB composition [22]. The lower angle peak, splits into (100),(001) and it may be due to the deformation and distortion created by BT.

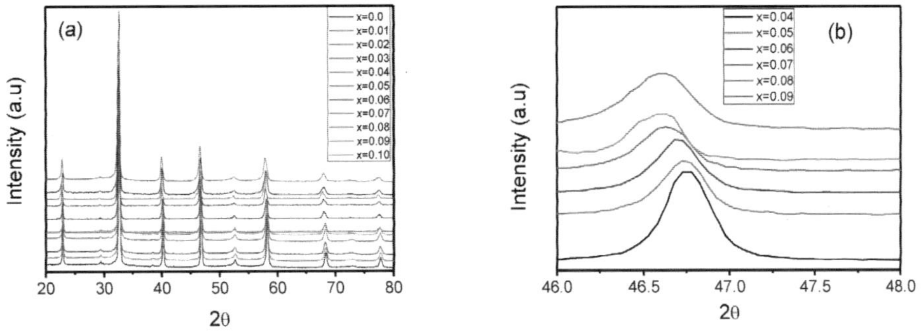

Fig. 1(a) X-Ray diffraction pattern of (1-x) $(Bi_{0.5}Na_{0.5})TiO_3$-x$BaTiO_3$ ceramics, (b) reduced X-Ray diffraction pattern of (1-x) $(Bi_{0.5}Na_{0.5})TiO_3$-x$BaTiO_3$ ceramics

3.2 Raman Study:

The Raman spectroscopy is an effective technique to investigate the presence of functional groups in solid solutions, which provides better insight into the structural study. Fig 2(a) represents the Raman spectroscopy study of (1-x) $(Bi_{0.5}Na_{0.5})Ti0_3$-$xBaTiO_3$ (x=0.0-0.1) to verify the existence of MPB region. The Raman peaks are relatively broad, which can be caused by the distorted octahedral $[BiO_6]$ and $[NaO_6]$ clusters or disorder in A-site of rhombohedral structureand the overlapping of Raman modes due to the lattice anharmonicity. The observed peaks are categorized into three wave number ranges. There are (1) low wavenumber range of 100–200 cm^{-1} which is believed to be related to Na–O vibrations; the peak splitting in this range indicates that there are two kinds of ABO_6interactions in the NBT system. (2) A mid wave number range of 200–400 cm^{-1}which is believed to be related to Ti–O vibrations; and (3) a high wave number range of 400– 800 cm^{-1}which is believed to be related to oxygen octahedral vibrations and rotations [23].The first Raman-active $A_1(TO_1)$ mode at around (146 cm^{-1}) is related to network modifiers or distorted octahedral $[BiO_6]$ and $[NaO_6]$ clusters. The Raman-active $E(TO_2)$ mode in the regions of 279 cm^{-1} is assigned at stretching arising from the bonds due tothe presence of octahedral $[TiO_6]$ clusters at short-range. The (TO_3) modes situated at around 542 cm^{-1} is described to the (←O←Ti→O→) stretching symmetric vibrations of the octahedral $[TiO_6]$ clusters. Finally, the (LO_3) mode found at 812 cm^{-1} is due to presence of the sites within the rhombohedral lattice pre containing octahedral distorted $[TiO_6]$ clusters. Fig. 2(b) shows the Raman spectroscopy plot for x=0.04-0.1 of (1-x) $(Bi_{0.5}Na_{0.5})Ti0_3$-$xBaTiO_3$solid solution. The mode around the region 542 cm^{-1}splitfor x≥0.05, shows a MPB region which also confirms the X-ray result. To find the most prominent composition among the MPB region which will be suitable for industrial application, we study the dielectric, ferroelectric and piezoelectric study.

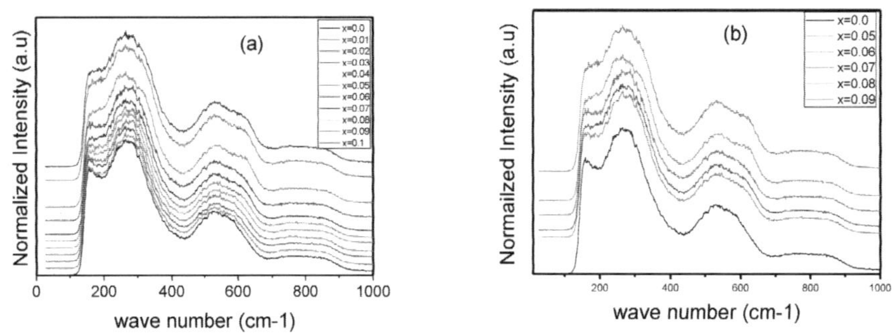

Fig. 2. Raman Spectroscopy study of (1-x) $(Ba_{0.5}Na_{0.5})TiO_3$-$xBaTiO_3$ceramic (a)x=0.0-0.1 (b) x=0.0,0.05,0.06,0.07,0.08,0.09

3.3 Dielectric behaviour:

The temperature dependence of dielectric constant ε' of (1-x)BNT- (x)BT system at various frequencies near the MPB region are shown in Fig.3. Two dielectric peaks have been observed in each composition as shown in Fig. 3. The observed two dielectric peaks can be attributed to the factors caused by the phase transitions from ferroelectric to anti-ferroelectric, which is called depolarization temperature (T_d) and from anti-ferroelectric to paraelectric phase, at which the maximum value of dielectric constant corresponding temperature is Curie temperature (T_m). The value of T_d and T_m are found to decrease with increasing the concentration of BTupto x=0.08 and then increases with further addition. The T_d plays an important role in the practical applications of the materials. The ferroelectric domains with lower T_d are less stable, hence T_d can be seen as the indication of the stability in ferroelectric domains. The MPB compositions exhibit a lower depolarization temperature, which implies a reduction of the stability of ferroelectric domains. The coexistence of a mixed rhombohedral–tetragonal phase could lead to more powerful stress in the MPB compositions, due to the incompatibility of their crystal lattices, resulting in a decrease of thermal stability in the long-range ferroelectric domains[24,25].This can be regarded as the origin of the lower depolarization temperature at the MPB compositions.It was also observed that T_m decreases with increasing BT content and reached a minimum value at 8 mol% BT and with further additions of BT increase in T_m value was observed. The enlargement of the unit cell and achange in the relative fractions of rhombohedral and tetragonal phases was thought to play an important role as the larger ionic radii of Ba^{2+} increase in lattice constant and consequently reduced the T_m value.It is again observed that the dielectric constant increases upto x=0.08 and then decreases with further addition.

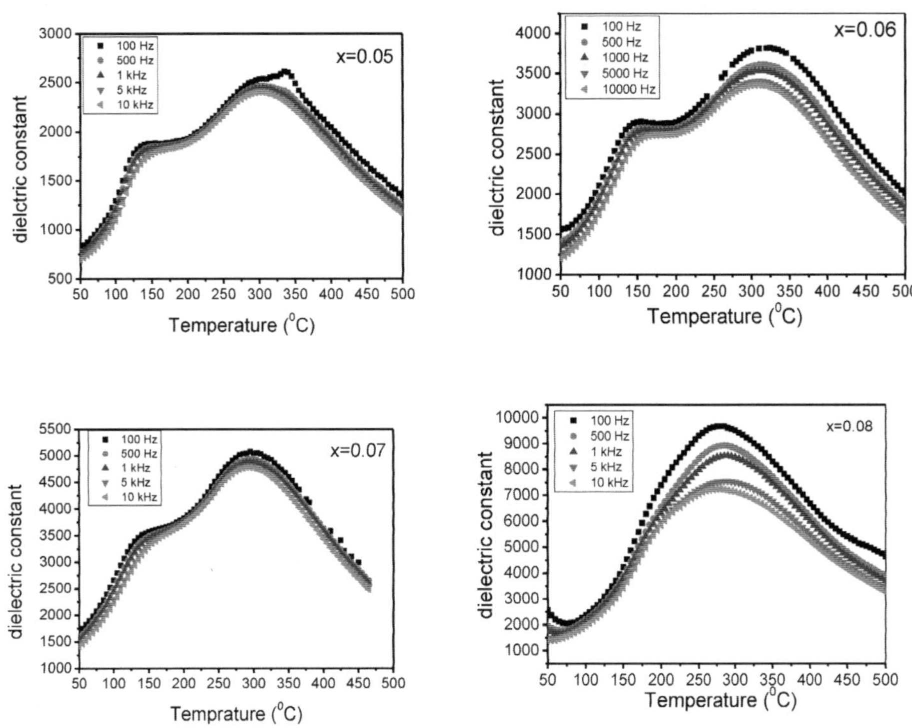

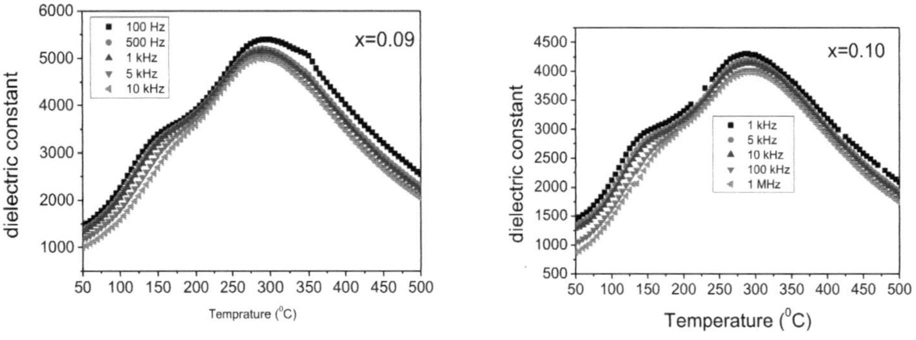

Fig 3. Temperature dependent dielectric study of (1-x) $(Ba_{0.5}Na_{0.5})TiO_3$-$xBaTiO_3$ ceramic

3.4 Ferroelectric and piezoelectric study:

At Room-temperature both unipolar and bipolar polarization–electric field (P–E) hysteresis data and strain–electric field (S–E) data for (1-x)BNT–xBT ceramics measured an electric field of 20 kV/cm at frequency 1Hz are shown in Figs. 4(a,b) and 5(a,b), respectively. Furthermore, the remnant polarization (P_r), the maximum polarization (P_{max}), the coercive field (Ec), the maximum strain (S_{max}), and the normalized strain coefficient ($d_{33} = S_{max}/E_{max}$) values are also listed in Table 1. It could be noted that the observed large strain seems to exist only over a narrow compositional region. However, the addition of a stable ferroelectric end-member BT in the solidsolution should result an increase in domain stability. Itcan be clearly seen from the table that the P_r and$E_c$increased considerably as the amount of BT increased upto 8 mol% and decreases with further addition of BT. The bipolar strain–electric field (S–E) loops also showed a normal butterflyshape with increased negative strain. The S_{max} and d_{33} values increases up to x=0.08 and then decreases. A slight asymmetry isobserved in the *(S-E)* loops at x=0.08, which may be due to the internal fieldinduced by the presence of defects [26]. It can be seen that low-field piezoelectricproperties exhibited a strong compositional and phase dependence within the MPB region of BNT–BT.Development of typical butterfly-shaped loops and the existence of remnant strain in the NBT-xBTceramics near MPB indicates their piezoelectric nature [27].

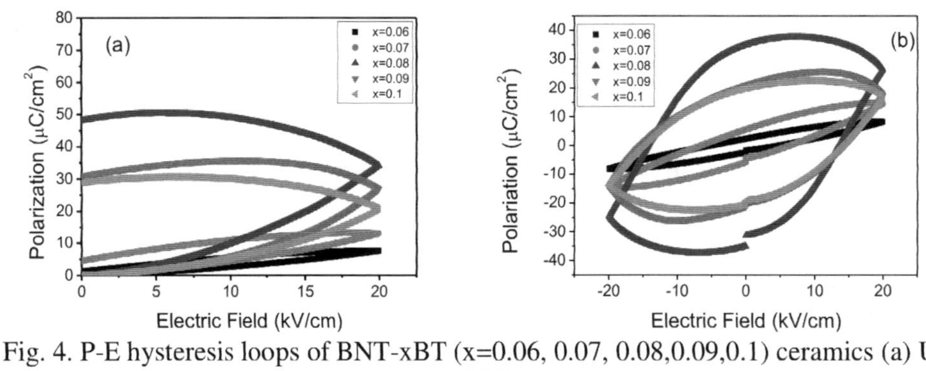

Fig. 4. P-E hysteresis loops of BNT-xBT (x=0.06, 0.07, 0.08,0.09,0.1) ceramics (a) Unipolar (b) Bipolar

IOP Conf. Series: Materials Science and Engineering **178** (2017) 012032 doi:10.1088/1757-899X/178/1/012032

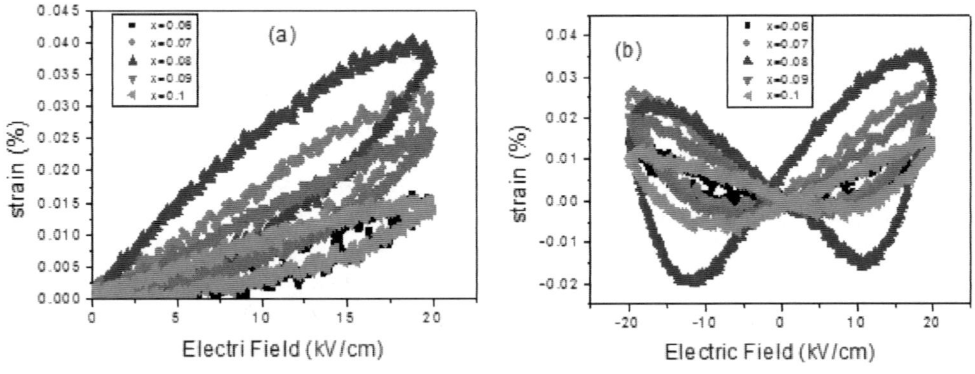

Fig. 5 Variation of induced strain% vs. electric field in the BNT-xBT ceramics with BNT-xBT (x=0.06, 0.07, 0.08,0.09, 0.1) ceramics (a) Unipolar (b) Bipolar

Parameters obtained from Dielectric, ferroelectric and piezoelectric study of (1-x) $(Bi_{0.5}Na_{0.5})TiO_3$-x$BaTiO_3$ ceramic

Composition	*Dielectric constant*	T_m	P_r (µC/cm^2)	S_{max} (%)	d_{33} (pm/V)
x=0.06	3821	325	1.98	0.015	75
x=0.07	5080	301	4.79	0.031	152
x=0.08	9672	280	48.16	0.038	194
x=0.09	5408	293	31.3	0.025	121
x=0.1	4307	296	29.13	0.014	72

4. Conclusion:

(1-x)BNT-xBTsolid solutions were successfully prepared by the solid state reactiontechnique. The X-ray diffraction and Raman spectroscopy study shows that there exist a MPB within the region x=0.05 -0.1. The dielectric data shows that the dielectric constant increases up to x=0.08 and then decreases with further addition of BT. The ferroelectric and piezoelectric study was performed for both unipolar and bipolar field and enhanced behaviour was observed at x=0.08. All the above study concludes that the highest MPB observed at x=0.08. High dielectric, large strain and highest remanent polarization was observed for x-0.08 which makes the composition suitable for memory, transducer and actuator application.

IOP Conf. Series: Materials Science and Engineering **178** (2017) 012032 doi:10.1088/1757-899X/178/1/012032

References:

[1] Jaffe. B, Cook.W, Jaffe.H, (1971) Piezoelectric Ceramics, ,Academic Press, New York P. 78.

[2] Haertling.H.G, (1999) Ferroelectric ceramics: history and technology, J. Am. Ceram. Soc., 82, 797–818.

[3] Smolenskii.A.G,Isupov.A.VAgranovskaya.I.A , Krainik.N.N, (1961)Sov,Phys.SolidState2(196) ,2651.

[4] Buhrer.F.C, (1962) Some propertiesof Bismuth perovskites,J.Chem.Phys.36 (3),798–803.

[5] Takanaka.T, Maruyama.K, Sakata.K, (1991) $(Bi_{1/2}Na_{1/2})TiO3$–BaTiO3 system for lead-freepiezoelectricceramics,Jpn.J.Appl.Phys. 30, 2236–2239 ,part19B .

[6] Pronin.P.I, Syrnikov.P.P, Isupov.A.V, Egorov.M.V, Zaitseva .V.N, (1980) Ferro 1971.

[7] Takenaka.T, Nagata.H,(2005) J. Eur. Ceram. Soc.25, 2693.

[8] Wang.X.X, Choy.H.S, Tang.G.X, W Chan.L.H, (2005) J. Appl. Phys.97, 104101.

[9] Chiang.M.Y, Gregory.F.W, Andrey.S.N, (1998) Appl. Phys.Lett. 73(25),3683–3685 .

[10] Hosono.Y, Harada.K Yamashita.Y, (2001) Jpn. J.Appl.Phys. 40,5722–5726 , part19B.

[11] Li.D.H, Feng.D.C,Xiang.H.P, (2003) Jpn.J.Appl.Phys.,427387–7391 ,part112.

[12] Gomah-Pettry.R.J, .Saïd.S, Marchet.P ,Mercurio.P, (2004) J.Eur.Ceram.Soc. 24(6) ,1165–1169 .

[13] Xu.Q, Chen.S, Chen.W, etal. (2004) J.Alloy. Compd. 381(1–2),221–225 .

[14] Xu.G, Duan.Z, Wang.X, Yang.D,(2005) J.Cryst.Growth,275 (1–2), 113–119 .

[15]Saburi.O , (1959)J. Phys. Soc.Japan 14, 1159–1174.

[16] Harman.G.G, (1957) Phys. Rev. 106, 1358–1359.

[17] Merz.J.W , (1949) Phys. Rev. 76, 1221–1225.

[18] Jona.F, Shirane.G, (1963) Ferroelectric Crystals, Elsevier, Amsterdam.

[19] Ge.Wenwei, Liu.Hong, Zhao.Xiangyong, Fang.Bijun , Li.Xiaobing, Wang.Feifei, Zhou.Dan, Yu.Ping, Pan.Xiaoming, DiLin ,HaosuLuo, (2008) J.Phys.D:Appl. Phys. 41(117) 1154035

[20] Zhang.Q, Zhang.Y, Wang.F, DiLin.X, Lin.Xiangyong,Zhao, HaosuLuo, (2010) J.Cryst.Growth, 312(30),457–460.

[21] Isupov, V.A., (2005) *Ferro- electrics*, 315(1), 123-147.

[22] Chen .Chao, JiangpingXiang, Li.Yueming, Wang.Feifei, Zhang .Q, Luo .Haosu, (2010), J. Appl.Phys., 108(12),124106 .

[23] Luo .Liang, Ge.Wenwei, Li.Jiefang, Viehland. D, Farely.Charles, Bodnar .Robert,.Zhang .Q, HaosuLuo, (2011,J.Appl.Phys. 109 (11),113507 .

[24] Dai.H.X , Digiovanni.A, Viehland.D, (1993) *J. Appl. Phys.*74, 3399–3405 .

[25] Yoon.S.M , Jang.M.H, Kim.S, (1995)*Jpn. J. Appl. Phys.* 4, 1916–1921.

[26] Zheng.J.X, Liu.Y.J, Peng.F.J, Liu.X, Gong.Q.Y, Zhou.S.K , Huang.H.D, (2013) Thin Solid Films,548,118-124 .

[27] Jaffe.B, Cook.R.W, Jaffe.H.Jr, (1971) Piezoelectric Ceramics , (Academic Press London and New York).

.

IOP Conf. Series: Materials Science and Engineering **178** (2017) 012033 doi:10.1088/1757-899X/178/1/012033

Selection of Levels of Dressing Process Parameters by Using TOPSIS Technique for Surface Roughness of En-31 Work piece in CNC Cylindrical Grinding Machine

SANJAY S. PATIL
PHD, Research Scholar, Department of Mechanical Engineering.
MIT, Kothrud, Pune , India.

e-mail: sspatil.scoe@sinhgad.edu

YOGESH J. BHALERAO
Principal
MIT Academyof engineering,
Alandi,Pune,India.

e-mail: ybhalerao@maepune.ac.in

Abstract: Grinding is metal cutting process used for mainly finishing the automobile components. The grinding wheel performance becomes dull by using it most of times. So it should be reshaping for consistent performance. It is necessary to remove dull grains of grinding wheel which is known as dressing process. The surface finish produced on the work piece is dependent on the dressing parameters in sub-sequent grinding operation. Multi-point diamond dresser has four important parameters such as the dressing cross feed rate, dressing depth of cut, width of the diamond dresser and drag angle of the dresser. The range of cross feed rate level is from 80-100 mm/min, depth of cut varies from 10 – 30 micron, width of diamond dresser is from 0.8 – 1.10mm and drag angle is from $40^{o} – 50^{0}$, The relative closeness to ideal levels of dressing parameters are found for surface finish produced on the En-31 work piece during sub-sequent grinding operation by using Technique of Order Preference by Similarity to Ideal Solution (TOPSIS).In the present work, closeness to ideal solution i.e. levels of dressing parameters are found for Computer Numerical Control (CNC) cylindrical angular grinding machine. After the TOPSIS technique, it is found that the value of Level I is 0.9738 which gives better surface finish on the En-31 work piece in sub-sequent grinding operation which helps the user to select the correct levels (combinations) of dressing parameters.

1. Introduction

Demand for high quality surface finish, less lead time, low cost and higher productivity are need of the hour in industries. Grinding is a metal cutting process which is used for finishing of hard stage parts. Grinding operation has numerous parameters which will decide it affect on the surface roughness of the work piece. Grinding wheel, work piece material characteristics, grinding wheel– work piece kinematics, dressing process and truing are parameters of grinding operations.

Xun Chen,.Brian Rowe established the relation for grinding performance by changing the dressing conditions for surface grinding using simulation. However it will be interesting to see the results for the same if operation is carried out in CNC cylindrical grinding machine. V.P.Darji and R.V. Rao (2014) used Multi Criteria Decision Making method for selection of material in sugar industries. Rajnish Kumar, Jagadish, and Amitava Ray (2014) presented TOPSIS method for selection of material for optimal design. R. Kumar, C. Bhomik and A. Ray (2013) demonstrated TOPSIS method for selection of cutting tool material. En-31 [13, 14] is more commonly used in manufacturing industries. Surface finish is an important parameter in manufacturing engineering and it can influence the performance of mechanical parts and the production cost.

The objective of the grinding process is to generate parts with high surface finish, with minimum tolerance. Dressing [5, 7] is the process of removing dull grains of grinding wheel in order to make it sharp. Topography of wheel is important aspect which has an effect on the surface finish produced of the work piece. Topography [12] of wheel is generated by dressing process. Selection of appropriate process parameters for producing good quality of surface finish on the specific material plays important role in manufacturing. Selection of correct process parameter reduces manufacturing time, increase in productivity, increase efficiency of process, less cost and it makes less confusion for the selecting the range of parameters. So it is necessary to select correct values (level) of dressing parameters for generating better surface finish of work piece. Improper selection of dressing parameters (level) causes poor surface finish on the En-31 work piece, high power loss and it is uneconomical.

2. Literature Review, Conclusion from Reviewed Literature and Need of Present Research Work

A. Fritsche and F. Bleicher (2015) focused on evaluating and influencing dressing results by changing the grain size distribution based on statistical and experimental investigation. They developed "Active Grain Stack" (AGS) model which give information of a structure of grinding wheel and how different layers interact with each other. Fritz K Locke et al. (2016) developed a method to simulate a detailed 3- D model of the grinding wheel structure. They used volumetric composition which includes no. of grins, their orientation; their distribution in grinding wheel. The 3D model has been developed to measure and transfer the shape of Cubic Boron Nitride (CBN) grains. In addition, a method has been created numerically for realistic CBN grains

After reviewing the literature and industrial survey, it is found that, dressing parameters are highly effective in producing good quality of surface finish on the work piece in CNC cylindrical angular grinding machine. Very few researchers have worked on EN31 material. Therefore there is a need to find out a solution which will be nearly equal to the ideal levels of these parameters to have a good surface finish of En31 work piece which would help the user to select the correct levels of dressing parameters which is the main objective of grinding process.

The objective of this work is to develop TOPSIS method for selection of closeness to the ideal dressing parameters for generating better surface finish on the work piece. A seven steps approach method can be used in finding closeness to ideal dressing parameters for generating better surface finish on the work piece.

3. TOPSIS Method

TOPSIS was first presented by Yoon (1980) and Hwang and Yoon (1981), for solving Multiple Criteria Decision Making (MCDM) problems based on the concept that the chosen alternative should have the shortest Euclidian distance from the Positive Ideal Solution (PIS) and the farthest from the Negative Ideal

Solution (NIS). For instance, PIS maximizes the benefit and minimizes the cost, whereas the NIS maximizes the cost and minimizes the benefit. It assumes that each criterion require to be maximized or minimized. TOPSIS is a simple and useful technique for ranking a number of possible alternatives according to closeness to the ideal solution. TOPSIS is one of the popular techniques so more discussion is not given.

4. TOPSIS Method for present research work

Referring the steps of the TOPSIS method, the following steps have been applied for the present research work.

Step 1- There are four input parameters namely dressing cross feed rate, dressing depth of cut, width of the diamond dresser and drag angle of the dresser. Levels of these parameters are given in table 1. Level I includes cross feed rate of 80 mm/minute, dressing depth of cut of 10 micron, width of dresser 0.8 mm and drag angle of dresser of 40^0. Similarly Level II and Level III are designed as shown in table 1. Decision matrix is given in the table I.

Table 1. Input parameters and levels

Name of parameters	Dressing cross feed rate (mm/min.)	Dressing depth of cut(micron)	Width of the diamond dresser(mm)	Drag angle of the dresser(degree)
Level I	80	10	0.8	40
Level II	90	20	0.9	45
Level III	100	30	1.1	50

Step 2- - Normalized decision matrix

Relative performance of normalized values of dressing process parameters namely dressing cross feed rate, dressing depth of cut, width of diamond dresser and drag angle of dresser are calculated and given in table 2.

Table 2. Normalized decision matrix

Name of parameters	Dressing cross feed rate (mm/min.)	Dressing depth of cut(micron)	Width of the diamond dresser(mm)	Drag angle of the dresser(degree)
Level I	0.511	0.267	0.490	0.5111
Level II	0.5750	0.5346	0.552	0.5750
Level III	0.6388	0.8019	0.6748	0.6516

Step 3- Weighted decision matrix

IOP Conf. Series: Materials Science and Engineering **178** (2017) 012033 doi:10.1088/1757-899X/178/1/012033

Each dressing process parameter is equally important after normalized dressing process parameters and Level I, Level II and Level III also. So Analytical Hierarchical Process (AHP) process is used for to quantify the relative importance of each dressing process parameters as shown in equation I. Dressing process parameters are critically assisted in term of surface finish produced on the EN-31 work piece in sub-sequent grinding operation. Order of ranking descends as follows.

1. Dressing cross feed rate.

2. Dressing depth of cut

3. Width of diamond dresser

4. Drag angle of dresser

Weighted decision matrix is formed by weight of parameter multiply by each element of column.

$$Wi = [0.6384 \quad 0.1866 \quad 0.1201 \quad 0.0544]$$ --- I (Found by AHP method)

Table 3. Weighted decision matrix

Name of parameters	Dressing cross feed rate (mm/min.)	Dressing depth of cut(micron)	Width of the diamond dresser(mm)	Drag angle of the dresser(degree)
Level I	0.3262	0.04982	0.0588	0.02780
Level II	0.3670	0.09975	0.0662	0.03128
Level III	0.4078	0.1496	0.0810	0.03544

Step 4-Identify positive and negative solutions

Positive ideal solution (V+) and negative ideal solution (V-) are determined and given in table 4

Table 4. Identify positive and negative solutions

V+	0.3262	0.04982	0.0588	0.03544
V-	0.4078	0.1496	0.0810	0.02780

Step 5- Separation measure for each alternative is calculated. The positive ideal solution and Negative ideal solution are defined as follows.

$$S+ = \{0.20358 \quad 0.10885 \quad 0.20385\}$$

$$S+ = \{0.20358 \quad 0.10885 \quad 0.00764\}$$

Step 6 - Relative closeness to each solution is given in step 6.

$$\text{Ci} = \{0.963829 \quad 0.515339 \quad 0.03617\}$$

If the values are less than one then it is acceptable. In step 6 all values are less than one hence solution as acceptable and maximum value is the best.

Step 7 –Ranking of preference order.

Level 1 has value of 0.963827, Level II has value of 0.515339, Level III has value of 0.0361 and Level I has maximum value and i.e. the best. Level III is worst option for surface roughness on En-31 work piece. Level II is intermediate solution for surface finish on En-31 work piece. User will produce good quality surface finish on En-31 by selecting option level I. Histogram of different levels of dressing parameters are shown in figure. 1

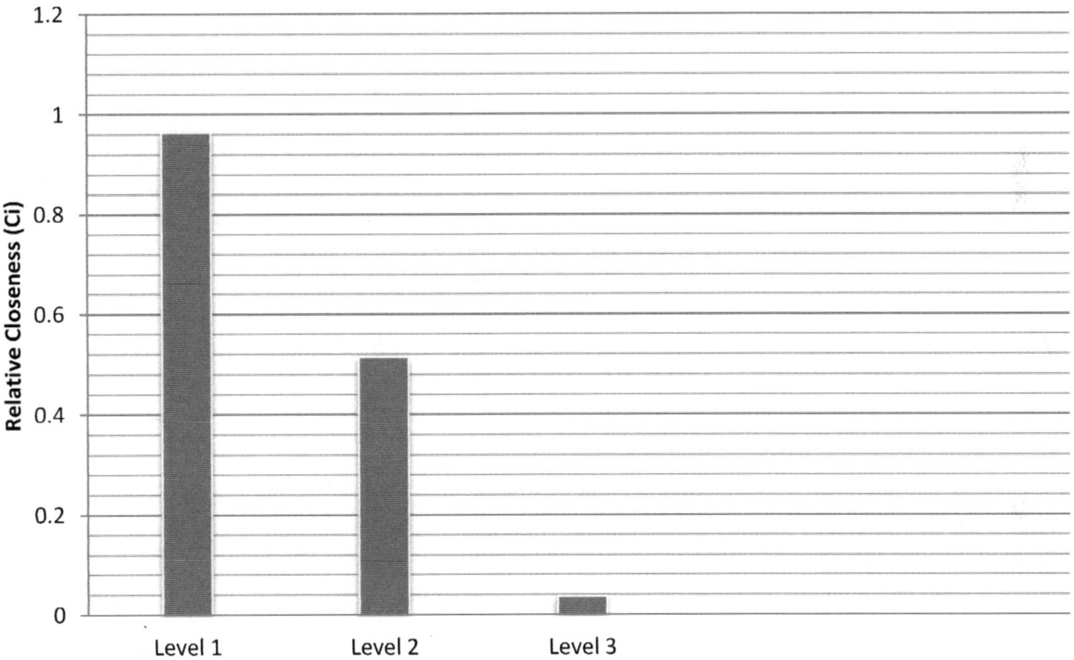

Figure 1. Histogram of different levels of dressing parameters.

5. Conclusions

TOPSIS method shows that closeness to the ideal solution of dressing parameters namely cross feed rate, depth of cut, width of diamond dresser and drag angle of multipoint diamond dresser is level I. The values of parameters are 80 mm/minute cross feed rate, 10micron depth of cut, 0.8mm width of diamond dresser and 40^0 drag angle of diamond dresser. These values are best for generating better surface finish on the En-31 work piece in sub-sequent grinding operation in CNC cylindrical angular grinding machine.

6. Acknowledgment

Authors would like to express their sincere gratitude to Mr. Prakash Patil, Senior Manager, Grindwell Norton Limited- Abrasive Solutions, Pune, India and all other domain experts for providing their

IOP Conf. Series: Materials Science and Engineering **178** (2017) 012033 doi:10.1088/1757-899X/178/1/012033

expertise in the development of proposed selection of dressing process parameter for EN-31 using TOPSIS for CNC cylindrical grinding machine.

7. References

[1] Hwang C.L.and Yoon K."Multiple attribute decision making: methods and applications" Berlin Heidelberg New York,Springer. 1981

[2] Ashish chauan, M.K. Pradhan,"Selection of non-conventional manufacturing process: a combine topsis-ahp approach"5th International and 26th Al India Manufacturing Technology, Design and Research Conference (AIMTDR 2014) December 12th-14th December 2014,IIT, Guwahati, Assam,India

[3] V.P.Darji, R.V.Rao "Intelligent Multi Criteria Decision Making Method for Material Selection in Sugar Industry." Procedia Material Science 5, 2585 – 2594, International Conference on Advances in Manufacturing and Materials Engineering, AMME 2014.

[4] Yu Novoselov, S. Bratan, V. Bogutsky, "Analysis of Relation between Grinding Wheel Wear and Abrasive Grain Wear." International Conference on Industrial Engineering, ICIE2016.Procedia Engineering 150(2016) 809-814.

[5] Xun Chen, W.Brian Rowe. "Analysis and Simulation of Grinding Process. part I." International Journal of Machine Tools & Manufacture, Vol.36 no. 8 (1996) pp 871-882.

[6] Fritz K Locke, Sebastian Barth, Christian Wrobel, Markus Weib, Patrick Mattfeld, Karl –Heinz Brakhage, Michael Room, "Modeling of the grinding wheel structure depending on volumetric composition" 7th HPC 2016 – CIRP Conference on High Performance . Cutting Procedia CIRP 46(2016) 276-280.

[7] Sanjay S. Patil,Yogesh J Bhalerao "Effect of Dressing Parameters on Surface Roughness and Grinding Ratio in Surface Grinding" Proceeding of International Conference on Industrial Tribology (ICIT-2012) Pune, December 2012

[8] Srikrishan S, Sreenivasulu Reddy. A, Vani S, " A New Car Selection in the Market using TOPSIS Technique" International Journal of Engineering Research and General Science, Volume 2, Issue- 4, June –July, 2014

[9] Y.J. Lai, T.Y.Liu, and L. Hwang, TOPSIS for MODM, Europen Journal of Operation Research. 76, 486-500,1994,

[10]A.Fritsche, F.Bleicher," Evaluating and Influencing Dressing Results by Changinize g the Grain Distribution Based on Statical and Experimental Investigations. 12th Global Conference on Sustainable Manufacturing. Procedia CIRP 26(2015) 718-723.

[11] Rajinish Kumar,Jagadish, Amitava Ray."Selection of Material for Optimal Design using Multi-Criteria Decision Making" Procedia Material Science 6 (2014) 590-596, 3rd International Conference on Materials Processing and Characterisation (ICMPC 2014)

[12] Rajnish Kumar, Chiranjib Bhomik and Amitava Ray,"Selection of Cutting tool Material by TOPSIS Method," National Conference on Recent Advancement in Mechanical Engineering (NCRAME 2013)

[13] Harsh Y Valera, Sanket N Bhavasr,"Experimental Invetsigation of Surface Roughness and Power Consumption in Turning operation of EN 31 Alloy Steel," Procedia Technology 14 (2014) 528-534. 2nd International Conference on Innovations in Automation and Mechetronics Engineering. ICIAME 2014.

[14] Lijohn P George, K. Varughese Job, I.M.Chandran," Study on Surfcae Roughness and its Prediction in Cylindrical Grinding Process based on Taguchi method of optimization," International Journal of Scientific and Research Publications, Volume 3, Issue 5, May 2013.

IOP Conf. Series: Materials Science and Engineering **178** (2017) 012034 doi:10.1088/1757-899X/178/1/012034

Effect of WC and Si additions on the mechanical properties of ZrB_2-WC-Si based cermet

R.R.Rout[a], B.Mishra[a],N.Sahoo[a], S.K. Karak[b*]

[a]Dalmia Institute of Scientific and Industrial Research, Rajgangpur
[b]National Institute of Technology, Rourkela,Odisha,India

Corresponding author E-mail: skkarak@gmail.com

Abstract: The present paper deals with the study of mechanical behavior of ZrB_2-WC-Si cermets with respect to variation in weight percentage of ZrB_2, WC and Si. Pellets are pressed in hydraulic press with a load of 6T followed by sintering though spark plasma at 1800^{0}C under inert atmosphere. Vickers hardness test of the cermet ZrB_2-60-WC-20-Si-20 showed hardness of 27.5 GPa where as it was quiet lower in case of cermets ZrB_2-60-WC-10-Si-30 and ZrB_2-60-WC-15-Si-25. Wear test showed that the mass loss became lower in case of ZrB_2-60-WC-20-Si-20 in comparison to other cermets. This was attributed to the formation of metastable phases of WSi_2 and SiC those are responsible for the increase in hardness of the cermet. Resulting matrix became stronger and also intergranular-intragranular region became densified. Furthermore, this effect was supposed to reduce matrix wear of the cermet having composition ZrB_2-60-WC20-Si-20 .

1. Introduction

Ultra High Temperature Ceramics (UHTC) like zirconium diboride (ZrB_2) have important properties which includes high melting point (>3000^{0}C), high strength (>500MPa) and electrical conductivity (10^7 S/m), as well as excellent chemical stability. Due to these useful high temperature advantages, it was applied in different critical areas such as thermal protection systems, cutting tools, high temperature electrodes, and molten metal crucibles [1–5]. But this application became restricted as ZrB_2 ceramics are very much difficult for producing dense ceramic parts with complex shapes. Many researchers have reported that ZrB_2 ceramics are densified by hot pressing [6,7], which are limited to produce specimens with simple shapes. Report showed that ZrB_2 was densified below 2000^{0}C by use of additives such as B_4C [8], $MoSi_2$ [9], and C [10]. Proper densification was attributed to activation of sintering by removal of surface oxide impurities from the ZrB_2 particle surfaces (B_4C and C) or by liquid phase sintering ($MoSi_2$). Microwave sintering was another alternative to hot pressing and conventional sintering. Microwave sintering has the advantages of uniform and rapid heating since the energy was directly coupled into the specimen rather than conducted into the specimen from an external heat source [11]. In addition to promote densification of ZrB_2 in pressureless sintering, B_4C was also reported to absorb microwave energy [12]. To aid sintering, a low melting metallic phase such as silicon was added and the effects of Si addition on the microstructure and mechanical properties of SiC–B_4C based cermets was studied [13]. In order to improve the toughness and high-temperature hardness or strength property, various transition metal carbides (WC, NbC, TaC, etc.) are added to improve cermet properties [14]. Wei-Ming Guo et al have done research work on high-temperature deformation of ZrB_2 ceramics with WC additive in four-point bending method through hot-pressing at 1900 °C for 0–60 min in argon atmosphere under a static load of 25 MPa. After 60 min bending deformation, around 8% strain was reached without apparent macroscopic cracks, which showed superplastic behaviour. This was attributed to the presence of a secondary W-containing phase formed by the addition of WC [15].

For low temperature sintering of composite materials, Spark plasma sintering (SPS) was one of the advance sintering technique that densified the composite body within a certain time period. It was also a promising technique for those materials that are difficult to sinter by conventional sintering route. Advantages of SPS over conventional pressureless sintering include faster heating rates, shorter dwell time to retain finer microstructure [16-18]. Due to localized high-temperature spark effect, adsorptive gas and impurities existing on the surface of powder particles can be eliminated. To promote sintering by diffusion, high-speed ion migration between contacting particles were possibly due to the applied electric field and then transferring material in micro and nano levels [19]. On the contrary, conventional sintering process was carried out at higher temperature with prolonged time, resulting in grain growth and concomitant inferior mechanical properties. Moreover, there are no reports available on the synthesis of ZrB_2-WC-Si cermets by spark plasma sintering (SPS). Hence, the present study attempts to prepare ZrB_2-WC-Si cermets by SPS technique followed by its characterization.

2. Experimental Technique

Commercially available ZrB_2 powder (98% purity, particle size~35µm,Esk Ceramics, Gmbh &Co, Germany),WC powder (98.5% purity, particle size~30µm,Hb metals, India) Si (99.5% pure, particle size~30µm,BAIDAO silicon metal ind. ltd China) are used as-received input raw materials. Different compositions containing ZrB_2, WC and Si are manually blended and ball milled for 50hrs.Optimum composition of different cermets are decided as per variation of metal content (Si) and metallic carbide content (WC) to study the effect on the mechanical properties of ZrB_2-WC-Si cermet. Optimal composition of ZrB_2-WC-Si cermets are given in Table-1.Toluene was taken as solvent for wet milling of above mix. After air drying, the milled powders are pressed in uniaxial acting hydraulic press under a load of 6 tons for a dwell time of 120sec. Green pellets were made with 5mm diameter and 3mm thickness. Pellets are dried in oven at temperature of 80^0C for overnight. Then dried pellets are sintered at 1800^0C for 3min in spark plasma method under inert gas atmosphere. The bulk densities and apparent porosity of sintered pellets are measured by Archimedes principle. Hardness of sintered pellets are measured by vickers hardness test (LM248AT, LECO Corporation, USA) and wear test (TR-208 M1, DUCOM Instrument, USA) was carried out through ball on disc wear method. Wear pattern of different cermets are analysed through optical microscope (SCOPE-A1,Carlzeiss India). Different phases of the ZrB_2-WC-Si cermet after ball milling as well as sintered cermets were analysed by X-ray diffraction analysis (XRD) (ULTIMA-IV, Rigaku, Japan). Morphological features of input powder and sintered sample are analysed by scanning electron microscope (SEM) (JSM 6480LV, JEOL, USA).

Table.1. Compositions of ZrB_2-WC-Si cermet

Composition	ZrB_2 Wt %	WC Wt %	Si Wt %
ZrB_2-60-WC-10-Si-30	60	10	30
ZrB_2-60-WC-15-Si-25	60	15	25
ZrB_2-60-WC-20-Si-20	60	20	20

3. Results and discussion

3.1. Morphology of composite powder(before & after milling)

Fig.1 shows the SEM micrographs of ZrB_2-WC-Si powder at milling time of 0hr and 50hrs for different compositions. Micrograph confirmed the size, shape and morphology of ZrB_2, WC, and Si powders in different forms. From the figure, it was observed that ZrB_2 and WC powder particles are irregular in shape with sharp edges, whereas Si powders are angular in nature.SEM analysis also confirmed that the size of as received powder of ZrB_2, WC, and Si are in the range of 30-35 µm and that for milled powders in the range 10-15 µm.

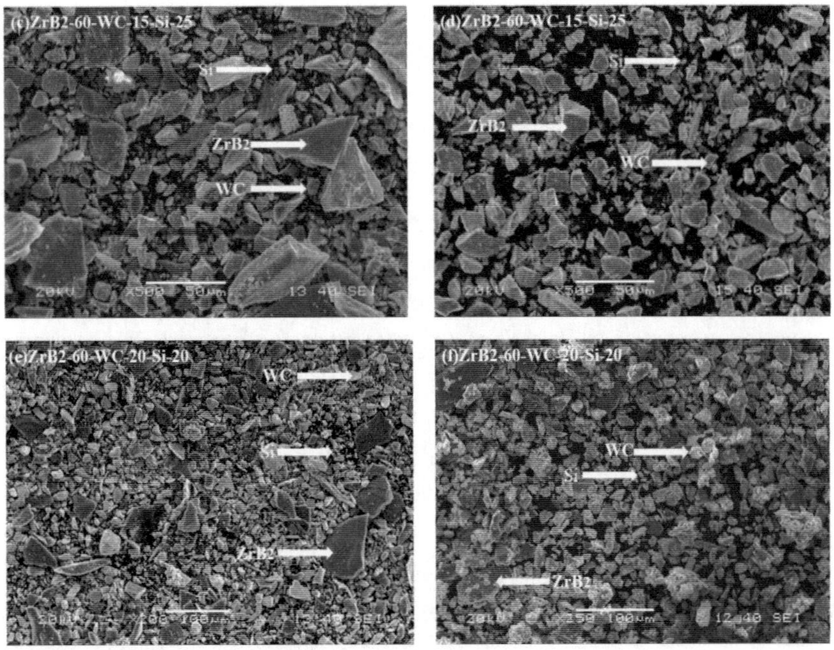

Figure 1. SEM micrographs of unmilled (left) and milled (right) powders: (a) and (b) ZrB_2-60-WC-10-Si-30, (c) and (d) ZrB_2-60-WC-15-Si-25 (e) and (f) ZrB_2-60-WC-20-Si-20

It is also cleared that the particle size of milled powder getting reduced with respect to milling period. Different phases like ZrB_2,WC and Si phases are uniformly distributed throughout whole structure after milling of 50hr. Confirmation of these phases are identified by use of EDS instrument. Same observation could be correlated with XRD pattern of the milled powder.

3.2. Phase analysis

Fig.2 shows the X-ray diffraction pattern of three different cermets having compositions ZrB_2-60-WC-10-Si-30, $ZrB_2$60-WC-15-Si-25 and $ZrB_2$60-WC-20-Si-20 taken at milling time of 0h and 50 h. It was observed that the peak intensity gradually in reducing manner with respect to milling time and accordingly peak broadened [20]. This was attributed to the reduction in particle size and introduction of lattice strain during milling. The reduction in particle size and lattice strain phenomenon arouse due to different collision took place between different parts of mill and also with particles of powder [21]. Further, even after 50 hrs of milling peaks of ZrB_2, WC, and Si are found with some minor impurity phases.

From the XRD patterns, the presence of ZrB_2, WC and Si phases at each stage of milling are also confirmed. In case of XRD analysis of milled powders at higher milling time (50 h), phenomenon like peak shifting as well broadening were observed upto minimum extent. This might be due to the fact gradual reduction in micron-level particle into sub-micron or nano-metric level fragmentation of particles caused by high energy planetary ball mill.

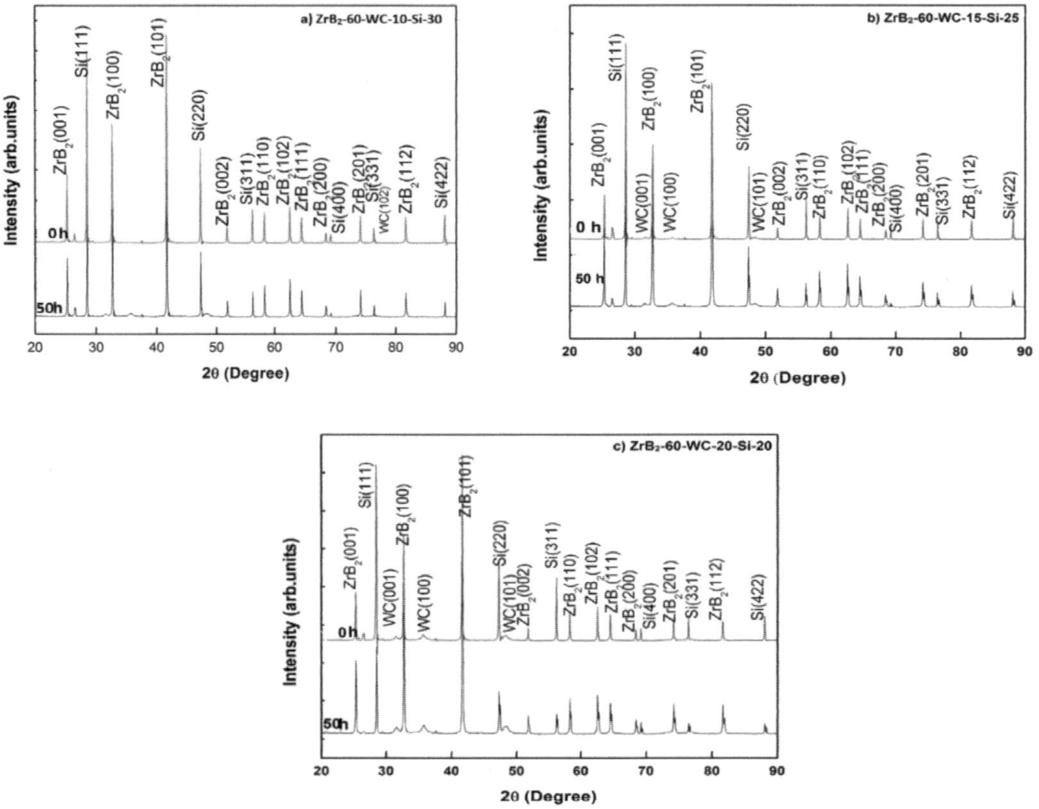

Figure 2. XRD spectra of milled powders of ZrB_2-WC-Si system at milling time 0h & 50h: (a) ZrB_2-60-WC-10-Si-30, (b)ZrB_2-60-WC-15-Si-25(c) ZrB_2-60-WC-20-Si-20

3.3. Characterization of different cermets (after SPS)

3.3.1. Morphology of sintered cermets

Fig.3 shows the microstructues of ZrB_2-WC-Si cermet sintered at 1800°C through spark plasma sintering method. Due to less holding time during sintering, grain growth was restrcited through out the process. However some localized plasma arc was accelerated the liquid phase sintering by the formation of glassy phases (perticulary seen in composition ZrB_2-60-WC-20-Si-20). Complete structure was densified by the formation of metastable phases (WSi_2) and metallic carbide (SiC) as identified by EDS in that corresponding area. Presence of these phases could also be confirmed from X-ray diffraction pattern of corresponding samples (fig.4).

National Conference on Processing and Characterization of Materials IOP Publishing
IOP Conf. Series: Materials Science and Engineering **178** (2017) 012034 doi:10.1088/1757-899X/178/1/012034

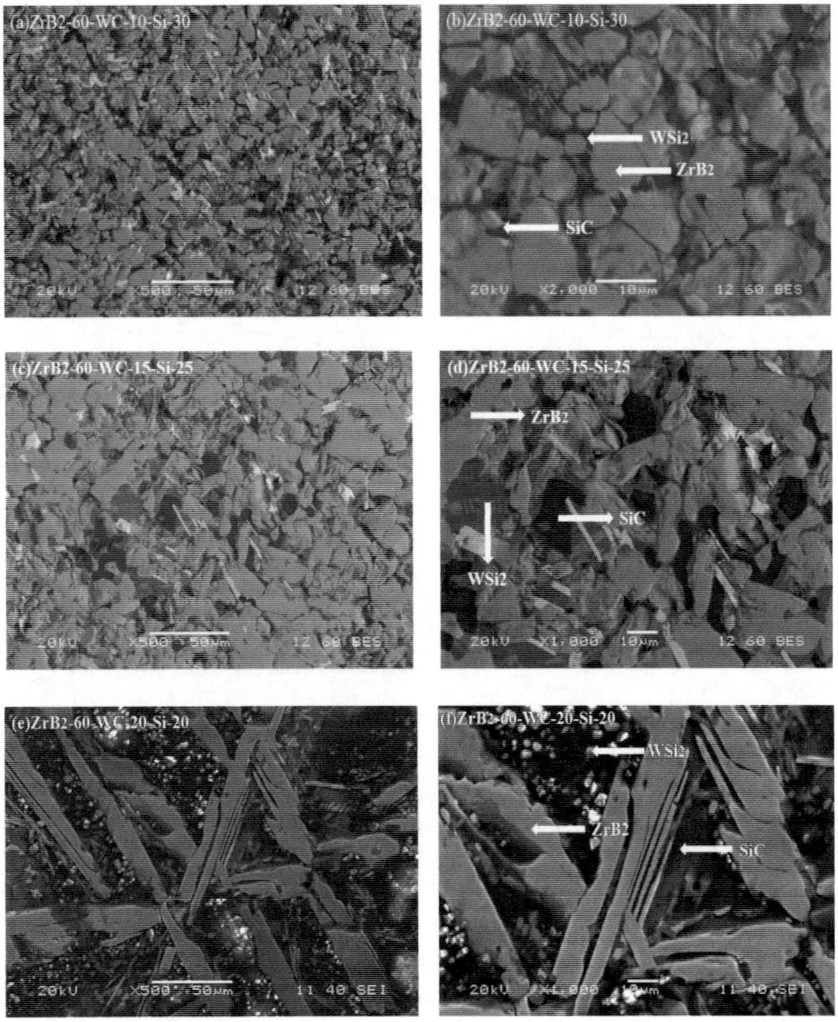

Figure 3. SEM micrographs of ZrB_2-WC-Si cermets sintered at 1800°C: (a) and (b) for ZrB_2-60-WC-10-Si-30, (c) and (d) ZrB_2-60-WC-15-Si-25 (e) and (f) ZrB_2-60-WC-20-Si-20

3.3.2. Phase analysis of sintered cermets

X-ray diffraction pattern of three different ZrB_2-WC-Si cermets sintered at 1800°C through plasma sintering technique was shown in fig.4. XRD analysis showed the presence of parent phases like ZrB_2, WC, Si and product phases of all samples after sintering. As sintering was carried out above the melting point of Si (at 1400^0C), hence reaction initiated with C (from WC partially) and finally formed SiC as confirmed from SEM micrograph of corresponding sample. XRD pattern showed the occurance of primary phases like ZrB_2, WC,Si and besides these phases some metastable phase like WSi_2 are also present as intermediates. It could also be observed that while increasing WC content from composition ZrB_2-60-WC-10-Si-30 to ZrB_2-60-WC-20-Si-20, the formation of SiC was very much effective.

IOP Conf. Series: Materials Science and Engineering **178** (2017) 012034 doi:10.1088/1757-899X/178/1/012034

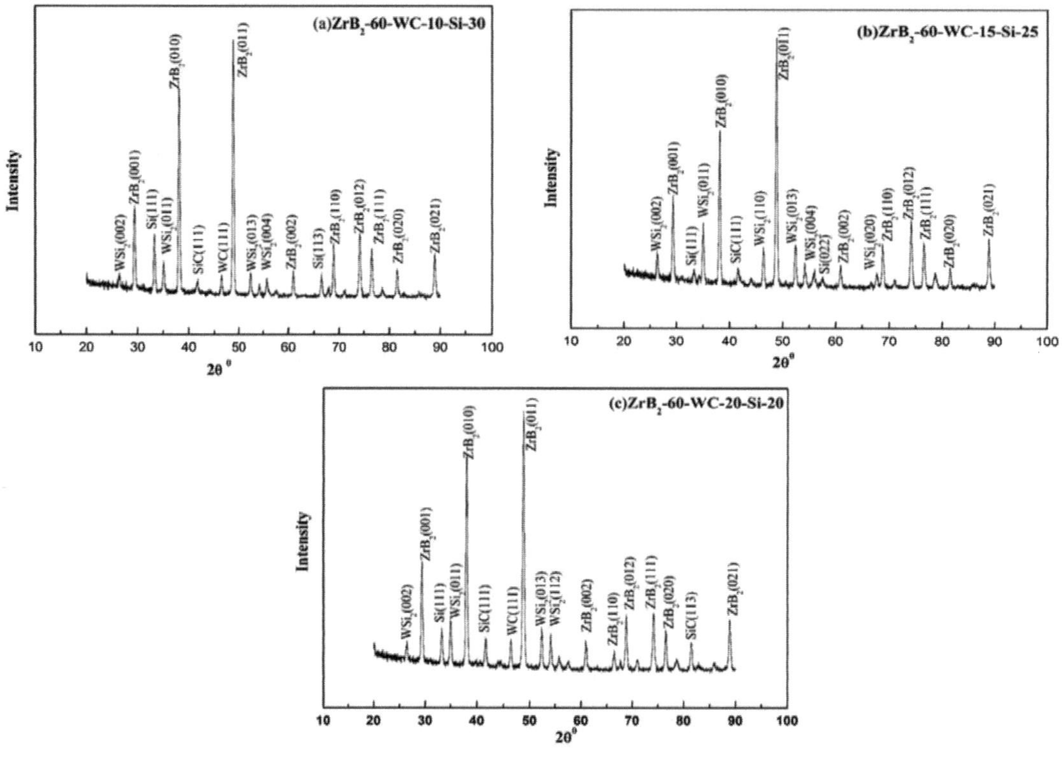

Figure 4. XRD patterns of ZrB_2-WC-Si cermet sintered at 1800°C (a) ZrB_2-60-WC10-Si-30, (b) ZrB_2-60-WC-15-Si-25 and (c) ZrB_2-60-WC-20-Si-20

3.3.3. Densification

Fig.5 shows the variation of apparent porosity and bulk density of different samples of ZrB_2-WC-Si cermets. It was cleared that the apparent porosity of the cermets decreases accordingly with increasing WC content upto 20% and also reducing metallic Si content upto 20%. Similarly bulk density was found higher in case of ZrB_2-60-WC-20-Si-20 in comparison to cermet ZrB_2-60-WC-10-Si-30 and ZrB_2-60-WC-15-Si-25. This might be attributed to the formation of metastable compounds likes WSi_2 and SiC in the matrix part by sealing the pores through liquid phase sintering and hence increasing the bulk density of the cermets even with lower metal content.

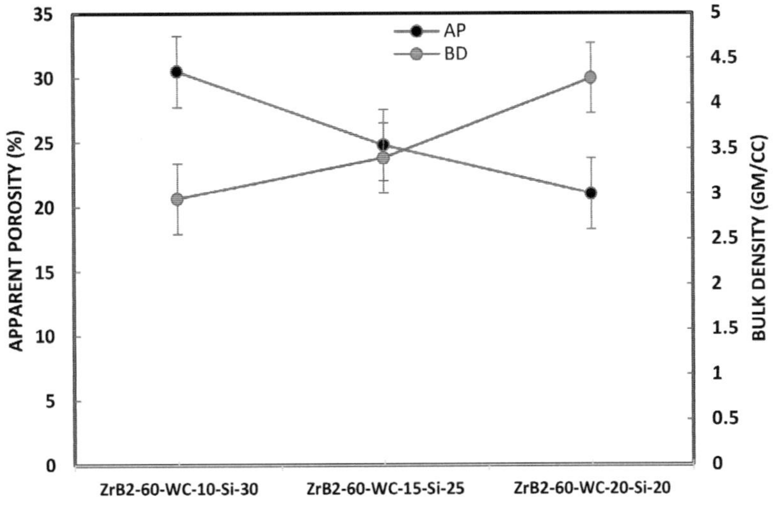

Figure 5. Apparent porosity and bulk density of sintered ZrB_2-WC-Si cermet

3.3.4. Micro-hardness

Fig.6 shows the variation of hardness of the ZrB_2-WC-Si cermets with respect to different compositions. The hardness values of sintered samples showed similar trends like density of corresponding cermets. It was observed that the amounts and distribution of ZrB_2, WC and Si phases along with product phases are affecting the hardness and density of the cermet with respect to sintering temperature. It was evident from the figure that the hardness value linearly increasing with the WC content of the material.The presence of residual porosity in case of cermets containing 10 wt. % and 15 wt. % WC, might be account for the part of decrease in measured hardness in comparison to cermet containing 20 wt. % WC. Residual stresses in sintered product was another factor affecting hardness of cermet. Due to increase in density and lower apparent posity of the cermet ZrB_2-60-WC-20-Si-20, the hardness value was increasing accordingly. Also due to formation of prominent SiC phases in the ZrB_2 matrix part of ZrB_2-60-WC-20-Si-20 cermet, the hardness value was found to be increasing drastically however same phenomenon observed with lower fraction in case of cermets ZrB_2-60-WC-10-Si30 and ZrB_2-60-WC-15-Si-25. S. Zhu etal.[11] have observed that the hardness value of ZrB_2 based composites found to be 17.5 GPa while sintering at 1820^{0}C through microwave sintering. Furthermore, the cermet having composition ZrB_2-60-WC-20-Si-20 recorded extraordinary level of improvement of the hardness (i.e. 27.5 GPa).

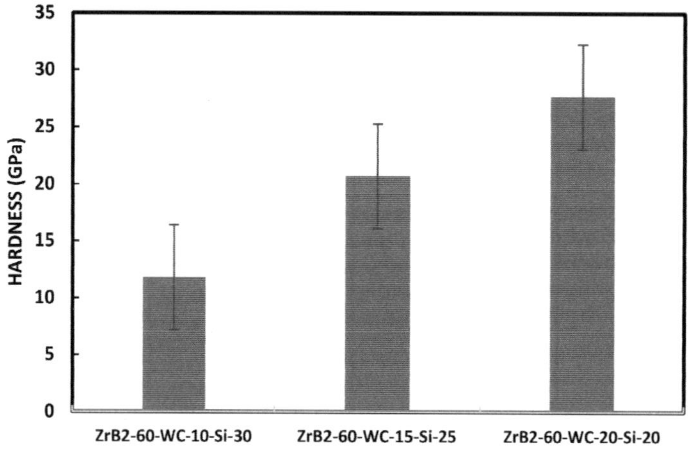

Figure 6. Hardness of ZrB_2-WC-Si cermet with varying compositions

3.3.5. Wear behaviour

Wear behavior of ZrB_2-WC-Si cermet for different compositions are studied by means of graphical representation as shown in fig.7 and fig.8. It was observed that the cermet ZrB_2-60-WC-20-Si-20 showed lower wear depth with respect to time and sliding distance. This might be due to liquid phase sintering of metallic Si in the matrix during SPS and formation of WSi_2 and SiC phases simultaneously enhancing the hardness of the cermet as confirmed from XRD and SEM analysis. As a result of which intergranular and intragranular region became densified resulting less wear of the matrix of the cermet.

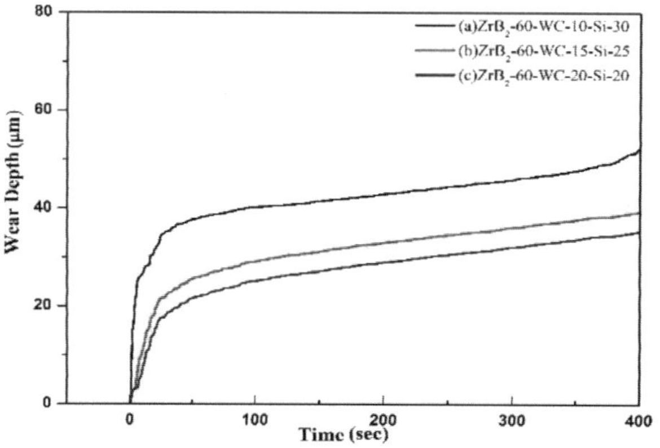

Figure 7. Variation of wear depth with sliding time for different ZrB_2-WC-Si cermets

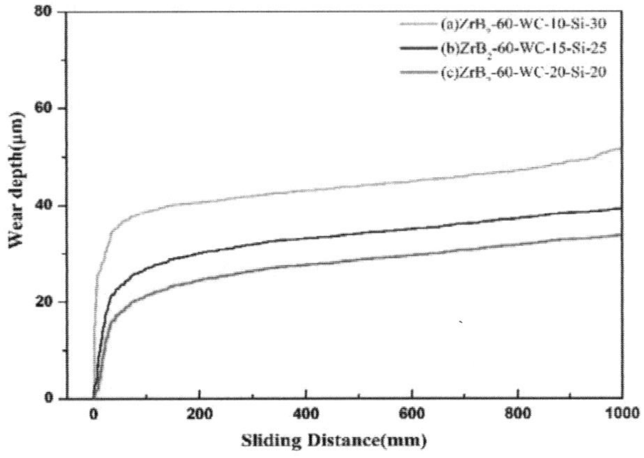

Figure 8. Variation of wear depth with sliding distance for different ZrB_2-WC-Si cermets

From the results discussed earlier, the denisty and hardness are reached at maximum value in case of cermet ZrB_2-60-WC-20-Si-20. Hence matrix part of corresponding cermet became strengthen and the wear pattern gradually in reducing nature.

4. Conclusion

ZrB_2-WC-Si cermet having different compositions in the form of ZrB_2-60-WC-10-Si-30, ZrB_2-60-WC-15-Si-25 and ZrB_2-60-WC-20-Si-20 are synthesized through SPS technique. XRD analysis of the sintered cermets shows the presence of parent phases along with metastable phase like WSi_2 and also SiC formation in different cermets, however it is most prominent in ZrB_2-60-WC-20-Si-20. It could be correlated with SEM analysis of sintered pellets that confirmed the presence of different phases like ZrB_2 , WC in the matrix along with SiC. Lower porosity and higher bulk density are observed in case of sintered ZrB_2-60-WC20-Si20 cermet. Hardness of ZrB_2-60-WC-20-Si-20 cermet shows higher value i.e. 27.5 GPa as compared to the other cermets ZrB_2-60-WC10-Si-30 and ZrB_2-60-WC15-Si-25. The wear pattern is improving gradually upto maximum level in case of cermet containing 20% WC and 20% Si.

Acknowledgement

Authors are thankful to research and development department of OCL INDIA Ltd. Rajgangpur for giving all major testing facility and input material collection during the research work.

References

[1] Cutler RA 1991 *ASM International* **4** 787–803.
[2] Mroz C. 1995 *American Ceram. Soc. Bull.***76** 164–5.
[3] Upadhya K, Yang J M and Hoffman W P 1997 *American Ceram. Soc. Bull.* **58** 51–6.
[4] Norasetthekul S, Eubank P T, Bradley W L, Bozkurt B and Stucker B 1999 *J. Mater. Sci.* **34** 1261–70.
[5] Levine S R, Opila E J, Halbig M C, Kiser J D, Singh M and Salem J A 2002 *J. Eur. Ceram. Soc.* **22** 2757–67.
[6] Chamberlain A L, Fahrenholtz W G, Hilmas G E and Ellerby D T 2004 *J. Am. Ceram. Soc.* **87(6)** 1170–2.
[7] Monteverde F, Bellosi A 2005 *Solid State Sci.* **5** 622–30.
[8] Wang H L, Wang C A, Yao X F and Fang D N 2007 *J. Am. Ceram. Soc.* **90** 1992–7.
[9] Zhang X H, Qu Q, Han J C, Han W B and Hong C Q 2008 *Scripta Mater.* **59** 753–6.

[10] Guo S Q, Kagawa Y and Nishimura T 2009 *J. Eur. Ceram. Soc.* **29(4)** 787–94.
[11] Zhu S, Fahrenholtz W, Hilmas G, Zhang S, Yadlowsky E and Keitz *Composites: Part A* **39** 449-53.
[12] Zhang H, Yan Y, Huang Z, Liu X and Jiang D 2009 *Scripta Materilia* **60** 559-62.
[13] Glasson D R, Jones J A 1969 *J. Appl. Chem.* **19** 125–37.
[14] Manoj Kumar B V, Bikramjit Basuw S, Kang J and Ramkumar 2006 *J. Am. Ceram. Soc.* **89 [12]** 3827–31.
[15] Guo W, Yang Z and Zhang G 2011 *Int. J. of Ref. Metals and Hard Mat.* **29** 705–9.
[16] Hulbert D M, Jiang D, Dudina D V and Mukherjee A K 2009 *Int. J. Refract. Met. and Hard Mater.* **27** 367–75.
[17] Hayun S, Kalabukhov S, Ezersky V, Dariel M P and Frage N 2010 *Ceram. Int.* **36** 451–57.
[18] Munir Z A, Anselmi U and Ohyanagi M 2006 *J. Mater. Sci.* **41** 763–77.
[19] Sairam K, Sonber J K , Murthy T S R, Subramanian C ,Fotedar R K, Nanekar P and Hubli R C 2014 *Int. J. Refract. Met. Hard Mater.* **42** 185–92.
[20] Zamoraa V, Ortiz A L, Guiberteau F, Nygren M and Shaw L L 2011 *J. of Euro.Cer. Soc.* **31** 2407–14.
[21] Sahani P, Karak S K, Mishra B, Chakravarty D and Chaira D 2016 *Int. J. of Ref. Metals and Hard Materials* **57** 31-41.